Birkhäuser Advanced Texts
Basler Lehrbücher

Edited by
Herbert Amman, University of Zürich
Ranee Brylinski, Penn State University

Steven G. Krantz
Harold R. Parks

The Geometry of Domains in Space

Birkhäuser
Boston • Basel • Berlin

Steven G. Krantz
Department of Mathematics
Washington University
St. Louis, MO 63130

Harold R. Parks
Department of Mathematics
Oregon State University
Corvallis, OR 97331

Library of Congress Cataloging-in-Publication Data
Krantz, Steven G. (Steven George), 1951-
  The geometry of domains in space / Steven G. Krantz, Harold R. Parks.
    p. cm. — (Birkhäuser Advanced Texts)
  Includes bibliographical references and index.
  ISBN 0-8176-4097-5 (hard sewn : alk paper). — ISBN 3-7643-4097-5 (hard sewn : alk. paper)
  1. Mathematical analysis. 2. Geometry. I. Parks, Harold R., 1949-   . II. Title.   III. Series: Birkhäuser Advanced Texts (Boston, Mass.)
QA300.K644   1999
515—dc21
                                                                     98-44619
                                                                     CIP

AMS Subject Classifications: 26-01, 26-02, 26A24, 26A45, 26A51, 26B05, 26B15, 26B20, 26B25, 26B30, 26B35, 28-01, 28A05, 25A12M, 28A15, 28A25, 28A75, 28A78, 30C62, 30C65, 33B15, 35J05, 46E20, 46E30, 46E35, 49K20, 49R05, 51M25, 52A20, 53-01, 53A04, 53A05, 53A10, 52A20, 53B20, 53C45

Printed on acid-free paper.
© 1999 Birkhäuser Boston

Birkhäuser

All rights reserved. This work may not be translated or copied in whole or in part without the written permission of the publisher (Birkhäuser Boston, c/o Springer-Verlag New York, Inc., 175 Fifth Avenue, New York, NY 10010, USA), except for brief excerpts in connection with reviews or scholarly analysis. Use in connection with any form of information storage and retrieval, electronic adaptation, computer software, or by similar or dissimilar methodology now known or hereafter developed is forbidden.
The use of general descriptive names, trade names, trademarks, etc., in this publication, even if the former are not especially identified, is not to be taken as a sign that such names, as understood by the Trade Marks and Merchandise Marks Act, may accordingly be used freely by anyone.

ISBN 0-8176-4097-5
ISBN 3-7643-4097-5

Typeset by the author in LATEX.
Printed and bound by Edwards Brothers, Inc., Ann Arbor, MI.

9 8 7 6 5 4 3 2 1

To Herbert Federer

Contents

Preface		ix
1.	**Elementary Topics**	**1**
	1.1 Smooth Functions	1
	1.2 The Concept of Defining Function	8
	1.3 Measure Theory	15
2.	**Domains with Smooth Boundaries**	**27**
	2.1 The Tangent Bundle and Normal Bundle of the Boundary	27
	2.2 The Second Fundamental Form and Curvature	34
	2.3 Surfaces with Constant Mean Curvature	45
3.	**Measures**	**57**
	3.1 The Carathéodory Construction	57
	3.2 Rectifiability	66
	3.3 Minkowski Content	74
	3.4 A Space-Filling Curve	81
	3.5 Covering Lemmas	83
	3.6 Functions of Bounded Variation	100
	3.7 Domains with Finite Perimeter	107
	3.8 The Area Formula	123
	3.9 The Co-Area Formula	136
4.	**Sobolev Spaces**	**143**
	4.1 Basic Definitions and Results	143
	4.2 Restriction and Trace Theorems for Sobolev Spaces	148
	4.3 Domain Extension Theorems for Sobolev Spaces	153

5. Smooth Mappings — 157
- 5.1 Sard's Theorem 157
- 5.2 Extension Theorems 162
- 5.3 Proof of the Whitney Extension Theorem 167
- 5.4 Application of the Whitney Extension Theorem 179
- 5.5 Multidimensional Versions of the Fundamental Theorem of Calculus 183

6. Convexity — 191
- 6.1 The Classical Notion of Convexity 191
- 6.2 Other Characterizations of Convexity 201
- 6.3 Exhaustion Functions 211
- 6.4 Convexity of Order k 217

7. Steiner Symmetrization — 223
- 7.1 Basic Properties 223
- 7.2 The Isodiametric, Isoperimetric and Brunn-Minkowski Inequalities 231
- 7.3 Equality in the Isoperimetric Inequality 243

8. Topics Related to Complex Analysis — 247
- 8.1 Quasiconformal Mappings 247
- 8.2 Weyl's Theorem on Eigenvalue Asymptotics of a Domain in Space 259

A.1. Metrics on the Collection of Subsets of Euclidean Space — 275

A.2. The Constants Associated to Euclidean Space — 281

Guide to Notation — 287

Bibliography — 291

Index — 305

Preface

The analysis of Euclidean space is well-developed. The classical Lie groups that act naturally on Euclidean space—the rotations, dilations, and translations—have both shaped and guided this development. In particular, the Fourier transform and the theory of translation invariant operators (convolution transforms) have played a central role in this analysis.

Much modern work in analysis takes place on a domain in space. In this context the tools, perforce, must be different. No longer can we expect there to be symmetries. Correspondingly, there is no longer any natural way to apply the Fourier transform. Pseudodifferential operators and Fourier integral operators can play a role in solving some of the problems, but other problems require new, more geometric, ideas.

At a more basic level, the analysis of a smoothly bounded domain in space requires a great deal of preliminary spadework. Tubular neighborhoods, the second fundamental form, the notion of "positive reach", and the implicit function theorem are just some of the tools that need to be invoked regularly to set up this analysis. The normal and tangent bundles become part of the language of classical analysis when that analysis is done on a domain. Many of the ideas in partial differential equations—such as Egorov's canonical transformation theorem—become rather natural when viewed in geometric language. Many of the questions that are natural to an analyst—such as extension theorems for various classes of functions—are most naturally formulated using ideas from geometry.

Some of the geometric tools that should be part of the working kit of every analyst are Sard's theorem, Sobolev spaces, geometric measures, barycentric coordinates, Steiner symmetrization, the isoperimetric inequality, Weyl's theorem, covering theorems, convex functions, and Stokes's theorem in all its many formulations. We cover these, and more. An effort has been made to organize topics logically, and to make the presentation self-contained—and to not require a great deal of background on the part of the reader.

An important special case of the preceding discussion is *convex domains*. The classical literature of convexity is highly non-analytic. It does not address the relationships between convex functions and convex domains. The use of the Minkowski functional, convex exhaustion functions, convexity with respect to a family of functions, strictly and strongly convex domains, and the attendant analytic tools, are ill-documented. And such documentation that exists is incomplete and is scattered in the literature. This book contains a thoroughgoing treatment of the basic analytic ideas!in the theory of convexity.

We hope that this book will serve as a text for graduate students intending to pursue advanced work in geometric analysis, and also as a useful reference for professional mathematicians and other mathematical scientists. The analysis of domains exemplifies a growing interaction between analysis and geometry—both classical differential. We hope that this symbiotic development will continue, perhaps with the aid of the present monograph.

Prerequisites for reading this book are minimal. A grounding in advanced calculus is perhaps the most important of these, more for the attainment of a certain amount of mathematical sophistication rather than for knowledge of any particular theorem. Familiarity with manifolds, or at least with the concept of doing calculus on a surface in space, will be helpful in several spots. All necessary ideas from differential geometry are developed here from first principles.

St. Louis, Missouri　　　　　　　　　　　　　　　　　　　　Steven G. Krantz
Corvallis, Oregon　　　　　　　　　　　　　　　　　　　　　Harold R. Parks

Chapter 1

Elementary Topics

1.1 Smooth Functions

Let $U \subset \mathbb{R}^N$ be any open set. A function $f : U \to \mathbb{R}^M$ is said to be **continuously differentiable of order** k, or C^k, if f possesses all partial derivatives of order not exceeding k and all of those partial derivatives are continuous; we write $f \in C^k$ or $f \in C^k(U)$ if U is not clear from context. If the range of f is also not clear from context, then we write $f \in C^k(U; \mathbb{R}^M)$. We also extend this notation to allow $f \in C^0$ to indicate that f is continuous.

The **support** of a continuous function $f : U \to \mathbb{R}^M$, denoted supp f, is the closure of the set of points where $f \neq 0$. We will use C_c^k to denote the C^k functions with compact support.

Let $\mathbb{Z}$ denote the integers, $\mathbb{Z}^+$ the non-negative integers, and $\mathbb{N}$ the positive integers or natural numbers. A **multi-index** α is an element of $(\mathbb{Z}^+)^N$, the product of N copies of $\mathbb{Z}^+$. If $\alpha = (\alpha_1, \alpha_2, \ldots, \alpha_N)$ is a multi-index and $x = (x_1, x_2, \ldots, x_N)$ is a point in $\mathbb{R}^N$, then we introduce the following standard notation

$$x^\alpha \equiv (x_1)^{\alpha_1}(x_2)^{\alpha_2}\ldots(x_N)^{\alpha_N}$$

$$|\alpha| \equiv \alpha_1 + \alpha_2 + \cdots + \alpha_N,$$

$$\frac{\partial^\alpha}{\partial x^\alpha} = D^\alpha \equiv \frac{\partial^{\alpha_1}}{\partial x_1^{\alpha_1}} \frac{\partial^{\alpha_2}}{\partial x_2^{\alpha_2}} \cdots \frac{\partial^{\alpha_N}}{\partial x_N^{\alpha_N}}$$

$$\alpha! \equiv (\alpha_1!)(\alpha_2!)\ldots(\alpha_N!).$$

With this notation, a function f on U is C^k if $D^\alpha f$ exists (in the classical sense of differentiation) and is continuous, for all multi-indices α with $|\alpha| \leq k$. The function is said to be C^∞ (**infinitely differentiable**) provided that $f \in C^k$ for every positive k. The notation $f \in C_c^\infty$ will indicate that f is C^∞ and has compact support. The function is said to be in C^ω

(that is, it is **real analytic**) provided that it has a convergent power series expansion about each point of U. We direct the reader to the companion reference Krantz and Parks [2] for matters related to real analytic functions.

Suppose that $U \subset \mathbb{R}^N$ is an open set, that M is a positive integer, and that $\kappa \in \mathbb{Z}^+ \cup \infty$. We will introduce two commonly used topologies on $C^\kappa(U; \mathbb{R}^M)$. Here it will be most convenient to define a topology by specifying what it means for a sequence of functions to converge in the topology.

Definition 1.1.1 *Let $f_1, f_2, \ldots$ be a sequence in $C^\kappa(U; \mathbb{R}^M)$.*

(i) *We say f_i converges to $f \in C^\kappa(U; \mathbb{R}^M)$ in the **compact-open C^κ topology** if, for every compact $K \subset U$ and every multi-index $\alpha = (\alpha_1, \alpha_2, \ldots, \alpha_N)$ with $|\alpha| - 1 < \kappa$, the sequence $D^\alpha f_i$ converges uniformly to $D^\alpha f$ on K.*

(ii) *We say f_i converges to $f \in C^\kappa(U; \mathbb{R}^M)$ in the **Whitney topology** if, for every multi-index $\alpha = (\alpha_1, \alpha_2, \ldots, \alpha_N)$ with $|\alpha| - 1 < \kappa$, the sequence $D^\alpha f_i$ converges uniformly to $D^\alpha f$.*

Remark 1.1.2 The compact-open C^κ topology is sometimes called the **weak topology on C^κ** and the Whitney topology is sometimes called the **strong topology** or **fine topology on C^κ**.

The usual vector operations can be applied to multi-indices, but certain additional notation is often useful: For multi-indices $\alpha = (\alpha_1, \alpha_2, \ldots, \alpha_N)$ and $\beta = (\beta_1, \beta_2, \ldots, \beta_N)$, we write

$$\beta < \alpha \quad \text{if} \quad \beta_i < \alpha_i \text{ for } i = 1, 2, \ldots, N$$
$$\beta \leq \alpha \quad \text{if} \quad \beta_i \leq \alpha_i \text{ for } i = 1, 2, \ldots, N$$
$$\alpha - \beta \equiv (\alpha_1 - \beta_1, \alpha_2 - \beta_2, \ldots, \alpha_N - \beta_N) \quad \text{if} \quad \beta \leq \alpha,$$
$$\binom{\alpha}{\beta} \equiv \frac{\alpha!}{\beta!(\alpha - \beta)!} \quad \text{if} \quad \beta \leq \alpha.$$

With this notation, the **generalized Leibniz's Rule** for derivatives of products of functions of N variables is written as

$$\frac{\partial^\alpha}{\partial x^\alpha}(fg) = \sum_{\beta \leq \alpha} \binom{\alpha}{\beta} \frac{\partial^\beta f}{\partial x^\beta} \frac{\partial^{\alpha-\beta} g}{\partial x^{\alpha-\beta}}, \tag{1.1}$$

provided $f, g \in C^k$ and $|\alpha| \leq k$. This is easily proved by induction on $|\alpha|$.

While a C^k function is most directly recognized by simply differentiating the function, it is important to realize that the smoothness of a function can be determined without actually differentiating it. One method for doing this is with the calculus of finite differences. Another is by determining the rate at which the function can be approximated by some regular family of

1.1. SMOOTH FUNCTIONS

testing functions (such as polynomials). We refer the reader to Krantz [1] for a thorough treatment of these matters.

In the nineteenth century it was widely held among mathematicians that a generic real-valued function—on the real line for instance—is smooth at most points and that its singularities are isolated. This point of view is carried on today in calculus classes. But Weierstrass taught us that a continuous function can be nowhere differentiable, and we now know that it is actually the nowhere differentiable functions that are generic among the continuous ones on a compact interval (see Krantz [2], Theorem 12.10).

Other naïvetés in elementary function theory are of more recent vintage. In the early part of this century, mathematicians wondered whether a smooth function could be compactly supported. This confusion stemmed from a widely held misunderstanding of the definition of "function" (see Langer [1] for a dramatic error of Euler stemming from a misunderstanding of the function concept). For instance in the 1930s there was general disagreement as to whether the expression

$$\phi(x) = \begin{cases} e^{-1/x^2} & \text{if } x > 0 \\ 0 & \text{if } x \leq 0 \end{cases}$$

is really a function (see Boas [1]). The issue was that ϕ is defined by two formulas rather than one.

Of course we now understand that a real-valued function on a set $U \subset \mathbb{R}^N$ is a subset f of $U \times \mathbb{R}$ such that

- if $u \in U$, then there is a number $y \in \mathbb{R}$ such that $(u, y) \in f$;
- if $(u, x) \in f$ and $(u, y) \in f$, then $x = y$.

With this definition of function, the expression ϕ certainly defines a function. One verifies easily, using for instance l'Hôpital's rule, that ϕ is C^∞. To illustrate this assertion, let us look at just the first derivative. The function ϕ is plainly differentiable at points other than $x = 0$. At the origin we calculate for $h > 0$ that

$$\frac{\phi(h) - \phi(0)}{h - 0} = \frac{e^{-1/h^2}}{h} = \frac{1/h}{e^{1/h^2}}.$$

We wish to calculate

$$\lim_{h \to 0} \frac{1/h}{e^{1/h^2}}.$$

We apply l'Hôpital's rule to obtain

$$\lim_{h \to 0^+} \frac{-1/h^2}{(-2/h^3)e^{1/h^2}} = \lim_{h \to 0^+} \frac{h}{2e^{1/h^2}}.$$

This last expression plainly equals 0. The limit of the difference quotient from the left is trivially zero. We conclude that $\phi'(0)$ exists and equals 0.

Thus ϕ' exists at every point and is continuous. Subsequent derivatives may be analyzed similarly.

The function ϕ is now a useful tool. For consider

$$\phi_{[0,1]}(x) = \phi(x) \cdot \phi(1-x).$$

This function has the properties

- $1 \geq \phi_{[0,1]} \geq 0$
- $\phi_{[0,1]}(x) > 0$ if and only if $0 < x < 1$
- $\phi_{[0,1]} \in C^\infty(\mathbb{R})$.

Thus we have constructed a compactly supported, C^∞ function. We denote the collection of such functions by $C_c^\infty(\mathbb{R})$.

It is easy to see, using translations and dilations, that *any* compact interval can be the support of a C^∞ function. Precisely, the function

$$\phi_{[a,b]} \equiv \phi_{[0,1]}\bigg((x-a)/(b-a)\bigg)$$

is C^∞ and has support $[a,b]$ when $a < b$.

Examples of smooth functions in $\mathbb{R}^N$ may be constructed just by using algebraic operations on smooth functions of one variable. For example,

$$F(x_1, x_2) \equiv \phi_{[0,1]}(x_1) \times \phi_{[3,5]}(x_2) \tag{1.2}$$

is a C^∞ function of two real variables whose support is the rectangle $[0,1] \times [3,5]$. However the collection of smooth functions of two or more variables is much richer than this example suggests. The next two theorems support this contention:

Theorem 1.1.3 (C^∞ Urysohn Lemma) *Let A and B be any two disjoint closed subsets of $\mathbb{R}^N$. There exists a function $f \in C^\infty(\mathbb{R}^N)$, taking values in $[0,1]$, such that f is identically equal to 1 on the set A and f is identically equal to 0 on the set B.*

Theorem 1.1.4 (C^∞ Varieties) *Let $V \subset \mathbb{R}^N$ be any closed set. There is a C^∞ function f on $\mathbb{R}^N$ such that*

$$V = \{x \in \mathbb{R}^N : f(x) = 0\}.$$

The first theorem says that any two closed sets may be separated, in a function-theoretic sense, by a C^∞ function (for the ontology of the notion of function-theoretic separation, refer to the Stone-Weierstrass theorem in Rudin [1]). The second theorem says that *any* closed set is the zero set of a C^∞ function. Contrast this with the rigid structure of zero sets for either polynomials or real analytic functions (see Krantz and Parks [2], Section 5.2). Both theorems can be seen to be corollaries of the following result:

1.1. SMOOTH FUNCTIONS

Lemma 1.1.5 (LSC Domination Principle)
If $f : \mathbb{R}^N \to [0, \infty]$ is lower-semi-continuous, then there exists $\phi : \mathbb{R}^N \to [0, \infty)$ such that

$$\phi \in C^\infty, \tag{1.3}$$
$$0 \leq \phi \leq f, \tag{1.4}$$
$$\{x : \phi(x) > 0\} = \{x : f(x) > 0\}. \tag{1.5}$$

Notation 1.1.6 *Throughout this book, we will use $\mathbb{B}(x,r)$, $r > 0$, to denote the open ball about x with radius r. We will also use $\overline{\mathbb{B}}(x,r)$ $r \geq 0$, to denote the closed ball about x with radius r.*

Proof: We use the characterization that the function f is lower-semi-continuous if and only if $\{x : f(x) > \alpha\}$ is open for every $\alpha \in \mathbb{R}$. By redefining f to equal 1 on $\{x : f(x) > 1\}$, we may and shall assume that f takes only finite values.

Exhausting $\{x : f(x) > 0\}$ by compact subsets, we can find a sequence of points $p_1, p_2, \ldots$ and positive reals $\sigma_1, \sigma_2, \ldots$ such that

$$\{x : f(x) > 0\} = \bigcup_{i=1}^\infty \mathbb{B}(p_i, \sigma_i).$$

For each $i = 1, 2, \ldots$, define the function $\phi_i \in C^\infty$ by setting

$$\phi_i(x) = \frac{1}{2} f(p_i) \, \phi_{[-1,1]}\bigl(|x - p_i|^2/\sigma_i^2\bigr).$$

Noting that $\phi_{[-1,1]}$ is everywhere less than 1, we see that each ϕ_i satisfies

$$0 \leq \phi_i \leq f.$$

For each $i = 1, 2, \ldots$, set

$$k_i = \sup\left\{ \left|\frac{\partial^\alpha \phi_i}{\partial x^\alpha}\right| : \alpha \text{ is a multi-index with } |\alpha| \leq i \right\}.$$

Then the series

$$\sum_{i=1}^\infty \frac{2^{-i}}{1 + k_i} \phi_i$$

converges uniformly, as do all its term-by-term derivatives of any order. The function defined by the series has the desired properties. ∎

Proof of the C^∞ Varieties Theorem 1.1.4: Let V be a closed subset of $\mathbb{R}^N$. We apply Lemma 1.1.5 with $f = \chi_{cV}$, the characteristic function of the complement of V. The resulting function ϕ is C^∞ and $V = \{x : \phi(x) = 0\}$. ∎

Notation 1.1.7 *Let A be a set in a topological space. We will denote the interior of A by $\mathring{A}$, the closure of A by $\overline{A}$, and the boundary of A by ∂A.*

Proof of the C^∞ Urysohn's Lemma 1.1.3: Let A and B be disjoint closed subsets of $\mathbb{R}^N$. First construct a closed set $\widetilde{A}$ with

$$A \subset (\widetilde{A})^\circ,$$
$$\widetilde{A} \cap B = \emptyset,$$

a task we leave to the reader.

Use the C^∞ Varieties Theorem 1.1.4 to obtain a non-negative C^∞ function ψ with $\widetilde{A} = \{x : \psi(x) = 0\}$. Set $c = \inf_{x \in B} \psi(x)$. Apply the LSC Domination Principle 1.1.5 with

$$f(x) = \begin{cases} 0 & \text{if } x \in A \\ c & \text{if } x \notin A \end{cases}$$

to obtain a function $\phi \in C^\infty$ that has A as its zero set and is less than or equal to ψ on B. We can then define

$$\Psi(x) = \begin{cases} \frac{\psi(x)}{\phi(x)} & \text{if } x \notin A \\ 0 & \text{if } x \in A. \end{cases}$$

The function Ψ is C^∞ because ψ vanishes in a neighborhood of A. Also Ψ equals 0 on A and is greater than or equal to 1 on B. To complete the proof we compose Ψ with any non-negative C^∞ function from $\mathbb{R}$ to $\mathbb{R}$ that vanishes on $(-\infty, 0]$ and equals 1 on $[1, \infty)$, for example

$$\frac{1}{\int_\mathbb{R} \phi_{[0,1]}} \int_{-\infty}^t \phi_{[0,1]}(s)\, ds. \qquad \blacksquare$$

Now let us look at C^∞ partitions of unity. Let $E \subset \mathbb{R}^N$ be a closed set. Let $\mathcal{U} = \{U_\alpha\}_{\alpha \in A}$ be an open covering of $\mathcal{U}$. The following result will be proved in Section 3.5.

Lebesgue Covering Theorem *There is an open refinement $\mathcal{V} = \{V_\beta\}$ of the covering $\mathcal{U}$ (that is, each $V_\beta \in \mathcal{V}$ is open and is a subset of some $U_\alpha \in \mathcal{U}$) such that*

(i) *$\mathcal{V}$ still covers E.*

(ii) *No point of E lies in more than $N+1$ of the sets V_β. (This assertion is summarized by saying that the covering $\mathcal{V}$ has **valence** not exceeding $N+1$.)*

Now we have

1.1. SMOOTH FUNCTIONS

Theorem 1.1.8 (Smooth partitions of unity) *Let $\mathcal{U} = \{U_\alpha\}_{\alpha \in A}$ be an open cover of a closed set E. Then there exist C^∞ functions ϕ_j, each ϕ_j compactly supported in some U_{α_j}, such that*

$$\sum_j \phi_j(x) = 1 \quad \text{for all} \quad x \in E.$$

Further, no point of E lies in the support of more than $N+1$ of the functions ϕ_j.

Definition 1.1.9 *The collection of functions $\{\phi_j\}$ is termed a **smooth partition of unity** subordinate to the covering $\mathcal{U}$.*

Proof of Theorem 1.1.8: Use Lebesgue's Covering Theorem to pass to an open refinement $\mathcal{V}$ of E having valence not exceeding $N + 1$. We can also assume that $\mathcal{V}$ is countable (either by writing E as a countable union of compact sets or by appealing to Lindelöf's Theorem, see for instance Dugundji [1], p. 174). For each j, apply Theorem 1.1.4 to the closed set ^cV_j to obtain a C^∞ function μ_j that is supported precisely on V_j. Replacing μ_j by μ_j^2 if necessary, but without changing notation, we can assume each μ_j is also a non-negative function. Notice that each point of E lies in at most $N + 1$ of the sets V_j and that for each $x \in E$ there is at least one j such that $\mu_j(x) > 0$.

Apply Theorem 1.1.4 to E itself to obtain a function μ^* supported exactly on ^cE and which we may also assume is non-negative. Define

$$\psi(x) = \mu^*(x) + \sum_j \mu_j(x).$$

For each x there is a neighborhood W_x such that at most $N + 2$ of the functions involved in the definition of ψ are non-vanishing on W_x. It follows that ψ is C^∞. Note that ψ is nowhere zero, and we have

$$\psi(x) = \sum_j \mu_j(x) \quad \text{for} \quad x \in E.$$

Thus the functions

$$\phi_j(x) \equiv \frac{\mu_j(x)}{\psi(x)}$$

have the desired properties. ∎

1.2 The Concept of Defining Function

Definition 1.2.1 *Let $\Omega \subset \mathbb{R}^N$ be a bounded domain. Let $\rho : \mathbb{R}^N \to \mathbb{R}$ be a C^k function, $k \geq 1$. We say that ρ is a **defining function** for Ω if the following properties obtain:*

$$\Omega = \{x \in \mathbb{R}^N : \rho(x) < 0\}; \tag{1.6}$$

$$^c\bar{\Omega} = \{x \in \mathbb{R}^N : \rho(x) > 0\}; \tag{1.7}$$

$$\mathbf{grad}\,\rho(x) = \nabla \rho(x) \neq 0 \quad \text{for all } x \in \partial\Omega. \tag{1.8}$$

Notice that, from freshman calculus, the vector field $\nabla\rho$ is normal to $\partial\Omega$ at each point of $\partial\Omega$. We often denote this vector field by ν and its value at a particular boundary point P by $\nu(P)$ or ν_P. Observe that, according to our definition of a defining function, ν_P will be an *outward normal* to $\partial\Omega$ at P. Thus our setup automatically orients $\partial\Omega$, the boundary of the domain Ω.

Definition 1.2.2 *A vector $\mathbf{v} = (v_1, \ldots, v_N) \in \mathbb{R}^N$ is a **tangent vector** to $\partial\Omega$ at $P \in \partial\Omega$ if $\mathbf{v} \perp \nabla\rho(P)$, where the orthogonality is determined by the usual Euclidean inner product. Another way to say this is that $\mathbf{v}$ is tangent to $\partial\Omega$ at P if*

$$\sum_{j=1}^{N} (\partial\rho/\partial x_j)(P) \cdot v_j = 0. \tag{1.9}$$

We write $\mathbf{v} \in \mathrm{Tan}(\partial\Omega, P)$.

In some contexts it is useful to replace the defining function ρ with the function $\tilde\rho(x) \equiv \rho(x)/|\nabla\rho(x)|$. This function is well-defined when x is near $\partial\Omega$ and can be suitably modified away from $\partial\Omega$ so that it, too, is a defining function. The attractive feature of $\tilde\rho$ is that $|\nabla\tilde\rho(P)| = 1$ for every $P \in \partial\Omega$. The process of replacing a given defining function ρ by another defining function $\tilde\rho$ raises the question of whether the geometric objects that we wish to study in this book are *dependent* on the choice of defining function. Fortunately they are not. A significant part of what we do in the present section is to establish this independence.

If $k \in \mathbb{N}$ and ρ is a C^k function that is a defining function for a domain $\Omega \subset \mathbb{R}^N$, then we say that Ω has C^k **boundary**. One might wonder whether other reasonable definitions of C^k boundary are equivalent to this one. What if the boundary is locally the graph of a C^k function? What if the boundary is given parametrically? What if, instead, we wish to think of the boundary as a regularly imbedded C^k manifold? Fortunately, all of these approaches are equivalent. We shall see the details of this claim in what follows.

Let us repeat that the condition defining tangent vectors simply mandates that $w \perp \nu_P$ at P. After all, we know from calculus that $\nabla\rho$, which is parallel to ν_P, points in a normal direction, and since the surface $\partial\Omega$

1.2. THE CONCEPT OF DEFINING FUNCTION

is $(N-1)$-dimensional the normal direction is uniquely determined; thus the normal direction, and hence the tangent vectors, are independent of the choice of ρ. In principle, this settles the issue of whether the tangent vectors to $\partial\Omega$ are well-defined. Nevertheless, this point is so important, and the point of view that we are considering so pervasive, that further discussion is warranted. The issue is this: If $\tilde{\rho}$ is another defining function for Ω, then it should give the same tangent vectors as ρ at any point $P \in \partial\Omega$. The key to seeing that this is so is to write $\tilde{\rho}(x) = h(x) \cdot \rho(x)$, for h a function that is non-vanishing near $\partial\Omega$. Then, for $P \in \partial\Omega$,

$$\begin{aligned}\sum_{j=1}^{N}(\partial\tilde{\rho}/\partial x_j)(P) \cdot w_j &= h(P) \cdot \left(\sum_{j=1}^{N}(\partial\rho/\partial x_j)(P) \cdot w_j\right) \\ &\quad + \rho(P) \cdot \left(\sum_{j=1}^{N}(\partial h/\partial x_j)(P) \cdot w_j\right) \\ &= h(P) \cdot \left(\sum_{j=1}^{N}(\partial\rho/\partial x_j)(P) \cdot w_j\right) \\ &\quad + 0, \end{aligned} \qquad (1.10)$$

because $\rho(P) = 0$. Thus w is a tangent vector at P vis-à-vis ρ if and only if w is a tangent vector vis-à-vis $\tilde{\rho}$. But why does h exist?

After a change of coordinates, it is enough to assume that we are dealing with a piece of $\partial\Omega$ that is a portion of flat, $(N-1)$-dimensional real hypersurface (just use the Implicit Function Theorem—see for example, Smith [1], Chapter 10, Theorem 8.2). That is, we may assume that the defining function ρ is given by

$$\rho(x_1, x_2, \ldots, x_N) = x_N.$$

If $\tilde{\rho}$ is another defining function for $\partial\Omega$ near P, then we set

$$h(x) = \begin{cases} \frac{1}{x_N}\tilde{\rho}(x) & \text{if } x_N \neq 0, \\ (\partial\tilde{\rho}/\partial x_N)(x) & \text{otherwise.} \end{cases} \qquad (1.11)$$

The Mean Value Theorem allows us to conclude that h is continuous along $\partial\Omega = \{x : x_N = 0\}$. The continuity of h and the vanishing of ρ on $\partial\Omega$ is sufficient for the product rule to hold for the partial derivatives of $\tilde{\rho} = h \cdot \rho$ (even if h is not differentiable at points of $\partial\Omega$). Thus (1.10) is verified.

It is natural to ask just how smooth the function h defined by (1.11) must be. Because of the presence of $\frac{\partial\tilde{\rho}}{\partial x_N}$ in its definition, the best that can be hoped is that h would be C^{k-1}, and that is in fact the case. Applying the generalized Leibniz's Rule (1.1) to

$$\frac{1}{x_N}\tilde{\rho}(x),$$

we see that, for any multi-index α with $|\alpha| \leq k-1$, if we write $\alpha = (\alpha', \alpha_N)$, then

$$\frac{\partial^\alpha}{\partial x^\alpha}\left(\frac{1}{x_N}\tilde{\rho}\right) = \sum_{i=0}^{\alpha_N}(-1)^i \frac{\alpha_N!}{(\alpha_N - i)!}\left(\frac{1}{x_N}\right)^{i+1} \frac{\partial^{(\alpha', \alpha_N - i)}\tilde{\rho}}{\partial x^{(\alpha', \alpha_N - i)}}. \quad (1.12)$$

We can evaluate the limit of

$$\left(\frac{1}{x_N}\right)^{i+1} \frac{\partial^{(\alpha', \alpha_N - i)}\tilde{\rho}}{\partial x^{(\alpha', \alpha_N - i)}}$$

as x_N approaches 0 by using l'Hôpital's Rule. The resulting limit is

$$\frac{1}{(i+1)\,i!} \frac{\partial^{(\alpha', \alpha_N - i)}\tilde{\rho}}{\partial x^{(\alpha', 1 + \alpha_N)}}.$$

Substituting this limit in the right-hand side of (1.12) and using the identity $\frac{1}{j+1} = \sum_{i=0}^{j} \frac{(-1)^i}{i+1}\binom{j}{i}$, we see that the limit as x_N approaches 0 of the right-hand side of (1.12) is

$$\frac{1}{1+\alpha_N} \frac{\partial^{(\alpha', 1+\alpha_N)}\tilde{\rho}}{\partial x^{(\alpha', 1+\alpha_N)}}. \quad (1.13)$$

The discussion above proves the following:

Proposition 1.2.3 *If $\rho, \tilde{\rho}$ are C^k defining functions for a domain Ω, then there is a C^{k-1} function h defined near $\partial\Omega$ such that $\rho = h \cdot \tilde{\rho}$.*

The result in Proposition 1.2.3 is sharp. In fact, the next lemma shows that if one defining function for the domain Ω is

$$\rho(x_1, x_2, \ldots, x_N) = x_N$$

and a second defining function, near the origin, is

$$\tilde{\rho}(x_1, x_2, \ldots, x_N) = x_N(1 + |x|^{k-1+\lambda}),$$

with $0 < \lambda < 1$, then $\tilde{\rho} \in C^k$ while $h = \tilde{\rho}/\rho \in C^{k-1} \setminus C^k$.

Lemma 1.2.4 *Let $P(x_1, x_2, \ldots, x_N) \not\equiv 0$ be a homogeneous polynomial of degree d, let ℓ be an integer, and let λ satisfy $0 < \lambda < 1$. If we set*

$$\Phi(x) = P(x)\,|x|^{\ell+\lambda},$$

then $\Phi \in C^{d+\ell} \setminus C^{d+\ell+1}$ if $d + \ell \geq 0$ and $\Phi \notin C^0$ if $d + \ell \leq -1$.

Proof: The fact that $d + \ell \leq -1$ implies $\Phi \notin C^0$ is seen immediately by using polar coordinates, so we may and shall assume that $d + \ell \geq 0$.

Since $P \not\equiv 0$, we can find $\gamma = (\gamma_1, \gamma_2, \ldots, \gamma_N)$ such that $P(\gamma) \neq 0$. Then we see that

$$f(t) = P(t\gamma)|t\gamma|^{\ell+\lambda} = \text{const} \cdot t^{d+\ell+\lambda}, \qquad t \in \mathbb{R},$$

1.2. THE CONCEPT OF DEFINING FUNCTION

which is clearly not in $C^{d+\ell+1}$, so we conclude that $\Phi \notin C^{d+\ell+1}$.

Using polar coordinates again, we see that if $d + \ell = 0$, then $\Phi \in C^0$. The rest of the lemma then follows by induction on $d+\ell$ using the fact that

$$\frac{\partial}{\partial x_i}|x|^m = m\, x_i\, |x|^{m-2},$$

which allows us to conclude the following:
A sum of functions of the form

$$\sum_{i=1}^{s} P_i(x)\, |x|^{\ell_i+\lambda}, \tag{1.14}$$

where each P_i is a homogeneous polynomial of degree d_i, each ℓ_i is an integer, and

$$D = d_1 + \ell_1 = d_2 + \ell_2 = \cdots = d_s + \ell_s$$

has as its partial derivatives functions also of the form (1.14), but with the value of D reduced by 1. ∎

This somewhat protracted discussion of a small technical point seems necessary because it is recorded incorrectly in many places in the literature.

Defining Functions with Unit Normal Vectors

Note that, while the ratio of any two C^k defining functions for a domain must be at least C^{k-1}, it is not the case that if a given function ρ is a C^k defining function for some domain, then it can be multiplied by *any* C^{k-1} function and have the resulting product be a C^k defining function. Generally, such a product can only be expected to be of class C^{k-1}. In particular, the defining function $\tilde{\rho}$ defined by $\tilde{\rho} = \rho/|\nabla\rho|$ is one order of differentiability less smooth than ρ, and that is the price to be paid in obtaining a unit normal vector by this change of defining function.

It is not necessary to sacrifice an order of differentiability to obtain a defining function with a unit length gradient vector. Provided the domain has a boundary that is C^2 or smoother, the signed distance to the boundary will do the job. This is not obvious; it is a somewhat subtle application of the Implicit Function Theorem as we will see below.

Lemma 1.2.5 *If Ω has a C^k $k \geq 2$, defining function ρ and $P \in \partial\Omega$, then there is a neighborhood U of P such that for all $Q \in U$ there is a unique point $Q' \in \partial\Omega \cap U$ with*

$$|Q - Q'| = \operatorname{dist}(Q, \partial\Omega). \tag{1.15}$$

Moreover, Q' is a C^{k-1} function of Q.

Proof: For fixed Q, the point Q' must be an extremum of the function $|Q - Q'|^2$ subject to the constraint $\rho(Q') = 0$. By the Lagrange Multiplier

Theorem (see for example Smith [1], Chapter 11, Theorem 4.7) there must exist a real number λ such that Q' satisfies

$$Q - Q' - \lambda \nabla \rho(Q') = 0, \qquad (1.16)$$
$$\rho(Q') = 0.$$

We will apply the Implicit Function Theorem to show that for Q near enough to P the equations (1.16) have a unique solution near P that is a C^{k-1} function of Q.

Consider the function $F : \mathbb{R}^{2N+1} \to \mathbb{R}^{N+1}$ defined by

$$\begin{pmatrix} Q \\ Q' \\ \lambda \end{pmatrix} \mapsto \begin{pmatrix} Q - Q' - \lambda \nabla \rho(Q') \\ \rho(Q') \end{pmatrix}.$$

The derivative (or Jacobian matrix) DF of F is represented by the matrix of size $(N+1) \times (2N+1)$ given by

$$\begin{pmatrix} I_{N \times N} & -I_{N \times N} - \lambda \operatorname{\mathbf{Hess}}[\rho(Q')] & \nabla \rho(Q') \\ 0_{1 \times N} & [\nabla \rho(Q')]^{\mathbf{t}} & 0 \end{pmatrix},$$

where $I_{\ell \times \ell}$ denotes the $\ell \times \ell$ identity matrix and $0_{\ell \times m}$ denotes the $\ell \times m$ zero matrix. So we have

$$DF(P, P, 0) = \begin{pmatrix} I_{N \times N} & -I_{N \times N} & \nabla \rho(P) \\ 0_{1 \times N} & [\nabla \rho(P)]^{\mathbf{t}} & 0 \end{pmatrix}.$$

Note that $F(P, P, 0) = 0 \in \mathbb{R}^{N+1}$. The Implicit Function Theorem will apply if

$$\det \begin{bmatrix} -I_{N \times N} & \nabla \rho(P) \\ [\nabla \rho(P)]^{\mathbf{t}} & 0 \end{bmatrix} \neq 0,$$

but this is true since one can readily verify by induction on ℓ that, for $v \in \mathbb{R}^\ell$,

$$\det \begin{bmatrix} -I_{\ell \times \ell} & v \\ v^{\mathbf{t}} & 0 \end{bmatrix} = (-1)^\ell |v|^2. \qquad \blacksquare$$

Theorem 1.2.6 *If Ω has a C^k, $k \geq 2$, defining function ρ, then the signed distance function*

$$\tilde{\rho}(Q) = \operatorname{sgn}(\rho(Q)) \operatorname{dist}(Q, \partial \Omega)$$

is a C^k function in a neighborhood of $\partial \Omega$.

1.2. THE CONCEPT OF DEFINING FUNCTION

Proof: Fix a point $P \in \partial \Omega$. Let us introduce the notation $g : U \to \mathbb{R}^N$ for the function mapping Q to Q', the nearest point of $\partial \Omega$, as in the preceding lemma. Equations (1.16) then become

$$x - g(x) = \lambda(x) \, (\nabla \rho) \circ g(x) \tag{1.17}$$
$$\rho \circ g(x) = 0. \tag{1.18}$$

It follows that also

$$\tilde{\rho}(x) = \lambda(x) \, |(\nabla \rho) \circ g(x)| \tag{1.19}$$
$$\tilde{\rho}^2(x) = |x - g(x)|^2 = \sum_{i=1}^{N} (x_i - g_i(x))^2. \tag{1.20}$$

Differentiating (1.18) with respect to x_j, we have

$$\sum_{i=1}^{N} \frac{\partial \rho}{\partial x_i} \circ g(x) \frac{\partial g_i}{\partial x_j} = 0, \tag{1.21}$$

for $j = 1, 2, \ldots, N$. Differentiating (1.20) with respect to x_j and using (1.17) and (1.21), we find that

$$\begin{aligned}
2 \tilde{\rho} \frac{\partial \tilde{\rho}}{\partial x_j} &= 2 \sum_{i=1}^{N} \left(\delta_{ij} - \frac{\partial g_i}{\partial x_j} \right) (x_i - g_i) \\
&= 2(x_j - g_j) - \sum_{i=1}^{N} \frac{\partial g_i}{\partial x_j} \lambda \frac{\partial \rho}{\partial x_i} \circ g(x) \\
&= 2(x_j - g_j) \\
&= 2 \lambda \frac{\partial \rho}{\partial x_j} \circ g.
\end{aligned}$$

Thus we have

$$\tilde{\rho} \nabla \tilde{\rho} = \lambda \, (\nabla \rho) \circ g.$$

By (1.19) we conclude that

$$\nabla \tilde{\rho} = \frac{(\nabla \rho) \circ g}{|(\nabla \rho) \circ g|} \tag{1.22}$$

must hold, at least on the complement of $\partial \Omega$. But the left-hand side of (1.22) is continuous and the right-hand side is C^{k-1}, so the equation must hold on $\partial \Omega$ as well. This shows that $\nabla \tilde{\rho}$ is C^{k-1}. Therefore, $\tilde{\rho}$ is C^k. ∎

Other proofs that the signed distance function is locally C^k, $k \geq 2$, if the domain is C^k can be found in Foote [1], Gilbarg and Trudinger [1], and Krantz and Parks [1].

Equivalent Notions of Smooth Boundaries

Let us now turn to the idea of the boundary being locally a graph. Fix a domain $\Omega \subset \mathbb{R}^N$ and a point $P \in \partial\Omega$. We say that $\partial\Omega$ is **locally the graph of a function near** P if there is an open N-dimensional neighborhood U of P, a Euclidean coordinate system $t_1, t_2, \ldots, t_N$ on U, and a function $u(t_1, t_2, \ldots, t_{N-1})$ such that

$$\partial\Omega \cap U = \{(t_1, t_2, \ldots, t_{N-1}, u(t_1, t_2, \ldots, t_{N-1})) \in U\}.$$

This definition is most convenient if the set U is a product region in the t-coordinate system.

Observe that if $\partial\Omega$ is locally the graph of a C^k function u near P, and if the positive t_N axis points *out* of the domain, then the function $\rho_U(t) \equiv t_N - u(t_1, t_2, \ldots, t_{N-1})$ behaves like a defining function for Ω near P. In other words, if we express ρ_U in the original spatial coordinates $x_1, x_2, \ldots, x_N$ on U, then $\rho_U(x) > 0$ on $U \cap {}^c\Omega$, $\rho_U(x) < 0$ on $U \cap \Omega$, and $\nabla \rho_U \neq 0$ on $\partial\Omega$. If $\partial\Omega$ is locally the graph of a C^k function near every point of $\partial\Omega$, then we cover $\partial\Omega$ by a family of open sets U_α, each with its associated defining function ρ_α. Now apply the Partition of Unity Theorem 1.1.8 to this open covering of the set $E = \partial\Omega$. We obtain compactly supported, C^∞ functions ϕ_j that sum to 1 on $\partial\Omega$. Setting

$$\rho(x) = \sum_{j=1}^{\infty} \phi_j(x) \rho_{\alpha_j}(x),$$

where ϕ_j is supported on U_{α_j}, we obtain a function defined on a full neighborhood of $\partial\Omega$ that satisfies the properties of a defining function on that neighborhood. It is then elementary to extend the defining function to all of space if needed.

The preceding machinery may be run in reverse. If ρ is a C^k defining function for Ω, then fix a point $P \in \partial\Omega$. Assume the coordinates have been rotated so that ν_P points in the direction of the positive x_N axis. The Implicit Function Theorem then implies that there is a neighborhood V of $(P_1, P_2, \ldots, P_{N-1})$ and a C^k function $u(x_1, x_2, \ldots, x_{N-1})$ such that

$$\rho(x_1, x_2, \ldots, x_{N-1}, u(x_1, x_2, \ldots, x_N)) = 0$$

on V. But this just says that $\partial\Omega$ is locally the graph of the function u near P.

Another common method for specifying a hypersurface in space is parametrically. Let $W \subset \mathbb{R}^{N-1}$ be an open set and let $\psi_1, \psi_2, \ldots, \psi_N$ be C^k, real-valued functions on W. Set $t' = (t_1, t_2, \ldots, t_{N-1})$ for the coordinate on W. If the mapping

$$\Psi : t' \mapsto (\psi_1(t'), \psi_2(t'), \ldots, \psi_N(t'))$$

is one-to-one and if the matrix $(\partial \psi_i/x_k)$ has rank $N-1$ at each point of W, then the image M of Ψ is called a C^k **hypersurface given parametrically**.

Now let us relate the notion of parametrically given hypersurface to some of the preceding ideas. With notation as in the last paragraph, fix a point $P = (P_1, P_2, \ldots, P_N)$ in the surface M. The matrix

$$\begin{pmatrix} \frac{\partial \psi_1}{\partial t_1} & \frac{\partial \psi_1}{\partial t_2} & \cdots & \frac{\partial \psi_1}{\partial t_{N-1}} \\ \frac{\partial \psi_2}{\partial t_1} & \frac{\partial \psi_2}{\partial t_2} & \cdots & \frac{\partial \psi_2}{\partial t_{N-1}} \\ \vdots & \vdots & & \vdots \\ \frac{\partial \psi_N}{\partial t_1} & \frac{\partial \psi_N}{\partial t_2} & \cdots & \frac{\partial \psi_N}{\partial t_{N-1}} \end{pmatrix} \quad (1.23)$$

has rank $N-1$. For convenience of notation, let us suppose that the submatrix consisting of the first $N-1$ rows of (1.23) is non-singular. Then the Inverse Function Theorem tells us that the mapping

$$t' \mapsto (\psi_1(t'), \psi_2(t'), \ldots, \psi_{N-1}(t'))$$

has a C^k inverse function near $(P_1, P_2, \ldots, P_{N-1})$. We will denote that inverse function by Φ. Then near P we can write M as the graph of a function:

$$(x_1, x_2, \ldots, x_{N-1}, \psi_N[\Phi(x_1, x_2, \ldots, x_{N-1})]),$$

$(x_1, x_2, \ldots, x_{N-1})$ near $(P_1, P_2, \ldots, P_{N-1})$.

Thus we have a third method for thinking of the boundary of a domain: as a locally parametrized hypersurface. We see that all three points of view are equivalent, and one passes back and forth among them using local coordinates and the Implicit Function Theorem. A fourth point of view is to consider the boundary as a locally regularly imbedded C^k manifold (see Hirsch [1]). Since that method will not be used in this book, we shall not discuss it in detail.

1.3 Measure Theory

In this section we review some basic measure theory, mainly to set our notation and terminology rather than to provide a complete treatment. The reader interested in an exhaustive presentation should see Federer [4].

Definition 1.3.1 *Let X be a set and let S be the set of all subsets of X. By a* **measure on a set** *X we mean a function*

$$m : S \to [0, +\infty]$$

satisfying the conditions

(i) $m(\emptyset) = 0$,

(ii) $m(A) \leq \sum_{B \in \mathcal{B}} m(B)$ if $\mathcal{B}$ is at most countable and $A \subset \bigcup_{B \in \mathcal{B}} B$.

If m is a measure on a set X, it is convenient to postulate, as we have, that every subset $A \subset X$ has associated a value $m(A)$, but some subsets are more important than others.

Definition 1.3.2 *Let m be a measure on the set X. A subset $A \subset X$ is said to be m-**measurable** if*

$$m(B) = m(B \cap A) + m(B \setminus A) \tag{1.24}$$

holds for every $B \subset X$.

It is obvious from this definition (and the definition of a measure) that any set having measure 0 is m-measurable. The sets of measure 0 are negligible in measure theory, and any statement $S(x)$ containing a free variable from a set X with a measure m is said to hold for m-**almost all** x or m-**almost everywhere** if

$$m(\{x \in X : S(x) \text{ is false}\}) = 0.$$

The main result concerning the m-measurable sets is expressed by the following definition and theorem (see Federer [4], §2.1.3).

Definition 1.3.3 *By a σ-**algebra** in a set X we mean a family $\mathcal{S}$ of subsets of X such that*

(i) $\emptyset \in \mathcal{S}$, $X \in \mathcal{S}$,

(ii) *if $A \in \mathcal{S}$, then $^cA \in \mathcal{S}$,*

(iii) *if $\mathcal{B} \subset \mathcal{S}$ is at most countable, then $\bigcup_{B \in \mathcal{B}} B$ and $\bigcap_{B \in \mathcal{B}} B$ are elements of $\mathcal{S}$.*

Theorem 1.3.4 *If m is a measure on X, then the family of m-measurable sets forms a σ-algebra in X and m is countably additive on pairwise disjoint collections of m-measurable sets.*

Note that many texts would use the term *outer measure* for what we have called a measure. Because of the preceding theorem, if m is a measure in our sense, then when restricted to the σ-algebra of m-measurable sets, m is a measure in the alternative sense of a function defined on a σ-algebra and countably additive on disjoint unions.

Example 1.3.5 *Let $X = \mathbb{R}$. For any non-empty open interval $(a, b) \subset \mathbb{R}$, set*

$$m((a,b)) = \begin{cases} b - a & \text{if } -\infty < a < b < +\infty, \\ +\infty & \text{if } a = -\infty \text{ or } b = +\infty. \end{cases}$$

Since any open subset of $\mathbb{R}$ is uniquely expressible as an at most countable union of pairwise disjoint intervals, m will be well-defined on open sets if it

1.3. MEASURE THEORY

is extended additively to pairwise disjoint unions of open intervals. Finally, for an arbitrary $A \subset \mathbb{R}$, we define

$$m(A) = \inf\{m(V) : V \text{ is open and } A \subset V\}.$$

We leave it to the reader to verify that this defines a measure. This measure is called **Lebesgue measure** and will be denoted by $\mathcal{L}$. Sets that are $\mathcal{L}$-measurable are referred to as **Lebesgue measurable sets**.

Measure theory becomes more interesting for the analyst when there is also a topology on the space. Recall that, if X is a topological space, then the **Borel sets** are the elements of the smallest σ-algebra that contains the open sets.

Definition 1.3.6

(i) *A measure m is* **regular** *if every set A is a subset of a measurable set B having the same measure, that is $m(A) = m(B)$.*

(ii) *Let X be a topological space and let m be a measure on X. We say that m is* **Borel regular** *if all Borel sets are m-measurable and for every subset $A \subset X$ there exits a Borel set B with*

$$A \subset B \quad \text{and} \quad m(A) = m(B). \tag{1.25}$$

(iii) *Let X be a locally compact Hausdorff space and let m be a measure on X. We say that m is a* **Radon measure** *if all Borel sets are m-measurable and the following conditions hold:*

- *If $K \subset X$ is compact, then $m(K) < \infty$.*

- *If $V \subset X$ is open, then*

$$m(V) = \sup\{m(K) : K \text{ is compact and } K \subset V\}.$$

- *For any $A \subset X$, $m(A) = \inf\{m(V) : V \text{ is open and } A \subset V\}.$*

The interested reader should consult Federer [4], §2.2.4, to see the proof that any sufficiently rich Borel regular measure on a complete separable metric space must have a non-measurable set. The proof in Federer [4] uses the Axiom of Choice in the form of the Well-Ordering Principle. In general, the question of measurability is inextricably tied to the study of abstract set theory. In a notable paper, R. M. Solovay [1] has shown that if there exists an inaccessible cardinal, then there is a model of set theory in which all subsets of $\mathbb{R}$ are Lebesgue measurable. Obviously, one would now want to know whether or not an inaccessible cardinal exists, but unfortunately it is impossible to show that the existence of an inaccessible cardinal is consistent with the usual axioms of set theory (see Jech [1], Theorem 27, p. 85).

Proposition 1.3.7 (Inner Regularity of Radon Measures)
Suppose m is a Radon measure on X. If A is m-measurable, $m(A) < \infty$, and $0 < \epsilon$, then there exists a compact set $K \subset A$ such that $m(A) < m(K) + \epsilon$.

Proof: There exists an open set U such that $A \subset U$ and $m(U \setminus A) < \epsilon/2$. With the choice of U now fixed, there exists another open set V such that $U \setminus A \subset V$ and $m(V) < \epsilon/2$. We can also find a compact set K' with $K' \subset U$ and $m(U \setminus K') < \epsilon/2$. Finally, we set $K = K' \setminus V$, so K is compact, $K \subset A$, and we have

$$m(A \setminus K) = m(A \setminus (K' \setminus V)) \leq m(U \setminus K') + m(V) < \epsilon. \qquad\blacksquare$$

The following result is a useful tool (see Federer [4], §2.3.2(9)).

Theorem 1.3.8 (Carathéodory's Criterion) *If m is a measure on the metric space X, then all Borel subsets of X are m-measurable if and only if*

$$m(A \cup B) \geq m(A) + m(B) \qquad (1.26)$$

holds whenever $A, B \subset X$ with $\mathrm{dist}(A, B) > 0$.

Suslin Sets

Carathéodory's Criterion shows us that all Borel subsets of $\mathbb{R}$ are Lebesgue measurable. It is natural to ask whether the Borel sets are the only Lebesgue measurable sets. Answering that and related questions leads one into the study of descriptive set theory and away from the topics in this book. We shall only briefly touch on these matters.

Definition 1.3.9 *A set $A \subset \mathbb{R}^M$ will be called a **Suslin set** if there exists a Borel set $B \subset \mathbb{R}^M \times \mathbb{R}^N$, for some N, such that A is the image of B under projection onto the first factor.*

Originally, Suslin sets were known as *analytic sets*[1] (and still are known as *ensembles analytiques* in French). The classic book on the subject is Lusin [1].[2] The main facts are contained in the following theorem (see Federer [4], §2.2.10–2.2.13).

---

[1] Folklore has it that as a graduate student Suslin was given the task of proofreading a paper by the already famous Lebesgue in which Lebesgue had stated as a lemma, the proof of which was omitted because it was obvious, the assertion that the projection of a Borel set is a Borel set. Suslin realized that the assertion was in fact false and produced a counterexample. Jech [1], page 593, confirms at least the broad outline of this story and cites the original papers.

[2] The author's name is also transliterated as Luzin.

1.3. MEASURE THEORY

Theorem 1.3.10 [3]

(i) *Every Borel set is a Suslin set.*

(ii) *If A is a Suslin set in $\mathbb{R}^{M_1}$ and $f : A \to \mathbb{R}^{M_2}$ is continuous, then $f(A)$ is a Suslin set.*

(iii) *There exists a Suslin set that is not a Borel set.*

(iv) *If m is a measure on $\mathbb{R}^M$ such that all Borel sets are m-measurable, then every Suslin set is m-measurable.*

Measurable Functions

Definition 1.3.11 *Suppose that m is a measure on X and that Y is a topological space. A function defined on m-almost all of X and having values in Y is said to be m-**measurable** if $f^{-1}(U)$ is m-measurable for every open $U \subset Y$. A pair of m-measurable functions f and g are considered to be equal (for purposes of measure theory) if they agree at m-almost all points of X.*

It is a series of elementary exercises to verify that the set of real-valued (or complex-valued or $\mathbb{R}^N$-valued) measurable functions is closed under the usual algebraic operations—of course, due care must be exercised in dealing with division by zero. It is another elementary exercise to verify that, for measurable functions taking values in a metric space, the pointwise almost everywhere limit of a sequence of measurable functions is measurable.

While the condition of measurability is not very restrictive, it is still sufficient to ensure some nice properties, as the next two theorems show (see Federer [4], §2.3.5 and §2.3.7).

Theorem 1.3.12 (Lusin's Theorem) *If m is a Radon measure on a locally compact Hausdorff space X, $f : X \to \mathbb{R}$ is m-measurable, A is a compact set with $m(A) < \infty <$, and $0 < \epsilon$, then there exists a compact set C such that $m(A \setminus C) < \epsilon$ and f restricted to C is continuous.*

Theorem 1.3.13 (Egoroff's Theorem) *If m is a measure on X with $m(X) < \infty$, the m-measurable functions $f_i : X \to \mathbb{R}$, $i = 1, 2, \ldots$, converge m-almost everywhere to $g : X \to \mathbb{R}$, and $0 < \epsilon$, then there exists an m-measurable set B such that $m(X \setminus B) < \epsilon$ and the convergence of f_i to g is uniform on B.*

---

[3]The reader comparing our definition to that in Federer [4], §2.2.10, should note that Federer's space $\mathcal{N}$ is homeomorphically mapped onto a Borel subset of $\mathbb{R}$ by the mapping

$$(n_1, n_2, \ldots) \mapsto \sum_{i=1}^{\infty} \exp\left(-\sum_{j=1}^{i} n_j\right).$$

One of the main reasons for the invention of measure theory is the extension of the notion of integration from the relatively simple Riemann integration taught in freshman calculus to a more general and flexible tool useful in modern analysis.

Definition 1.3.14 *Let m be a measure on X.*

(i) *By an m-**step function** we mean an m-measurable function, s, taking at most countably many real values such that*

$$\sum_{y \in \mathbb{R}} y \cdot m(s^{-1}\{y\}) \qquad (1.27)$$

is well-defined, using the convention $0 \cdot \infty = \infty \cdot 0 = 0$. That is, the series in (1.27) either converges, diverges to $+\infty$, or diverges to $-\infty$.

(ii) *If f is an m-measurable function defined on m-almost all of X and taking values in $[-\infty, +\infty]$, we define the **upper integral of** f to be the infimum of the expressions in (1.27) as s varies over all step functions that are greater than or equal to f at m-almost all points of X. The upper integral of f is denoted by*

$$\int^* f\, dm.$$

(iii) *Likewise, if f is an m-measurable function defined on m-almost all of X and taking values in $[-\infty, +\infty]$, we define the **lower integral of** f to be the supremum of the expressions in (1.27) as s varies over all step functions that are less than or equal to f at m-almost all points of X. The lower integral of f is denoted by*

$$\int_* f\, dm.$$

(iv) *If $\int^* f\, dm = \int_* f\, dm$, then we denote the common value of the upper and lower integrals by*

$$\int f\, dm.$$

(v) *We say that f is m-**integrable** if the upper and lower integrals of f are equal and finite. Often the measure m is clear from context and we will say simply that f is **integrable***

(vi) *Suppose X is a topological space and f is an m-measurable function on X. If for each $p \in X$ there exists an open set U with $p \in U$ and such that $f \cdot \chi_U$ is m-integrable, then we say f is **locally m-integrable** or **locally integrable** when the measure is clear from the context.*

1.3. MEASURE THEORY

The preceding integration of m-measurable functions behaves well with respect to limits as the next three classical theorems tell us.

Theorem 1.3.15 (Fatou's Lemma) *Suppose m is a measure on X. If $f_1, f_2, \ldots$ are non-negative m-measurable functions on X, then*

$$\liminf_{i \to \infty} \int_X f_i \, dm \leq \int_X \liminf_{i \to \infty} f_i \, dm. \tag{1.28}$$

Theorem 1.3.16 (Lebesgue's Monotone Convergence Theorem) *Suppose m is a measure on X. If $f_1, f_2, \ldots$ are non-negative m-measurable functions on X such that*

$$f_i(x) \leq f_{i+1}(x) \quad \text{holds for } m\text{-almost all } x \in X,$$

then

$$\lim_{i \to \infty} \int_X f_i \, dm = \int_X \lim_{i \to \infty} f_i \, dm. \tag{1.29}$$

Theorem 1.3.17 (Lebesgue's Dominated Convergence Theorem) *Suppose m is a measure on X. If $f_1, f_2, \ldots$ and g m-measurable functions on X such that*

$$f_i(x) \quad \text{converges for } m\text{-almost all } x \in X,$$

$$|f_i(x)| \leq g(x) \quad \text{holds for } m\text{-almost all } x \in X,$$

and

$$\int_X g \, dm < \infty,$$

then

$$\lim_{i \to \infty} \int_X f_i \, dm = \int_X \lim_{i \to \infty} f_i \, dm. \tag{1.30}$$

Product Measures and Fubini's Theorem

Definition 1.3.18 *Suppose m is a measure on X and n is a measure on Y. The **product measure** $m \times n$ is the measure on $X \times Y$ defined for $S \subset X \times Y$ by setting $(m \times n)(S)$ equal to the infimum of the set of numbers of the form*

$$\sum_{i=1}^{\infty} m(A_i) \cdot n(B_i),$$

where $A_1, A_2, \ldots$ are m-measurable subsets of X, $B_1, B_2, \ldots$ are n-measurable subsets of Y, and

$$S \subset \bigcup_{i=1}^{\infty} A_i \times B_i. \tag{1.31}$$

In (1.31) we use the convention that $0 \cdot \infty = \infty \cdot 0 = 0$.

The reader can find the proof of the following theorem in Federer [4], §2.6.2.

Theorem 1.3.19 (Fubini) *Suppose m is a measure on X and n is a measure on Y.*

(i) *$m \times n$ is a regular measure on $X \times Y$.*

(ii) *The cartesian product of measurable sets is $m \times n$ measurable, and the measure of a cartesian product is the product of the measures.*

(iii) *If $S \subset X \times Y$ is a countable union of sets of finite measure, then*

(a) *$\{x : (x,y) \in S\}$ is m-measurable for n-almost every $y \in Y$,*

(b) *$\{y : (x,y) \in S\}$ is n-measurable for m-almost every $x \in X$,*

(c)

$$(m \times n)(S) = \int m(\{x : (x,y) \in S\})\, dn$$
$$= \int n(\{y : (x,y) \in S\})\, dm.$$

(iv) *If f is an $m \times n$ integrable function such that $\{(x,y) : f(x,y) \neq 0\}$ is a countable union of sets of finite measure, then*

$$\int f\, d(m \times n) = \int \left(\int f\, dm \right) dn = \int \left(\int f\, dn \right) dm.$$

Definition 1.3.20 *Lebesgue measure on $\mathbb{R}^N$ will be denoted by $\mathcal{L}^N$ and is defined inductively by setting*

$$\mathcal{L}^N = \mathcal{L} \times \mathcal{L}^{N-1}.$$

Integrals with respect to Lebesgue measure will often be written using the notation

$$\int f(x_1, x_2, \ldots, x_N)\, dx_1\, dx_2 \ldots dx_N = \int f\, d\mathcal{L}^N.$$

Lebesgue Spaces

Definition 1.3.21 *Let m be a measure on X. Suppose $0 < p < \infty$. For a measurable $f : X \to \mathbb{C}$ we define*

$$\|f\|_p = \int_X |f|^p\, dm. \tag{1.32}$$

*The **Lebesgue space** $L^p(m)$ is the set of measurable $f : X \to \mathbb{C}$ for which $\|f\|_p < \infty$. We call $\|\cdot\|_p$ the L^p-**norm**.*

   In case $X = \mathbb{R}^N$ and $m = \mathcal{L}^N$, it is customary to write $L^p(\mathbb{R}^N)$, instead of $L^p(\mathcal{L}^N)$.

1.3. MEASURE THEORY

In the preceding definition, the value $p = \infty$ is conspicuously absent. A separate definition is required:

Definition 1.3.22 *Let m be a measure on X. For a measurable $f : X \to \mathbb{C}$ we define*

$$\|f\|_\infty = \inf \{t : 0 < t < \infty \text{ and } m(\{x : t < |f(x)|\}) = 0\}. \tag{1.33}$$

We set $L^\infty(m)$ equal to the family of measurable $f : X \to \mathbb{C}$ for which $\|f\|_\infty < \infty$, we call $\|\cdot\|_\infty$ the L^∞-norm, and in case $X = \mathbb{R}^N$, $m = \mathcal{L}^N$, we write $L^\infty(\mathbb{R}^N)$ instead of $L^\infty(\mathcal{L}^N)$.

For future reference we state some of the classic results concerning the Lebesgue spaces.

Theorem 1.3.23 *Let m be a measure on X. For $1 \leq p \leq \infty$, the Lebesgue space $L^p(m)$ is a complete metric space with the distance between f and g given by $\|f - g\|_p$.*

Theorem 1.3.24 *Let X be a locally compact Hausdorff space and let m be a Radon measure on X. For $1 \leq p < \infty$, the set of continuous functions with compact support $C_c(X)$ is dense in the Lebesgue space $L^p(m)$.*

For $1 < p < \infty$, it is not obvious that the function $d(f, g) = \|f - g\|_p$ on $L^p(m)$ satisfies the triangle inequality and thus defines a metric. The underlying result is the following:

Theorem 1.3.25 (Minkowski's Inequality) *Let m be a measure on X and let f and g be m-measurable functions taking values in $[0, \infty]$. For $1 < p < \infty$, it holds that*

$$\left(\int_X (f+g)^p \, dm\right)^{1/p} \leq \left(\int_X f^p \, dm\right)^{1/p} + \left(\int_X g^p \, dm\right)^{1/p}. \tag{1.34}$$

Another of the important inequalities in analysis is that of Hölder which applies to pairs of specially related exponents as in the next definition.

Definition 1.3.26 *Numbers p and q with $1 < p, q < \infty$ are called **conjugate exponents** if*

$$\frac{1}{p} + \frac{1}{q} = 1. \tag{1.35}$$

Theorem 1.3.27 (Hölder's Inequality) *Let m be a measure on X and let f and g be m-measurable functions taking values in $[0, \infty]$. If p and q are conjugate exponents with $1 < p, q < \infty$, then it holds that*

$$\int_X fg \, dm \leq \left(\int_X f^p \, dm\right)^{1/p} \times \left(\int_X g^q \, dm\right)^{1/q}. \tag{1.36}$$

Definition 1.3.28 *It is also customary to say that p and q are conjugate exponents if $p = 1$ and $q = \infty$ or if $p = \infty$ and $q = 1$.*

Smoothing

The goal of a smoothing operation is to achieve the ability to approximate the objects in a category of interest by objects that have better properties than generally are expected in the category. The prototype of such a theorem is the Weierstrass Approximation Theorem (see for example Krantz [2], Theorem 9.8) in which the category in question is that of continuous functions on a compact interval and the approximating objects are the polynomials. The impetus to develop smoothing operations is the hope that one will be able to extend proofs based on intuition concerning smooth objects to the entire category under consideration.

The main tool used in smoothing is the operation of convolution:

Definition 1.3.29 *For Lebesgue measurable functions $f, g : \mathbb{R}^N \to \mathbb{R}$ the* **convolution** *of f and g is the function $f * g$ given by*

$$(f * g)(x) = \int f(x - y)g(y) \, dy_1 dy_2 \ldots dy_N \tag{1.37}$$

for all x such that the integral is defined.

It is not immediately apparent what the domain of the convolution of f and g will be, nor what estimates the convolution might satisfy. It clearly follows from Hölder's Inequality (1.36) that if $f \in L^p(\mathbb{R}^N)$ and $g \in L^q(\mathbb{R}^N)$, for a pair of conjugate exponents p and q, then $(f * g)(x)$ is defined for all $x \in \mathbb{R}^N$. The situation is more subtle than that however, and we have the following result as a consequence of Fubini's Theorem (see Rudin [2], Chapter 7):

Theorem 1.3.30 *If $f, g \in L^1(\mathbb{R}^N)$, then $(f * g)(x)$ is defined for almost every $x \in \mathbb{R}^N$ and $f * g \in L^1(\mathbb{R}^N)$ with*

$$\|f * g\|_1 \leq \|f\|_1 \|g\|_1 \tag{1.38}$$

To apply convolution to smoothing, recall the function $\phi_{[-1,1]}$ constructed in Section 1.1. The important properties are

$$\phi_{[-1,1]} \geq 0, \tag{1.39}$$

$$\phi_{[-1,1]}(x) > 0 \text{ if and only if } |x| < 1, \tag{1.40}$$

$$\phi_{[-1,1]} \in C^\infty(\mathbb{R}). \tag{1.41}$$

By multiplying $\phi_{[-1,1]}$ by a constant, but without changing notation, we can insure the following additional property holds:

$$\int_{-\infty}^{\infty} \phi_{[-1,1]}(x) \, dx = 1. \tag{1.42}$$

1.3. MEASURE THEORY

For each positive integer N and each $\sigma > 0$, we define the function ψ_σ called a **Friedrichs mollifier** by setting

$$\psi_\sigma(x_1,\ldots,x_N) = \sigma^{-N}\prod_{i=1}^{N}\phi_{[-1,1]}(x_i/\sigma).$$

Such a family of functions is used to produce smooth approximations to continuous functions.

The main result on smoothing is the following (see Smith [1], Chapter 13, Section 12):

Theorem 1.3.31 *Let $f: \mathbb{R}^N \to \mathbb{R}$ be locally integrable.*

(i) *For each $\sigma > 0$, the function*

$$(f*\psi_\sigma)(x) = \int_{-\infty}^{\infty}\cdots\int_{-\infty}^{\infty} f(y)\psi_\sigma(x-y)\,dy_1\ldots dy_N$$

is C^∞.

(ii)
$$\operatorname{supp}(f*\psi_\sigma) \subset \left\{x : \operatorname{dist}(x,\operatorname{supp}(f)) \leq \sqrt{N}\sigma\right\}.$$

(iii) *If f is in L^p, then $f*\psi_\sigma$ converges to f in the L^p-norm as σ decreases to 0.*

(iv) *If f is continuous and C is a compact set, then $f*\psi_\sigma$ converges uniformly to f as σ decreases to 0.*

(v) *If f is uniformly continuous on $\mathbb{R}^N$, then $f*\psi_\sigma$ converges uniformly to f as σ decreases to 0.*

Part (iii) is proved using the following **generalized Minkowski Inequality** (see Zygmund [1], Chapter I, Equation (9.12)):

$$\left\{\int\left|\int h(x,y)\,dy\right|^p dx\right\}^{\frac{1}{p}} \leq \int\left\{\int|h(x,y)|^p\,dx\right\}^{\frac{1}{p}}dy, \qquad (1.43)$$

which expresses in a very general way the fact that the norm of a sum is less than or equal to the sum of the norms.

Chapter 2

Domains with Smooth Boundaries

2.1 The Tangent Bundle and Normal Bundle of the Boundary

Notation 2.1.1 *We will denote the sphere of radius $r > 0$ about $x \in \mathbb{R}^N$ by $\mathbb{S}(x,r)$. The unit sphere in $\mathbb{R}^N$ will be denoted by $\mathbb{S}^{N-1}$ so that $\mathbb{S}^{N-1} = \mathbb{S}(0,1)$.*

Definition 2.1.2 *Let ρ be a defining function for the domain Ω. The **Gauss map** is the function $\mathcal{G} : \partial\Omega \to \mathbb{S}^{N-1}$ defined by*

$$\mathcal{G}(x) = \frac{\nabla\rho(x)}{|\nabla\rho(x)|} \tag{2.1}$$

for $x \in \partial\Omega$. It is clear that $\mathcal{G}$ extends via the same formula to a neighborhood of $\partial\Omega$. We will use the same notation $\mathcal{G}$ for the extended function. The components of the Gauss map will be denoted by

$$\mathcal{G}(x) = (G_1(x), G_2(x), \ldots, G_N(x)). \tag{2.2}$$

Clearly, $\mathcal{G}(x)$ is a unit vector normal to $\partial\Omega$ at $x \in \partial\Omega$. Note that if ρ, and thus Ω, is in smoothness class C^k, then $\mathcal{G}$ is in smoothness class C^{k-1}.

In general differential topology (see Hirsch [1] for example), the normal bundle of a submanifold of Euclidean space consists of all pairs (p, w) where p is a point of the submanifold and w is a normal vector to the submanifold at the point p. For the boundary of a domain in Euclidean space, the normal bundle has the following natural definition.

Definition 2.1.3 *Let Ω be a domain with defining function $\rho \in C^k$, $k \geq 1$. The **normal bundle** of $\partial\Omega$, denoted $\mathrm{Nor}(\partial\Omega)$, is the subset of $\mathbb{R}^N \times \mathbb{R}^N$*

consisting of the points

$$\{(x, t\mathcal{G}(x)) : x \in \mathbb{R}^N \text{ with } \rho(x) = 0,\ t \in \mathbb{R}\}.$$

For $x \in \partial\Omega$,
$$\mathrm{Nor}(\partial\Omega, x) = \{t\mathcal{G}(x) : t \in \mathbb{R}\}$$
is the **normal space** *to* $\partial\Omega$ *at* x.

If $x(u_1, u_2, \ldots, u_{N-1}) = x(u)$ is a local C^k parametrization of the points with $\rho(x) = 0$, then
$$(x(u), t\mathcal{G}(x(u)))$$
is a local C^{k-1} parametrization of the normal bundle. It is easy to see that, if the differential of the parametrization $u \mapsto x(u)$ is of rank $N - 1$, then the differential of the parametrization $(u, t) \mapsto (x(u), t\mathcal{G}(x(u)))$ is of rank N.

Recall that the tangent vectors to the boundary of a domain were defined in Definition 1.2.2. The following definition is consistent with that earlier usage.

Definition 2.1.4 *Let Ω be a domain with defining function $\rho \in C^k$, $k \geq 1$. The* **tangent bundle** *of $\partial\Omega$, denoted* $\mathrm{Tan}(\partial\Omega)$, *is the subset of $\mathbb{R}^N \times \mathbb{R}^N$ consisting of the points*

$$\{(x, v) : x \in \mathbb{R}^N \text{ with } \rho(x) = 0,\ v \in \mathbb{R}^N \text{ with } v \cdot \mathcal{G}(x) = 0\}.$$

The tangent bundle of a manifold consists of pairs (p, v) where p is a point of the manifold and v is a tangent vector to the manifold at the point p. In an abstract manifold it is not assumed *a priori* that there is an imbedding into Euclidean space, so the tangent space must be described in terms of differentiation operators. For the boundary of a domain, there is a natural set of N tangential differentiation operators associated with a given defining function. We will see that they span the $(N-1)$-dimensional space of tangential differentiation operators.

Definition 2.1.5 *Let Ω be a domain with defining function $\rho \in C^k$, $k \geq 1$. For $i = 1, 2, \ldots, N$, we define the linear differential operators*

$$\delta_i = \frac{\partial}{\partial x_i} - G_i(x) \sum_{j=1}^N G_j(x) \frac{\partial}{\partial x_j}. \tag{2.3}$$

That the operators δ_i are tangential differentiation operators on
$$\partial\Omega = \{x : \rho(x) = 0\}$$
is shown by the next lemma.

2.1. THE BUNDLES ASSOCIATED WITH A BOUNDARY

Lemma 2.1.6 *Suppose $x_0 \in \partial\Omega$ and f and g are C^1 functions on some neighborhood U of x_0 such that*

$$f(x) = g(x) \quad \text{for } x \in U \cap \partial\Omega,$$

then

$$\delta_i f(x_0) = \delta_i g(x_0), \quad i = 1, 2, \ldots, N.$$

Proof: Set $h = f - g$ in U. Since $h(x) = 0$ for $x \in U \cap \partial\Omega$, we have $\nabla h(x_0) = \lambda \mathcal{G}(x)$, for some $\lambda \in \mathbb{R}$, that is,

$$\frac{\partial h}{\partial x_i}(x_0) = \lambda G_i(x_0),$$

for $i = 1, 2, \ldots, N$. We compute

$$\delta_i h = \lambda G_i(x_0) \left[1 - \sum_{j=1}^{N} |G_j(x_0)|^2 \right] = 0. \qquad \blacksquare$$

The next proposition tells us that the set of δ_i is $(N-1)$-dimensional and tells us which of the δ_i form a basis.

Proposition 2.1.7 *For x such that $\nabla \rho(x) \neq 0$, the set of differentiation operators $\{\delta_1, \delta_2, \ldots, \delta_N\}$ is linearly dependent, while if $G_{i_0} \neq 0$, then*

$$\delta_1, \delta_2, \ldots, \delta_{i_0-1}, \delta_{i_0+1}, \delta_{i_0+2}, \ldots, \delta_N \qquad (2.4)$$

are linearly independent.

Proof: Since $\nabla \rho(x) \neq 0$, it follows that $\mathcal{G}(x)$ is defined. Not all $G_i(x)$ are zero, but the linear combination

$$\sum_{i=1}^{N} G_i \delta_i.$$

is easily seen to vanish. Thus $\{\delta_1, \delta_2, \ldots, \delta_N\}$ is a linearly dependent set.

To see that the second part of the proposition is true, we suppose for convenience that $i_0 = N$, so $G_N \neq 0$. If

$$0 = \sum_{i=1}^{N-1} c_i \delta_i = \sum_{i=1}^{N-1} c_i \frac{\partial}{\partial x_i} - \sum_{i=1}^{N-1} c_i G_i \sum_{j=1}^{N} G_j(x) \frac{\partial}{\partial x_j},$$

then the coefficient of $\partial/\partial x_N$ is $-(\sum_{i=1}^{N-1} c_i G_i) G_N$, so $G_N \neq 0$ implies

$$\sum_{i=1}^{N-1} c_i G_i = 0.$$

But then $\sum_{i=1}^{N-1} c_i \delta_i$ simplifies to just $\sum_{i=1}^{N-1} c_i \frac{\partial}{\partial x_i}$. We conclude that $c_1 = c_2 = \cdots = c_{N-1} = 0$. Thus $\delta_1, \delta_2, \ldots, \delta_N$ are linearly independent. ∎

In the next section we will see the utility of the operators δ_i in giving some simple formulas for the curvature of the boundary of a domain.

The previous discussion has shown that the differentiation operators can be used to represent the abstract tangent space of the boundary of a domain. The correspondence to the more concrete Definition 2.1.4 is demonstrated by replacing the differentiation operators $\frac{\partial}{\partial x_j}$ in the definition of δ_i by the standard basis vectors $\mathbf{e}_i$.

Definition 2.1.8 *Let Ω be a domain with defining function $\rho \in C^k$, $k \geq 1$. We define the tangent vectors*

$$\begin{aligned} \mathcal{T}_i &= \mathbf{e}_i - G_i(x) \sum_{j=1}^{N} G_j(x)\mathbf{e}_j \\ &= \mathbf{e}_i - \left[\mathbf{e}_i \cdot \mathcal{G}(x)\right] \mathcal{G}(x). \end{aligned}$$

One easily checks that

$$\mathcal{T}_i \cdot \mathcal{G} = 0, \quad \text{for } i = 1, 2, \ldots, N.$$

As in the proof of the preceding proposition,

$$\sum_{i=1}^{N} G_i \mathcal{T}_i = 0,$$

and, if $G_{i_0} \neq 0$, then

$$\mathcal{T}_1, \mathcal{T}_2, \ldots, \mathcal{T}_{i_0-1}, \mathcal{T}_{i_0+1}, \mathcal{T}_{i_0+2}, \ldots, \mathcal{T}_N$$

are linearly independent. It may seem wrong to the reader to consider a set of N tangent vectors rather than $N-1$, but it is useful to have a set of vectors that always contains a basis, even as the point under consideration moves over $\partial \Omega$. We have the following theorem:

Theorem 2.1.9 *Suppose Ω is a C^k, $k \geq 1$, domain in $\mathbb{R}^N$ with defining function ρ. Then the tangent bundle can be written*

$$\left\{ \left(x, \sum_{i=1}^{N} v_i \mathcal{T}_i(x)\right) : x \in \mathbb{R}^N \text{ with } \rho(x) = 0, \ v = (v_1, v_2, \ldots, v_N) \in \mathbb{R}^N \right\}.$$

If $x(u_1, u_2, \ldots, u_{N-1}) = x(u)$ is a local C^k parametrization of the set of points with $\rho(x) = 0$, for u in a neighborhood of u_0, and if $G_N(x(u_0)) \neq 0$, then

$$\left(x(u), \sum_{i=1}^{N-1} v_i \mathcal{T}_i(x(u)) \right)$$

2.1. THE BUNDLES ASSOCIATED WITH A BOUNDARY

is a local C^{k-1} parametrization of the tangent bundle. It is easy to see that, if the differential of the parametrization $u \mapsto x(u)$ is of rank $N-1$, then the differential of the parametrization $(u,v) \mapsto (x(u), \sum v_i T_i(x(u)))$ is of rank $2(N-1)$. If $G_N(x(u_0)) = 0$, then there is some i_0 such that $G_{i_0}(x(u_0)) \neq 0$ and a similar parametrization is used in this case.

Alternative Definitions of Tangent Structures

The preceding discussion of the tangent and normal bundles has been based on calculus, but one can give more geometric definitions such as the next.

Definition 2.1.10 *For any set $S \subset \mathbb{R}^N$ and any point $p \in \overline{S}$ we define the **tangent cone** of S at p to be the set of $v \in \mathbb{R}^N$ such that for each $\epsilon > 0$ there exist $q \in S$ and $r > 0$ so that*

$$|q - p| < \epsilon, \quad |r(q-p) - v| < \epsilon.$$

We denote the tangent cone of S at p by

$$\mathrm{Tan}(S, p).$$

It is easy to verify that if Ω is a C^1 domain and $p \in \partial\Omega$, then $\mathrm{Tan}(\partial\Omega, p)$ as defined by Definition 2.1.10 coincides with the set $\mathrm{Tan}(\partial\Omega, p)$ defined in Definition 1.2.2, so our use of the notation $\mathrm{Tan}(S, p)$ is justified. Other easy observations are the following: The tangent cone is a closed set. If $v \in \mathrm{Tan}(S, p)$ and $r \geq 0$, then $rv \in \mathrm{Tan}(S, p)$. If $p \in \mathring{S}$, then $\mathrm{Tan}(S, p) = \mathbb{R}^n$.

It will also be useful to be able to speak of the possible "continuity" of the tangent cone $\mathrm{Tan}(S, p)$ as p varies. For this we need a topology. The Hausdorff distance (see Section A.1) defines a useful metric on the set of non-empty compact sets in Euclidean space. The Hausdorff distance between C_1 and C_2 will be denoted by $\mathrm{HD}(C_1, C_2)$. We will use the Hausdorff distance to define a metric on the space of non-trivial closed cones, and thus to define a topology:

Definition 2.1.11

(i) *A set $C \subset \mathbb{R}^N$ is a **cone** if*

$$C = \{rx : 0 \leq r < \infty, \ x \in C \cap \mathbb{S}(0,1)\},$$

*and C is a **non-trivial cone** if $C \cap \mathbb{S}(0,1) \neq \emptyset$. We do not require that a cone be closed under the operation of vector addition.*

(ii) *For C_1 and C_2 non-trivial closed cones in $\mathbb{R}^N$ we define the distance between them to be*

$$\mathrm{HD}\left(C_1 \cap \mathbb{S}(0,1), C_2 \cap \mathbb{S}(0,1)\right).$$

Another way to look at the tangent cone is provided by the following result.

Lemma 2.1.12 *For a set $S \subset \mathbb{R}^N$ and a point $p \in \overline{S}$, $\mathrm{Tan}(S,p)$ is the translation to the origin of the union of all half-lines*

$$\mathrm{r}(p,q) := \{p + t(q-p) : t \geq 0\} \tag{2.5}$$

originating at p such that a sequence $\{q_i\}$ of points in $\overline{S} \setminus \{p\}$ exists with q_i converging to p and with $\mathrm{r}(p,q_i)$ converging to $\mathrm{r}(p,q)$.

We leave it to the reader to define an appropriate metric on the set of half-lines in $\mathbb{R}^N$ and to prove the lemma.

Another geometric tangent structure suggested by the lemma is the following:

Definition 2.1.13 *For any set $S \subset \mathbb{R}^N$ and any point $p \in \overline{S}$ we define*

$$\mathrm{Tan}_2(S,p)$$

to be the the translation to the origin of the union of all lines

$$\mathrm{l}(q,q') := \{q + t(q'-q) : t \in \mathbb{R}\} \tag{2.6}$$

passing through p such that sequences $\{q_i\}$ and $\{q'_i\}$ of points in $\overline{S}$ exist with both q_i and q'_i converging to p and with $\mathrm{l}(q_i, q'_i)$ converging to $\mathrm{l}(q,q')$.

Again, the reader should define an appropriate metric on the spaces of lines in $\mathbb{R}^N$.

Example 2.1.14 *Let $S \subset \mathbb{R}^2$ be the graph of the function*

$$f(x) = \begin{cases} x^2 \sin(1/x) & \text{if } x \neq 0, \\ 0 & \text{if } x = 0. \end{cases}$$

Then $\mathrm{Tan}(S, (0,0))$ is the x-axis, but we have

$$\mathrm{Tan}_2(S, (0,0)) = \mathbb{R}^2.$$

This second notion of a tangent structure has a surprising and not well-known utility as evidenced by the next result.

Theorem 2.1.15 *If $\Omega \subset \mathbb{R}^N$ is a non-empty open set, then Ω is a C^1 domain if and only if at each point $p \in \partial \Omega$ the following hold:*

$$\mathrm{Tan}(S,p) = \mathrm{Tan}_2(S,p), \tag{2.7}$$

$$\mathrm{Tan}(S,p) \text{ is an } (N-1)\text{-dimensional plane.} \tag{2.8}$$

As far as we know, Theorem 2.1.15 is due to K. R. Lucas [1]. It is easy to see that, if $\Omega \subset \mathbb{R}^N$ is a C^1 domain, then the conditions (2.7) and (2.8) must hold. The interesting part of the result is the converse, which is a consequence of the following sequence of lemmas.

2.1. THE BUNDLES ASSOCIATED WITH A BOUNDARY

Lemma 2.1.16 *Suppose S is a closed subset of $\mathbb{R}^N$ and $\mathrm{Tan}_2(S,p_0)$ is a vector subspace of $\mathbb{R}^N$. Then the orthogonal projection onto the translation of $\mathrm{Tan}_2(S,p_0)$ to p_0 is a one-to-one function on $S \cap U$ for some neighborhood U of p_0.*

Proof: Assume that there exists a sequence of pairs of distinct points q_i and q_i', both converging to p_0, such that q_i and q_i' project orthogonally onto the same point of $\mathrm{Tan}_2(S,p_0)$. Considering the limit of $\mathrm{l}(q_i, q_i')$, we are led to the contradiction that there is a line in $\mathrm{Tan}_2(S,p_0)$ that is orthogonal to $\mathrm{Tan}_2(S,p_0)$. ■

Lemma 2.1.17 *Suppose that S is a closed subset of $\mathbb{R}^N$ and that p_0 is a point of S with*
$$\mathrm{Tan}(S,p_0) = \mathrm{Tan}_2(S,p_0).$$
If there is a neighborhood U of p_0 such that $\mathrm{Tan}(S,p)$ is a vector subspace of $\mathbb{R}^N$, for all $p \in S \cap U$, and the dimension of $\mathrm{Tan}(S,\cdot)$ is constant on $S \cap U$, then $\mathrm{Tan}(S,\cdot)$ is a continuous function at p_0.

Proof: Assume that there is a sequence of points p_i converging to p_0 such that $\mathrm{Tan}(S,p_i)$ does not converge to $\mathrm{Tan}(S,p_0)$. Then, for large enough i, we can choose q_i in S so that the half-line $\mathrm{r}(p_i,q_i)$ makes an angle with $\mathrm{Tan}(S,p_0)$ exceeding some positive ϵ, independent of i. But then the pairs of points p_i and q_i both converge to p_0 and the lines $\mathrm{l}(p_i,q_i)$ they determine make angles with $\mathrm{Tan}(S,p_0)$ exceeding the same ϵ. Passing to a subsequence if necessary, but without changing notation, we may assume that the lines $\mathrm{l}(p_i,q_i)$ converge to a line. That line is in $\mathrm{Tan}_2(S,p_0)$ by definition, but makes angle greater than ϵ with $\mathrm{Tan}_2(S,p_0)$ by the choice of p_i and q_i. That is a contradiction. ■

Lemma 2.1.18 *Suppose that S is a closed subset of $\mathbb{R}^N$ and that p_0 is a point of S with*
$$\mathrm{Tan}(S,p_0) = \mathrm{Tan}_2(S,p_0).$$
If there is a neighborhood U of p_0 such that $\mathrm{Tan}(S,p)$ is a vector subspace of $\mathbb{R}^N$ for all $p \in S \cap U$, and the dimension of $\mathrm{Tan}(S,\cdot)$ is constant on $S \cap U$, then p_0 is a relative interior point of the orthogonal projection of $S \cap U$ onto the translation of $\mathrm{Tan}_2(S,p_0)$ to p_0.

Proof: Let Π denote the translation of $\mathrm{Tan}_2(S,p_0)$ to p_0. Assume that there exists a sequence of points $q_i \in \Pi$ that converge to p_0 and are not images of $S \cap U$ under the orthogonal projection onto Π.

Suppose $r > 0$ is such that $\overline{\mathbb{B}}(p_0, r) \subset U$. For each i, let q_i' be the nearest point to q_i in Π that is in the image of $S \cap \overline{\mathbb{B}}(p_0, r)$ under orthogonal projection onto Π. Let p_i' be a point of $S \cap \overline{\mathbb{B}}(p_0, r)$ that projects to q_i'.

If for infinitely many i we had $|p_0 - p'_i| = r$, then a subsequence of the p'_i would converge to a point $p(r) \in S$ that projects orthogonally onto p_0 and satisfies $|p_0 - p(r)| = r$. Were this the situation for arbitrarily small choices of $r > 0$ with $\overline{\mathbb{B}}(p_0, r) \subset U$, we would be able to construct a tangent vector to S at p_0 that is orthogonal to Π. Plainly, this would be contradictory, and thus we may assume that, for all sufficiently large i, we have $|p_0 - p'_i| < r$.

For i such that p'_i is an interior point of $\overline{\mathbb{B}}(p_0, r_i)$, the right circular cylinder over $\Pi \cap \overline{\mathbb{B}}(q_i, |q'_i - q_i|)$ does not intersect $S \cap \overline{\mathbb{B}}(p_0, r)$ in its interior, but does have p'_i in its boundary. Since $\text{Tan}(S, p'_i)$ is of the same dimension as Π, there must be a vector in $\text{Tan}(S, p'_i)$ that is orthogonal to Π. Since this happens for infinitely many i, we contradict the continuity of $\text{Tan}(S, \cdot)$ proved in Lemma 2.1.17. ∎

Proof of Theorem 2.1.15: The hypotheses of the preceding three lemmas are satisfied. Thus, near each point of S, we can use the inverse of orthogonal projection onto the tangent plane to define a function onto S. The definition of $\text{Tan}(S, \cdot)$ guarantees that this function is differentiable, and we know that the derivative is continuous because of the continuity of the tangent planes proved in Lemma 2.1.17. ∎

2.2 The Second Fundamental Form and Curvature

Recall that the Gauss map for the boundary of a domain Ω with defining function ρ is given by

$$\mathcal{G}(x) = \frac{\nabla \rho(x)}{|\nabla \rho(x)|}.$$

Since $\mathcal{G}(x)$ is a unit vector normal to $\partial \Omega$ at x, then it must be independent of the choice of defining function. Thus we have the the next result.

Proposition 2.2.1 *If $\tilde{\rho}$ is another defining function for Ω and if $\widetilde{\mathcal{G}}$ is the Gauss map defined using $\tilde{\rho}$, then*

$$\mathcal{G}(x) = \widetilde{\mathcal{G}}(x) \quad \text{for } x \in \partial \Omega. \tag{2.9}$$

This is readily verified using Proposition 1.2.3.

Assume that ρ is at least C^2. Notice that, for any vector v, the directional derivative of $|\mathcal{G}|^2 \equiv 1$ in the direction of v is zero, so we have

$$\mathcal{G} \cdot \langle v, D\mathcal{G} \rangle \equiv 0. \tag{2.10}$$

Accordingly, if x is a point of $\partial \Omega$ and v is any vector, then $\langle v, D\mathcal{G}(x) \rangle$ is tangent to $\partial \Omega$.

2.2. THE SECOND FUNDAMENTAL FORM AND CURVATURE

Definition 2.2.2 *Suppose Ω is C^k, $k \geq 2$, and $x \in \partial\Omega$. The **Weingarten map** $\mathcal{W}_x : \mathrm{Tan}(\partial\Omega, x) \to \mathrm{Tan}(\partial\Omega, x)$ is defined by setting $\mathcal{W}_x(v)$ equal to the directional derivative of the Gauss map in the direction of v, that is*

$$\mathcal{W}_x(v) = \langle v, D\mathcal{G}(x)\rangle, \tag{2.11}$$

where v is tangent to $\partial\Omega$ at x.

Remark 2.2.3 Since it has been noted that $\langle v, D\mathcal{G}(x)\rangle$ is always tangent to $\partial\Omega$ for $x \in \partial\Omega$, even when v is not a tangent vector, one might hope or assume that if v is normal to $\partial\Omega$, then $\langle v, D\mathcal{G}(x)\rangle$ would vanish. Such is not the case. Perhaps the simplest example to illustrate this is to take

$$\rho(x, y) = y - xy$$

to be the defining function for $\{(x, y) : y < 0\}$ in some small neighborhood of the origin. We leave it to the reader to verify by direct calculation that, with $v = \begin{pmatrix} 0 \\ 1 \end{pmatrix}$ the upward pointing unit vector,

$$\langle v, D\mathcal{G}(0,0)\rangle = \begin{pmatrix} -1 \\ 0 \end{pmatrix}.$$

Proposition 2.2.4 *Suppose Ω is C^k, $k \geq 2$. If $\tilde{\rho}$ is another defining function for Ω and if $\widetilde{\mathcal{W}}$ is the Weingarten map defined using $\tilde{\rho}$, then*

$$\mathcal{W}_x(v) = \widetilde{\mathcal{W}}_x(v), \tag{2.12}$$

whenever v is tangent to $\partial\Omega$ at $x \in \partial\Omega$.

The proposition is most simply proved by a computation using coordinates. While such computations can be tedious, they are invaluable for understanding examples. For this purpose, we introduce some notation.

Definition 2.2.5

(i) *The standard basis vectors in $\mathbb{R}^N$ will be denoted by*

$$\mathbf{e}_1, \mathbf{e}_2, \ldots, \mathbf{e}_N.$$

(ii) *For a defining function ρ, we will write*

$$\rho_{,i} = \frac{\partial\rho}{\partial x_i}, \quad i \in \{1, 2, \ldots, N\} \tag{2.13}$$

$$\rho_{,ij} = \frac{\partial^2\rho}{\partial x_i \partial x_j}, \quad i, j \in \{1, 2, \ldots, N\}. \tag{2.14}$$

Lemma 2.2.6 *Suppose that ρ is C^k, $k \geq 2$.*

(i) For any pair of vectors $v = \sum_{i=1}^{N} v_i \mathbf{e}_i$ and $w = \sum_{i=1}^{N} w_i \mathbf{e}_i$ in $\mathbb{R}^N$, we have

$$\langle v, D\mathcal{G}(x)\rangle = \left(\sum_{i,j=1}^{N} \rho_{,ij} v_j \mathbf{e}_i\right) \left(\sum_{k=1}^{N} (\rho_{,k})^2\right)^{-1/2}$$
$$- \left(\sum_{i=1}^{N} \rho_{,i} \mathbf{e}_i\right) \left(\sum_{j,k=1}^{N} \rho_{,jk} \rho_{,k} v_j\right) \left(\sum_{k=1}^{N} (\rho_{,k})^2\right)^{-3/2} \quad (2.15)$$

and

$$w \cdot \langle v, D\mathcal{G}\rangle = \left(\sum_{i,j=1}^{N} \rho_{,ij} v_j w_i\right) \left(\sum_{k=1}^{N} (\rho_{,k})^2\right)^{-1/2}$$
$$- \left(\sum_{i=1}^{N} \rho_{,i} w_i\right) \left(\sum_{j,k=1}^{N} \rho_{,jk} \rho_{,k} v_j\right) \left(\sum_{k=1}^{N} (\rho_{,k})^2\right)^{-3/2}. \quad (2.16)$$

(ii) If, in addition, w is a tangent vector, then

$$w \cdot \langle v, D\mathcal{G}\rangle = \left(\sum_{i,j=1}^{N} \rho_{,ij} v_j w_i\right) \left(\sum_{k=1}^{N} (\rho_{,k})^2\right)^{-1/2}. \quad (2.17)$$

Proof: In terms of components we can write

$$\mathcal{G}(x) = \sum_{i=1}^{N} G_i \mathbf{e}_i,$$

with

$$G_i = \rho_{,i} \left(\sum_{k=1}^{N} (\rho_{,k})^2\right)^{-1/2}. \quad (2.18)$$

We have

$$\langle v, D\mathcal{G}(x)\rangle = \sum_{i,j=1}^{N} \frac{\partial G_i}{\partial x_j} v_j \mathbf{e}_i$$

with

$$\frac{\partial G_i}{\partial x_j} = \rho_{,ij} \left(\sum_{k=1}^{N} (\rho_{,k})^2\right)^{-1/2} - \rho_{,i} \left(\sum_{k=1}^{N} \rho_{,jk} \rho_{,k}\right) \left(\sum_{k=1}^{N} (\rho_{,k})^2\right)^{-3/2}, \quad (2.19)$$

2.2. THE SECOND FUNDAMENTAL FORM AND CURVATURE

so (2.15) follows. Equation (2.16) follows readily from (2.15), and then (2.17) follows from (2.16). ∎

Proof of Proposition 2.2.4: By Proposition 1.2.3, there exists a positive function h of class C^{k-1} such that

$$\begin{aligned} \rho &= h\tilde{\rho} \\ \rho_{,i} &= \frac{\partial h}{\partial x_i}\tilde{\rho} + h\tilde{\rho}_{,i} \\ \rho_{,ij} &= h_{,ij}\tilde{\rho} + \frac{\partial h}{\partial x_i}\tilde{\rho}_{,j} + \frac{\partial h}{\partial x_j}\tilde{\rho}_{,i} + h\tilde{\rho}_{,ij}. \end{aligned}$$

Now the last equation only makes sense if h is twice differentiable, *i.e.* if $k \geq 3$, but even in case $k = 2$ we have

$$\begin{aligned} \rho_{,i} &= h\tilde{\rho}_{,i} \\ \rho_{,ij} &= \frac{\partial h}{\partial x_i}\tilde{\rho}_{,j} + \frac{\partial h}{\partial x_j}\tilde{\rho}_{,i} + h\tilde{\rho}_{,ij} \end{aligned}$$

for points in $\partial\Omega$, in particular, at x.

Since v is tangent to $\partial\Omega$ at x, we have

$$\sum_{i=1}^N \rho_{,i} v_i = \sum_{i=1}^N \tilde{\rho}_{,i} v_i = 0.$$

Thus

$$\begin{aligned} \sum_{j=1}^N \rho_{,ij} v_j &= \sum_{j=1}^N \left(\frac{\partial h}{\partial x_i}\tilde{\rho}_{,j} + \frac{\partial h}{\partial x_j}\tilde{\rho}_{,i} + h\tilde{\rho}_{,ij} \right) v_j \\ &= \sum_{j=1}^N \frac{\partial h}{\partial x_j}\tilde{\rho}_{,i} v_j + \sum_{i,j=1}^N h\tilde{\rho}_{,ij} v_j \end{aligned} \quad (2.20)$$

and

$$\begin{aligned} \sum_{j,k=1}^N \rho_{,jk}\rho_{,k} v_j &= \sum_{j,k=1}^N \left(\frac{\partial h}{\partial x_j}\tilde{\rho}_{,k} + \frac{\partial h}{\partial x_k}\tilde{\rho}_{,j} + h\tilde{\rho}_{,jk} \right) h\tilde{\rho}_{,k} v_j \\ &= \sum_{j,k=1}^N \frac{\partial h}{\partial x_j} h \tilde{\rho}_{,k}^2 v_j + \sum_{j,k=1}^N h^2 \tilde{\rho}_{,jk} \tilde{\rho}_{,k} v_j. \end{aligned} \quad (2.21)$$

Substituting (2.20) and (2.21) into (2.15) and doing a bit of algebraic manipulation, we find that

$$\langle v, D\mathcal{G}(x) \rangle = \left(\sum_{i,j=1}^N \tilde{\rho}_{,ij} v_j e_i \right) \left(\sum_{k=1}^N (\tilde{\rho}_{,k})^2 \right)^{-1/2}$$

$$-\left(\sum_{i=1}^{N}\tilde{\rho}_{,i}\mathbf{e}_i\right)\left(\sum_{j,k=1}^{N}\tilde{\rho}_{,jk}\tilde{\rho}_{,k}v_j\right)\left(\sum_{k=1}^{N}(\tilde{\rho}_{,k})^2\right)^{-3/2}$$
$$=\langle v, D\widetilde{\mathcal{G}}(x)\rangle. \qquad\blacksquare$$

Corollary 2.2.7 *Let G_i, $i = 1, 2, \ldots, N$, be the components of the Gauss map.*

(i) *For any pair of vectors $v = \sum_{i=1}^{N} v_i\mathbf{e}_i$ and $w = \sum_{i=1}^{N} w_i\mathbf{e}_i$ in $\mathbb{R}^N$, we have*

$$w \cdot \langle v, D\mathcal{G}(x)\rangle = \sum_{i,j=1}^{N} \frac{\partial G_i}{\partial x_j} w_i v_j \qquad (2.22)$$

(ii) *For $j = 1, 2, \ldots, N$, we have*

$$\sum_{i=1}^{N} G_i \frac{\partial G_i}{\partial x_j} = 0. \qquad (2.23)$$

Proof: Part (i) follows from the proof of Lemma 2.2.6. Part (ii) then follows from (2.10) with $v = \mathbf{e}_j$. $\blacksquare$

Definition 2.2.8 *Suppose that Ω is C^k, $k \geq 2$, and that $x \in \partial\Omega$. The* **second fundamental form** *is the bilinear form II on $\text{Tan}(\partial\Omega, x)$ given by*

$$\text{II}(v, w) = w \cdot \mathcal{W}_x(v) \qquad (2.24)$$

where v and w are tangent to $\partial\Omega$ at x.

Proposition 2.2.9 *The second fundamental form is symmetric.*

Proof: This follows from the symmetry of the second derivative and (2.17). $\blacksquare$

The second fundamental form is the quadratic form associated with the Weingarten map. Since the second fundamental form is symmetric, we know from linear algebra that the Weingarten map will have only real eigenvalues and that there will exist a set of $N-1$ orthogonal eigenvectors. Also recall from linear algebra that a linear map from an $N-1$ dimensional vector space to itself has associated with it a set of $N-1$ invariants. The two most familiar and most commonly used of these invariants are the trace and the determinant. One definition of the other invariants is given next.

2.2. THE SECOND FUNDAMENTAL FORM AND CURVATURE

Definition 2.2.10 *If a linear map from an M dimensional real vector space to itself is represented by the $M \times M$ matrix A, then the **trace of order K of A** is denoted by $\mathrm{tr}_K(A)$ and equals the sum of all the $K \times K$ determinants that can be formed by intersecting any K rows of A with the same K columns. That is*

$$\mathrm{tr}_K(A) = \sum \begin{vmatrix} a_{i_1 i_1} & a_{i_1 i_2} & \cdots & a_{i_1 i_K} \\ a_{i_2 i_1} & a_{i_2 i_2} & \cdots & a_{i_2 i_K} \\ \vdots & \vdots & \cdots & \vdots \\ a_{i_K i_1} & a_{i_K i_2} & \cdots & a_{i_K i_K} \end{vmatrix},$$

where the sum extends over all choices of $1 \leq i_1 < i_2 < \ldots < i_K \leq M$.

Note that the usual trace of A equals the trace of order 1 of A and the determinant of A equals the trace of order M of A if A is $M \times M$.

Definition 2.2.11 *For $x \in \partial\Omega$,*

(i) *the $(N-1)$ eigenvalues of $\mathcal{W}_x$ are called the **principal curvatures** of $\partial\Omega$,*

(ii) *the unit eigenvectors of $\mathcal{W}_x$ are called the **directions of curvature** or **principal directions** of $\partial\Omega$,*

(iii) *the determinant of $\mathcal{W}_x$ is called the **Gaussian curvature** or **total curvature** of $\partial\Omega$ and will be denoted by $\mathbf{K}(x)$,*

(iv) *the number $\frac{1}{N-1}$ times the trace of $\mathcal{W}_x$ is called the **mean curvature** of $\partial\Omega$ at x and will be denoted by $\mathbf{H}(x)$,*

(v) *the number $\frac{1}{(N-1)(N-2)}$ times the trace of order 2 of $\mathcal{W}_x$ is called the **scalar curvature** of $\partial\Omega$ at x and will be denoted by $\mathbf{S}(x)$.*

By (2.2.7), we know that the second fundamental form is the restriction to the tangent plane of the quadratic form associated with the matrix $\frac{\partial G_i}{\partial x_j}$ given in (2.19). Additionally, we know that the range of that matrix is the tangent plane. The following lemma from linear algebra will allow us to use these facts to compute the mean curvature, the Gaussian curvature, and the sum of the squares of the principal curvatures.

Lemma 2.2.12 *Suppose A is an $N \times N$ matrix of rank $N-1$ with range Π. Let $\mathcal{A}$ denote the linear map from Π to Π that is the restriction to Π of the linear map determined by A.*

(i) For $K = 1, 2, \ldots, N-1$, the trace of order K of $\mathcal{A}$ equals the trace of order K of A, that is,
$$\operatorname{tr}_K[\mathcal{A}] = \operatorname{tr}_K[A].$$

(ii) The trace of $\mathcal{A}^2$ equals the trace of A^2, that is,
$$\operatorname{tr}[\mathcal{A}^2] = \operatorname{tr}[A^2].$$

Proof: Since the left-hand and right-hand sides of both equations are independent of choice of basis, we may use any bases that we find convenient. Choose an arbitrary set of $N - 1$ independent vectors in Π and extend it to a basis for $\mathbb{R}^N$ by adding a null vector of A as the Nth basis vector. With such a basis, A is represented by an $N \times N$ matrix with Nth row 0 and Nth column 0, while $\mathcal{A}$ is represented by the $(N-1) \times (N-1)$ matrix obtained by omitting the Nth row and Nth column of the matrix representing A. Part (i) is now evident. Part (ii) follows by considering the matrices representing A^2 and $\mathcal{A}^2$. ∎

Definition 2.2.13 For a C^2 function f defined on a domain in $\mathbb{R}^N$, we define the **Laplacian** of f, denoted $\triangle f$, by
$$\triangle f = \sum_{i=1}^{N} \frac{\partial^2 f}{\partial x_i^2}.$$

Theorem 2.2.14 Let Ω be a domain with defining function $\rho \in C^k$ $k \geq 2$. Let the tangential differentiation operators δ_i be as in (2.3). For $x \in \partial\Omega$, the mean curvature at x is given by
$$\begin{aligned}
\mathbf{H}(x) &= (N-1)^{-1} \sum_{i=1}^{N} \frac{\partial G_i}{\partial x_i} \\
&= (N-1)^{-1} \sum_{i=1}^{N} \delta_i G_i \\
&= (N-1)^{-1} |\nabla \rho|^{-3} \left(|\nabla \rho|^2 \triangle \rho(x) - \sum_{i,j=1}^{N} \rho_{,i}(x)\, \rho_{,j}(x)\, \rho_{,ij}(x) \right)
\end{aligned}$$
where the tangential differentiation operators δ_i are as in (2.3).

Proof: Fix a point $x_0 \in \partial\Omega$. Applying the preceding lemma to the Weingarten map, we see that
$$(N-1)\, \mathbf{H}(x) = \sum_{i=1}^{N} \frac{\partial G_i}{\partial x_i}.$$

2.2. THE SECOND FUNDAMENTAL FORM AND CURVATURE 41

Multiplying by $|\nabla\rho(x)|^3$, replacing $\frac{\partial G_i}{\partial x_i}$ by the expression given in (2.19), and simplifying, we obtain the result. ∎

Corollary 2.2.15 *Let Ω be a domain with defining function $\rho \in C^k$ $k \geq 2$. If*
$$|\nabla\rho| \equiv 1$$
holds in a spatial neighborhood of a point $x \in \partial\Omega$, then
$$\mathbf{H}(x) = \tfrac{1}{N-1}\,\triangle\,\rho(x).$$

Proof. We have
$$\sum_{i=1}^{N} \rho_{,i}^2 \equiv 1$$
in a neighborhood of x. Differentiating with respect to x_j, we find that
$$\sum_{i=1}^{N} \rho_{,i}\rho_{,ij} = 0$$
holds for each j. Multiplying by $\rho_{,j}$ and summing over j, we find that
$$\sum_{i,j=1}^{N} \rho_{,i}\rho_{,j}\rho_{,ij} = 0.$$
The result then follows from the last line in Theorem 2.2.14. ∎

Lemma 2.2.16 *Let the tangential differentiation operators δ_i be as in (2.3). For any $1 \leq i,j \leq N$, we have*
$$\delta_i G_j = \delta_j G_i. \tag{2.25}$$

Proof: Using (2.19), we see that

$$\begin{aligned}
\delta_j G_i &= \frac{\partial}{\partial x_j} G_i - G_j \sum_{k=1}^{N} G_k \frac{\partial}{\partial x_k} G_i \\
&= |\nabla\rho|^{-1}\rho_{,ij} - |\nabla\rho|^{-3}\rho_{,i}\sum_{k=1}^{N}\rho_{,k}\rho_{,jk} \\
&\quad - |\nabla\rho|^{-2}\rho_{,j}\sum_{k=1}^{N}\rho_{,k}\left(|\nabla\rho|^{-1}\rho_{,ik} - |\nabla\rho|^{-3}\rho_{,i}\sum_{\ell=1}^{N}\rho_{,\ell}\rho_{,k\ell}\right) \\
&= |\nabla\rho|^{-1}\rho_{,ij} - |\nabla\rho|^{-3}\rho_{,i}\sum_{k=1}^{N}\rho_{,k}\rho_{,jk} - |\nabla\rho|^{-3}\rho_{,j}\sum_{k=1}^{N}\rho_{,k}\rho_{,ik} \\
&\quad + |\nabla\rho|^{-5}\rho_{,i}\rho_{,j}\sum_{k,\ell=1}^{N}\rho_{,k}\rho_{,k\ell}\rho_{,\ell}.
\end{aligned}$$

Since the last expression is clearly symmetric in i and j, the result follows. ∎

Theorem 2.2.17 *Let Ω be a domain with defining function $\rho \in C^k$ $k \geq 2$. Let the tangential differentiation operators δ_i be as in (2.3). For $x \in \partial\Omega$, the sum of the squares of the principal curvatures is given by*

$$\sum_{i=1}^{N} j \frac{\partial G_i}{\partial x_j} \frac{\partial G_j}{\partial x_i} = \sum_{i,j=1}^{N} |\delta_i G_j|^2.$$

Proof: Let G be the matrix with entries $\frac{\partial G_i}{\partial x_j}$. By Lemma 2.2.12, we know that the sum of the squares of the principal curvatures equals $\operatorname{tr}[G^2]$. But the $(i,k)^{\text{th}}$ entry in G^2 is

$$\sum_{j=1}^{N} \frac{\partial G_i}{\partial x_j} \frac{\partial G_j}{\partial x_k},$$

so

$$\operatorname{tr}[G^2] = \sum_{i=1}^{N} \sum_{j=1}^{N} \frac{\partial G_i}{\partial x_j} \frac{\partial G_j}{\partial x_i}$$

as desired.

To see that $\sum_{i,j=1}^{N} |\delta_i G_j|^2$ also equals the sum of the squares of the principal curvatures, we use the preceding lemma to compute

$$\begin{aligned}
\sum_{i,j=1}^{N} |\delta_i G_j|^2 &= \sum_{i,j=1}^{N} \delta_i G_j \delta_j G_i \\
&= \sum_{i,j=1}^{N} \left(\frac{\partial G_j}{\partial x_i} - G_i \sum_{k=1}^{N} G_k \frac{\partial G_j}{\partial x_k} \right) \left(\frac{\partial G_i}{\partial x_j} - G_j \sum_{\ell=1}^{N} G_\ell \frac{\partial G_i}{\partial x_\ell} \right) \\
&= \sum_{i,j=1}^{N} \frac{\partial G_j}{\partial x_i} \frac{\partial G_i}{\partial x_j} - \sum_{i,j,\ell=1}^{N} \frac{\partial G_j}{\partial x_i} G_j G_\ell \frac{\partial G_i}{\partial x_\ell} \\
&\quad - \sum_{i,j,k=1}^{N} \frac{\partial G_i}{\partial x_j} G_i G_k \frac{\partial G_j}{\partial x_k} - \sum_{i,j,k\ell=1}^{N} G_i G_j G_k G_\ell \frac{\partial G_j}{\partial x_k} \frac{\partial G_i}{\partial x_\ell}.
\end{aligned}$$

Now, by part (ii) of Corollary 2.2.7, we see that

$$\sum_{i,j,\ell=1}^{N} \frac{\partial G_j}{\partial x_i} G_j G_\ell \frac{\partial G_i}{\partial x_\ell} = \sum_{i,\ell=1}^{N} G_\ell \frac{\partial G_i}{\partial x_\ell} \left(\sum_{j=1}^{N} G_j \frac{\partial G_j}{\partial x_i} \right) = 0$$

$$\sum_{i,j,k=1}^{N} \frac{\partial G_i}{\partial x_j} G_i G_k \frac{\partial G_j}{\partial x_k} = \sum_{j,k=1}^{N} G_k \frac{\partial G_j}{\partial x_k} \left(\sum_{i=1}^{N} G_i \frac{\partial G_i}{\partial x_j} \right) = 0$$

2.2. THE SECOND FUNDAMENTAL FORM AND CURVATURE

$$\sum_{i,j,k\ell=1}^{N} G_i G_j G_k G_\ell \frac{\partial G_j}{\partial x_k} \frac{\partial G_i}{\partial x_\ell} = \sum_{k,\ell=1}^{N} G_k G_\ell \left(\sum_{i=1}^{N} G_i \frac{\partial G_i}{\partial x_\ell}\right) \left(\sum_{j=1}^{N} G_j \frac{\partial G_j}{\partial x_k}\right)$$
$$= 0,$$

proving the result. ∎

The following theorem is an immediate consequence of Lemma 2.2.12. It makes use of the trace of order $N-1$ which is a matrix invariant defined in Definition 2.2.10.

Theorem 2.2.18 *Let Ω be a domain with defining function $\rho \in C^k$ $k \geq 2$. Let x be a point in $\partial\Omega$.*

(i) *The Gaussian curvature at x is given by*

$$\mathbf{K}(x) = \mathrm{tr}_{N-1} \left[\frac{\partial G_i}{\partial x_j}\right].$$

(ii) *The scalar curvature at x is given by*

$$\mathbf{S}(x) = \frac{1}{(N-1)(N-2)} \mathrm{tr}_2 \left[\frac{\partial G_i}{\partial x_j}\right].$$

Example 2.2.19 Let Ω be the region with defining function

$$\rho(x_1, x_2, \ldots, x_N) = \left(\sum_{i=1}^{N} x_i^2\right) - R^2,$$

so that Ω is the interior of the ball of radius R centered at the origin. One computes that

$$\rho_{,i} = 2x_i,$$
$$\rho_{,ij} = 2\delta_{ij},$$

where δ_{ij} is the Kronecker delta. It follows that at any point $x \in \partial\Omega$

$$\frac{\partial G_i}{\partial x_j} = \frac{\delta_{ij}}{R} - \frac{x_i x_j}{R^3}$$

holds. It is then routine to compute that

$$\sum_{i=1}^{N} \frac{\partial G_i}{\partial x_i} = \frac{N-1}{R} \quad \text{and} \quad \sum_{i,j=1}^{N} \frac{\partial G_i}{\partial x_j} \frac{\partial G_j}{\partial x_i} = \frac{N-1}{R^2},$$

so at every point of $\partial\Omega$ the mean curvature is $\frac{1}{R}$ and the sum of the squares of the principal curvatures is $\frac{N-1}{R^2}$.

The computation of the Gaussian curvature and the scalar curvature is a bit less straightforward. One can use the fact that for $y = (y_1, y_2, \ldots, y_M) \in \mathbb{R}^M$, the $M \times M$ matrix

$$I_{M \times M} - yy^{\mathrm{t}}$$

has the eigenvalues 1 and $1 - |y|^2$, with the multiplicity of the eigenvalue 1 being $M - 1$, where yy^{t} is the matrix with $(i,j)^{\mathrm{th}}$ entry $y_i y_j$,

to verify that the Gaussian curvature of $\partial\Omega$ has the expected value of $\frac{1}{R^{N-1}}$ and that the scalar curvature is $\frac{1}{R^2}$, at every point.

Plane Curves

Consider a region Ω in the plane. Suppose that ρ is a defining function for Ω of class C^k, $k \geq 2$, satisfying

$$|\nabla \rho| \equiv 1$$

in a neighborhood of $\partial\Omega$, and suppose $r(s)$ is a parametrization by arclength of $\partial\Omega$.

Definition 2.2.20 *The* **curvature** *of $\partial\Omega$ is the non-negative function κ defined by setting*

$$\kappa = \left|\frac{d^2 r}{ds^2}\right|. \tag{2.26}$$

Writing the components of r as $x(s)$ and $y(s)$, we can differentiate the equation $\rho[r(s)] = 0$ with respect to s to find

$$0 = \rho_{,1} x' + \rho_{,2} y', \tag{2.27}$$
$$0 = \rho_{,11} x'^2 + 2\rho_{,12} x' y' + \rho_{,22} y'^2 + \rho_{,1} x'' + \rho_{,2} y''. \tag{2.28}$$

Likewise, we differentiate the equation $x'^2 + y'^2 = 1$ to obtain

$$0 = x' x'' + y' y''. \tag{2.29}$$

Now, (2.27) and (2.29) can be thought of as a pair of linear equations simultaneously satisfied by (x', y'), so the determinant of the coefficients of that system vanishes, that is,

$$0 = \rho_{,2} x'' - \rho_{,1} y''. \tag{2.30}$$

Equation (2.30) tells us that (x'', y'') is orthogonal to $(\rho_{,2}, \rho_{,1})$ which in turn is orthogonal to the unit vector $(\rho_{,1}, \rho_{,2})$. Since we are in a two-dimensional space, we must have $|(\rho_{,1}, \rho_{,2}) \cdot (x'', y'')| = |(x'', y'')|$ or

$$|\rho_{,1} x'' + \rho_{,2} y''| = \sqrt{(x'')^2 + (y'')^2}. \tag{2.31}$$

2.3. SURFACES WITH CONSTANT MEAN CURVATURE

Similarly, we can differentiate the equation $\rho_{,1}^2 + \rho_{,2}^2 = 1$ with respect to the first and second variables to find

$$0 = \rho_{,1}\rho_{,11} + \rho_{,2}\rho_{,12}, \tag{2.32}$$
$$0 = \rho_{,1}\rho_{,12} + \rho_{,2}\rho_{,22}. \tag{2.33}$$

Equations (2.27) and (2.32) together imply

$$0 = \rho_{,11}y' - \rho_{,12}x', \tag{2.34}$$

while (2.27) and (2.33) imply

$$0 = \rho_{,12}y' - \rho_{,22}x'. \tag{2.35}$$

Multiplying (2.34) by y' and (2.35) by x' and adding the results to (2.28), we obtain

$$0 = \rho_{,11} + \rho_{,22} + \rho_{,1}x'' + \rho_{,2}y'', \tag{2.36}$$

and from (2.31) and Corollary 2.2.15 it follows that the mean curvature satisfies the equation

$$|\mathbf{H}| = \sqrt{(x'')^2 + (y'')^2}. \tag{2.37}$$

We can summarize the above calculations in the next theorem, which also embodies the part of the Frenet formulas appropriate to curves in the plane. Of course, there is a more general set of Frenet formulas applicable to curves in space[1] (see for example Struik [1]), but that will not be presented here.

Theorem 2.2.21 *For a C^k, $k \geq 2$, domain $\Omega \subset \mathbb{R}^2$ with $\partial\Omega$ parametrized by arc-length by $r(s)$, the following equations hold*

$$\kappa = |\mathbf{H}|, \tag{2.38}$$
$$\frac{d^2 r}{ds^2} = \pm \mathbf{H}\mathcal{G}. \tag{2.39}$$

2.3 Surfaces with Constant Mean Curvature

From Theorem 2.2.14 of the previous section, we know that if ρ is a function satisfying $\nabla\rho \neq 0$ on the surface $S = \{x : \rho(x) = 0\}$, then the mean curvature of S is given by the differential expression

$$(N-1)^{-1}|\nabla\rho|^{-3}\left(|\nabla\rho|^2 \Delta \rho - \sum_{i,j=1}^{N} \rho_{,i}\rho_{,j}\rho_{,ij}\right) \tag{2.40}$$

---

[1] More general Frenet formulas for curves in spaces of arbitrary dimension can be found in Shilov [1] or in Green [1].

evaluated on S. It is interesting to consider what happens when we reverse this process by first specifying the mean curvature and then seeing what the consequences are for the surface. The most natural specification for the mean curvature is to set it equal to a constant. Thus we will be considering the equation

$$|\nabla \rho|^2 \triangle \rho - \sum_{i,j=1}^{N} \rho_{,i}\, \rho_{,j}\, \rho_{,ij} = C(N-1)\,|\nabla \rho|^3, \qquad (2.41)$$

where C is a constant.

We saw in the previous section that a sphere of radius $1/C$ is a surface with mean curvature C at every point of the surface. Thus the ball provides an example of a domain with positive constant mean curvature on the boundary. In Example 2.3.6 we will see that the right circular cylinder is another example of a surface with constant mean curvature. The serious study of constant mean curvature surfaces dates back to the 1841 paper of Delaunay [1], in which he studied the rotationally symmetric constant mean curvature surfaces in $\mathbb{R}^3$, now called **Delaunay surfaces**. Clearly, the sphere and the cylinder are examples of rotationally symmetric constant mean curvature surfaces, but we will use the term "Delaunay surface" to mean those constant mean curvature surfaces other than the sphere and the cylinder; this choice of terminology is a matter of taste.

Definition 2.3.1 *By a* **surface** *in $\mathbb{R}^3$ we mean a set $S \subset \mathbb{R}^3$ such that for each point $p \in S$ there is an open set $U \subset \mathbb{R}^3$ such that $S \cap U$ with the relative topology is homeomorphic to an open subset of $\mathbb{R}^2$.*

The topological type of a surface in $\mathbb{R}^3$ that bounds a domain is characterized by two numbers: The genus and the number of ends. A useful heuristic definition of the genus is that a surface is of genus g if it is homeomorphic to a sphere with g handles attached. A more precise definition is as follows:

Definition 2.3.2 *If a surface is homeomorphic to a sphere, then its* **genus** *is 0; if a surface is homeomorphic to a torus, then its* **genus** *is 1; and if a surface is homeomorphic to a connected sum of $g \geq 2$ tori, then its* **genus** *is g.*

Remark 2.3.3 It takes some effort to make precise the notion of "connected sum" used in Definition 2.3.2; the interested reader may consult Bloch [1], Chapter 2, for example. The Classification Theorem for Surfaces tells us that every oriented, compact, connected 2-dimensional surface has a genus as in Definition 2.3.2.

Definition 2.3.4 *Let S be a surface. If there exist a compact, connected 2-dimensional surface S' and m distinct points $p_1, p_2, \ldots, p_m \in S'$ such that S is homeomorphic to*

$$S' \setminus \{p_1, p_2, \ldots, p_m\},$$

then S is said have m **ends**.

2.3. SURFACES WITH CONSTANT MEAN CURVATURE

	genus 0	genus$\geq$ 1
compact	Round spheres	None
1 end	None	None
2 ends	Cylinders Delaunay surfaces	None
3 or more ends	Kapouleas surfaces	

Imbedded Surfaces with
Constant Positive Mean Curvature
Table 2.1: Existence of Surfaces

Recently N. Kapouleas [1], [2], [3] has developed methods for constructing many examples of constant mean curvature surfaces, and Korevaar, Kusner and Solomon [1] have found important conditions that constant mean curvature surfaces must satisfy. William H. Meeks, III [1] has shown there are no constant mean curvature imbedded surfaces with one end. That recent work combined with earlier results of of A. D. Aleksandrov[2] [1], [2] has provided the answers to all existence questions concerning properly imbedded constant mean curvature surfaces of finite topological type in $\mathbb{R}^3$. The state of affairs is summarized in Table 2.1.

In the next lemma we give a fundamental construction based on the Inverse Function Theorem that allows us to express our constant mean curvature surfaces as local graphs. This will allow us to develop the classical examples.

Lemma 2.3.5 *Suppose Ω is a C^k, $k \geq 2$, domain with C^k defining function ρ. If $P = (p_1, p_2, \ldots, p_N) \in \partial \Omega$ and $\partial \rho / \partial x_N(P) > 0$, then there exists a C^k function f, defined on a ball in $\mathbb{R}^{N-1}$ centered at $(p_1, p_2, \ldots, p_{N-1})$ such that*

$$(x, f(x)) \in \partial\Omega \text{ for all } x \in B, \qquad (2.42)$$

$$f(p_1, p_2, \ldots, p_{N-1}) = p_N \qquad (2.43)$$

and

$$(1 + |\nabla f|^2) \Delta f - \sum_{j,k=1}^{N-1} \frac{\partial f}{\partial x_j} \frac{\partial f}{\partial x_k} \frac{\partial^2 f}{\partial x_j \partial x_k}$$
$$= -(N-1)(1 + |\nabla f|^2)^{3/2} \mathbf{H}(x, f(x)). \qquad (2.44)$$

Proof: It is immediate from the Inverse Function Theorem that locally there exists a C^k function, f, satisfying

$$\rho(x_1, x_2, \ldots, x_{N-1}, f(x_1, x_2, \ldots, x_{N-1})) = 0 \qquad (2.45)$$

---
[2]The author's name is also transliterated as Alexandrov.

and satisfying (2.43). Of course (2.45) is equivalent to (2.42). Now we set

$$u(x_1, x_2, \ldots, x_N) = x_N - f(x_1, x_2, \ldots, x_{N-1}).$$

The level set $u = 0$ gives $\partial\Omega$ near P, and every other level set of u is simply a translate of $\partial\Omega$ in the x_N direction, with the same normal direction as the original surface. Thus the mean curvature operator applied to u will produce the value of the mean curvature at the corresponding point on $\partial\Omega$. That is, we have

$$|\nabla u|^2 \Delta u - \sum_{j,k=1}^{N} \frac{\partial u}{\partial x_j} \frac{\partial u}{\partial x_k} \frac{\partial^2 u}{\partial x_j \partial x_k} = (N-1)|\nabla u|^3 \, \mathbf{H}(x, f(x)). \quad (2.46)$$

Noting that

$$\frac{\partial u}{\partial x_j} = -\frac{\partial f}{\partial x_j} \quad \text{for } 1 \le j \le N-1,$$

$$\frac{\partial u}{\partial x_N} = 1,$$

$$\frac{\partial^2 u}{\partial x_j \partial x_k} = -\frac{\partial f}{\partial x_j \partial x_k} \quad \text{for } 1 \le j, k \le N-1,$$

$$\frac{\partial^2 u}{\partial x_j \partial x_N} = 0 \quad \text{for } 1 \le j \le N,$$

we observe that (2.46) implies (2.45). ∎

Example 2.3.6 The right circular cylinder of radius R with axis in the x_1-direction has the defining function

$$\rho(x_1, x_2, x_3) = x_2^2 + x_3^2 - R^2.$$

The condition $\frac{\partial \rho}{\partial x_3} > 0$ is satisfied on the top of the cylinder, that is, for $x_3 > 0$. The appropriate function $f(x_1, x_2)$ is

$$f(x_1, x_2) = \sqrt{R^2 - x_2^2}.$$

Computing with (2.44), we find the mean curvature of the surface is $\frac{1}{2R}$.

Now let us apply Lemma 2.3.5 to a surface of constant mean curvature in $\mathbb{R}^3$ that is rotationally symmetric about an axis. We will write $x = x_1$, $y = x_2$, and $z = x_3$. Suppose that the surface is symmetric about the x-axis and that the surface has constant mean curvature $C \ge 0$ when considered as the surface bounding the region containing the x-axis. If $r(x)$ is the function describing the radius of the cross section of the surface perpendicular to the x-axis, then we have

$$z = \sqrt{r(x)^2 - y^2}.$$

2.3. SURFACES WITH CONSTANT MEAN CURVATURE

We compute

$$\frac{\partial z}{\partial x} = r r' (r^2 - y^2)^{-1/2},$$

$$\frac{\partial z}{\partial y} = -y (r^2 - y^2)^{-1/2},$$

$$\frac{\partial^2 z}{\partial x^2} = r r'' (r^2 - y^2)^{-1/2} - y^2 (r')^2 (r^2 - y^2)^{-3/2},$$

$$\frac{\partial^2 z}{\partial x \partial y} = y r r' (r^2 - y^2)^{-3/2},$$

$$\frac{\partial^2 z}{\partial y^2} = -r^2 (r^2 - y^2)^{-3/2}.$$

Substituting these equations in (2.44), we are led to the equation

$$r'' = \frac{1 + (r')^2}{r} - 2C[1 + (r')^2]^{3/2}, \tag{2.47}$$

which must be satisfied by the radius of a surface of constant mean curvature symmetric about an axis. We have proved the following proposition.

Proposition 2.3.7 *If $r(x) > 0$ is a C^k, $k \geq 2$, function such that the boundary surface of the region*

$$S = \{(x, y, z) : y^2 + z^2 < r(x)^2\}$$

has constant mean curvature $C \geq 0$, then $r(x)$ satisfies the equation (2.47).

While the existence and uniqueness theorem for ordinary differential equations applies to (2.47) for all values of C, it is unfortunately the case that only when $C = 0$ can a solution be found relatively easily. We summarize the facts for the case $C = 0$ in the following corollary, which the reader can verify.

Corollary 2.3.8 *If $r(x) > 0$ is a C^2 function such that the boundary surface of the region*

$$S = \{(x, y, z) : y^2 + z^2 < r(x)^2\}$$

has zero mean curvature, then $r(x)$ is given by

$$r(x) = \alpha \cosh(\gamma x) + \beta \sinh(\gamma x),$$

where

$$\gamma^2 (\alpha^2 - \beta^2) = 1.$$

If, additionally,

$$r_0 = r(x_0) \quad \text{and} \quad r_1 = r'(x_0),$$

then
$$\gamma = \frac{\sqrt{1+r_1^2}}{r_0},$$
$$\alpha = r_0 \cosh(\gamma x_0) - \frac{r_1}{\gamma} \sinh(\gamma x_0),$$
$$\beta = \frac{r_1}{\gamma} \cosh(\gamma x_0) - r_0 \sinh(\gamma x_0).$$

Remark 2.3.9 A surface having mean curvature zero is called a **minimal surface** because any smooth surface that locally minimizes area must have mean curvature zero. There are many interesting results and problems involving minimal surfaces, but we will not pursue them in this book. We refer the reader to Osserman [2] and Nitsche [1].

Delaunay Surfaces

For our further consideration of rotationally symmetric constant mean curvature surfaces, we will assume that $C > 0$ and that r' in (2.47) is not identically zero. A first integral of (2.47) can be found by multiplying both sides of (2.47) by r' and computing

$$2C\,r\,r' = r'\,[1+(r')^2]^{-1/2} - r\,r'\,r''\,[1+(r')^2]^{-3/2}$$
$$= \frac{d}{dx}\left\{r\,[1+(r')^2]^{-1/2}\right\},$$

so

$$\sqrt{1+(r')^2} = \frac{r}{C\,r^2 + \alpha}. \qquad (2.48)$$

While one might anticipate eventually finding a solution for (2.48), it would seem unlikely that the solution would be more revealing than the equation. Delaunay's amazing discovery is that the solution can be described nicely. Pursuing this assertion requires that we treat an apparently unrelated topic.

A **roulette** is a curve generated by a point on a rolling curve (the curve may or may not be closed). The best known example of a roulette is the **cycloid** which is the curve generated by a point on the circumference of a circle as the circle rolls without slipping along a straight line. The **epicycloid** is the curve generated by a point in the interior of a circular disc as the disc rolls without slipping along a straight line. The **hypocycloid** is similarly generated by a point exterior to the circle as the circle rolls without slipping along a straight line. We will also consider the epicycloid and the hypocycloid to be roulettes of a circle.

To consider a general roulette, let Γ be a plane curve and let q be the point which is to generate the roulette. Thus, the curve Γ is assumed to be in contact with a line L, and Γ is to be acted on by rigid motions, say parametrized by s, so that Γ and L remain in contact with the point of contact between Γ and L, as a function of s, giving an arc-length parametrization of both Γ and L. The rigid motions constitute the "rolling" and the

2.3. SURFACES WITH CONSTANT MEAN CURVATURE

requirement that the contact point gives an arc-length parametrization of *both* the curve and the line is what is meant by rolling "without slipping." The roulette is the path of q under the same rigid motions associated with rolling without slipping. The point q need not be a point on Γ.

Assume that $p(s)$ is the vector from q to a point on the curve giving an arc-length parametrization of Γ. The easiest approach in deriving the equation of the roulette is to consider a line, L, rolling without slipping along the curve Γ and to then give the coordinates of q relative to the line. Assume the initial point of contact between Γ and L is $p(0)$. We also associate with L a right-handed cartesian coordinate system with coordinates x and y as follows: The coordinate system associated with L will have L as its x-axis, the point $x = 0$ will be at the initial point of contact, $p(0)$, and the positive x-direction will be the unit tangent direction at the initial point of contact, that is, $p'(0)$. The y-axis is then chosen to give a right-handed coordinate system.

As the line L rolls along Γ without slipping, the point of contact between the line and the curve at $p(s)$ will have x-coordinate s. The positive x-direction will agree with $p'(s)$, the unit tangent vector to Γ at s, denoted $T(s)$. The vector from the point of contact of Γ and L to the point q is $-p(s)$, so the x-coordinate of q is given by

$$x(s) = s - p(s) \cdot T(s).$$

Recall from Definition 2.2.20 that the curvature κ of Γ is

$$\kappa = \left|\frac{d^2p}{ds^2}\right| = \left|\frac{dT}{ds}\right|.$$

The behavior of the y-coordinate of q is simplified by assuming that the curvature of Γ does not vanish at any point and that the unit tangent vector T and the unit normal vector N given by

$$N = \kappa^{-1}\frac{dT}{ds},$$

form a right-handed system. In that case, we have

$$y(s) = -p(s) \cdot N(s).$$

We have proved the following proposition.

Proposition 2.3.10 *If the C^k, $k \geq 2$, curve Γ has non-vanishing curvature and the arc-length parametrization $p(s)$ is such that the unit tangent vector $T(s)$ and unit normal vector $N(s)$ form a right-handed system, then the roulette of the point q at the origin is given parametrically by*

$$x(s) = s - p(s) \cdot T(s), \qquad (2.49)$$

$$y(s) = -p(s) \cdot N(s). \qquad (2.50)$$

Figure 2.1: **The Cycloid**

Example 2.3.11 Let a circle of radius r be parametrized by

$$p(s) = \begin{pmatrix} -a + r\cos(s/r) \\ r\sin(s/r) \end{pmatrix},$$

where $a > 0$. The center of this circle is at the point with coordinates $(-a, 0)$, so the point q that will generate the roulette is a units to the right of the center of the circle. We compute

$$T(s) = p'(s) = \begin{pmatrix} -\sin(s/r) \\ \cos(s/r) \end{pmatrix}, \quad N(s) = \begin{pmatrix} -\cos(s/r) \\ -\sin(s/r) \end{pmatrix}.$$

So the parametrization of the roulette is

$$x(s) = s - a\sin(s/r),$$
$$y(s) = r - a\cos(s/r).$$

When $0 < a < r$ this roulette is an epicycloid, when $a = r$ it is a cycloid (see Figure 2.1), and when $r < a$ it is a hypocycloid.

In practice, it is usually not particularly advantageous to eliminate the parameter s from the pair of equations (2.49) and (2.50) for $x(s)$ and $y(s)$, the previous example being a case in point.

Notice that, since we are working in the plane, the direction of the y-axis is determined by the direction of the x-axis. By making a convexity assumption, we can obtain a form for the parametrization of a roulette that avoids the requirement that the curve Γ be parametrized by arc-length and that even applies if Γ is only C^1. One should visualize the curve Γ *and* the region it encloses rolling along a straight line. We consider the curve traced out by a chosen interior point q, and for convenience of notation we suppose that that interior point is initially at the origin.

Proposition 2.3.12 *If the C^k, $k \geq 1$, curve Γ encloses a convex region with the origin in its interior, and is parametrized in the positively oriented direction by $p(t)$, then the roulette of the point q at the origin is given parametrically by*

$$x(t) = s(t) - |p'(t)|^{-1} p(t) \cdot p'(t), \tag{2.51}$$

$$y(t) = \sqrt{|p(t)|^2 - |p'(t)|^{-2}(p(t) \cdot p'(t))^2}, \tag{2.52}$$

2.3. SURFACES WITH CONSTANT MEAN CURVATURE

where $s(t)$ is the arc-length from $p(0)$ to $p(t)$.

Proof: We know that the unit tangent vector is given by $|p'(t)|^{-1}p'(t)$, so the formula for $x(t)$ is immediate. We also know that

$$|p|^2 = (p \cdot T)^2 + (p \cdot N)^2,$$

where N is such that T and N form a right-handed basis. Since the origin is interior to the region enclosed by Γ, we know that $p \cdot N$ must be negative, and the formula for $y(t)$ follows. ∎

Looking at Propositions 2.3.10 and 2.3.12, we may think that the roulette has lost one order of differentiablity compared to the original curve, but that is an illusion. In fact, we will see that a roulette is as smooth as the given parametrization of Γ:

Proposition 2.3.13 *If the curve Γ has non-vanishing curvature, encloses a convex region with the origin in its interior, and is parametrized by arclength in the positively oriented direction by the C^k, $k \geq 3$, function $p(s)$, then the roulette of the point q at the origin satisfies*

$$\frac{dx}{ds} = -\kappa p \cdot N > 0, \qquad (2.53)$$

$$\frac{dy}{ds} = \kappa p \cdot T, \qquad (2.54)$$

$$\frac{d^2x}{ds^2} = \kappa^2 p \cdot T - \frac{d\kappa}{ds} p \cdot N, \qquad (2.55)$$

$$\frac{d^2y}{ds^2} = \kappa + \frac{d\kappa}{ds} p \cdot T + \kappa^2 p \cdot N. \qquad (2.56)$$

Moreover, the roulette is the graph over the x-axis of a positive C^k function satisfying the differential equation

$$y \frac{dy}{dx} = F(x),$$

where $F(x)$ is a C^{k-1} function.

Proof: Equations (2.53) – (2.56) follow from (2.49) and (2.50) and the Frenet formula (see Theorem 2.2.21)

$$\frac{dN}{ds} = -\kappa T.$$

From (2.53) and (2.54) it follows that

$$\frac{dy}{dx} = -\frac{p(s) \cdot T(s)}{p(s) \cdot N(s)},$$

so from (2.50) we have
$$y \frac{dy}{dx} = p(s) \cdot T(s). \tag{2.57}$$

By (2.53) we can apply the Inverse Function Theorem to (2.49) to express s as a C^{k-1} function of x and consequently $p(s) \cdot T(s)$ is a C^{k-1} function of x, which we denote by F. ∎

From the proof of the theorem we obtain the following useful corollary.

Corollary 2.3.14 *Under the conditions of Theorem 2.3.13 the function $y(x)$ describing the roulette satisfies*

$$y \sqrt{1 + \left(\frac{dy}{dx}\right)^2} = |p|.$$

Proof: We use (2.50) and (2.57) to compute
$$\begin{aligned} |p|^2 &= (p \cdot N)^2 + (p \cdot T)^2 \\ &= y^2 + \left(y \frac{dy}{dx}\right)^2. \end{aligned}$$
∎

Example 2.3.15 We consider the roulette of an ellipse about one of its foci (see Figure 2.2). Let an ellipse with major axis of length $2a$ and minor axis of length $2b$ be parametrized by

$$p(\theta) = \begin{pmatrix} -c + a \cos \theta \\ b \sin \theta \end{pmatrix},$$

where $a^2 = b^2 + c^2$, thus the foci are $(-2c, 0)$ and $(0, 0)$. We compute

$$p'(\theta) = \begin{pmatrix} -a \sin \theta \\ b \cos \theta \end{pmatrix}.$$

A little manipulation then gives us
$$|p| = a - c \cos \theta, \tag{2.58}$$
$$|p'| = [a^2 - c^2 \cos^2(\theta)]^{1/2}, \tag{2.59}$$
$$p \cdot p' = c \sin \theta \, [a - c \cos \theta]. \tag{2.60}$$

Using (2.52), we find that
$$y = b \sqrt{\frac{a - c \cos \theta}{a + c \cos \theta}},$$

or, if we solve for $\cos \theta$,
$$\cos \theta = \frac{a}{c} \frac{b^2 - y^2}{b^2 + y^2}. \tag{2.61}$$

2.3. SURFACES WITH CONSTANT MEAN CURVATURE

Figure 2.2: **Roulette of the Focus of an Ellipse**

Since (2.51) will produce an equation for x in terms of $\cos\theta$ and $\sin\theta$, we could use (2.61) to eliminate θ and obtain a solution. Unfortunately, the solution obtained involves an elliptic integral. An alternative approach is to use Corollary 2.3.14 to write

$$\cos\theta = \frac{a - |p|}{c} = \frac{1}{c}\left(a - y\sqrt{1 + \left(\frac{dy}{dx}\right)^2}\right), \qquad (2.62)$$

so we can eliminate $\cos\theta$ from equations (2.61) and (2.62) and see that *the roulette of an ellipse generated by a focus of the ellipse satisfies the differential equation*

$$\sqrt{1 + \left(\frac{dy}{dx}\right)^2} = \frac{2ay}{y^2 + b^2} = \frac{y}{\frac{1}{2a}y^2 + \frac{b^2}{2a}}. \qquad (2.63)$$

Notice that the differential equation satisfied by the roulette of an ellipse (2.63) is the same as the differential equation (2.48) satisfied by the function $r(x)$ describing a rotationally symmetric surface with constant mean curvature $\frac{1}{2a}$. Thus we have obtained an aesthetically pleasing description of the solution of (2.48). In general, Delaunay discovered that a roulette formed by the focus of any conic section gives the longitudinal cross-section of a rotationally symmetric surface of constant mean curvature. The interested reader should also consult the article Eells [1].

Chapter 3

Measures

3.1 The Carathéodory Construction

Let $\mathcal{F}$ be a collection of subsets of $\mathbb{R}^N$. Let

$$\sigma : \mathcal{F} \to \{x \in \mathbb{R} : x \geq 0\}.$$

We refer to the function σ as a **gauge**. We need to make the following two assumptions concerning $\mathcal{F}$ and σ :

> For $\delta > 0$, each $E \subset \mathbb{R}^N$ has a covering by a set of at most countably many members of $\mathcal{F}$ having diameter less than δ. (3.1)

$$0 = \lim_{\delta \downarrow 0} \left(\sup\{\sigma(S) : S \in \mathcal{F},\ \text{diam}(S) < \delta\} \right). \tag{3.2}$$

Carathéodory's idea for constructing a measure from $\mathcal{F}$ and σ is as follows:

Definition 3.1.1 (Carathéodory [1])

(i) *Let $E \subset \mathbb{R}^N$ be any set. Let $\delta > 0$. Call $\mathcal{O} \subset \mathcal{F}$ a δ-admissible cover of E if $\mathcal{O}$ has at most countably many elements, each element of $\mathcal{O}$ has diameter less than δ, and finally*

$$E \subset \bigcup \mathcal{O}. \tag{3.3}$$

Let $\mathcal{A}_\delta = \mathcal{A}_\delta(E)$ denote the collection of all δ-admissible covers of E.

(ii) *Set*

$$\mathcal{H}_\delta(E) = \inf_{\mathcal{O} \in \mathcal{A}_\delta} \sum_{S \in \mathcal{O}} \sigma(S). \tag{3.4}$$

We call $\mathcal{H}_\delta$ the size δ approximating measure associated with $\mathcal{F}$ and σ.

(iii) Set
$$\mathcal{H}(E) = \lim_{\delta \to 0^+} \mathcal{H}_\delta(E). \tag{3.5}$$
We call $\mathcal{H}$ the **result of Carathéodory's construction from** $\mathcal{F}$ **and** σ.

By (3.1), δ-admissible covers will exist for any $\delta > 0$ and any $E \subset \mathbb{R}^N$. Also, notice that, by definition, if $0 < \delta_1 < \delta_2$, then $\mathcal{H}_{\delta_1}(E) \geq \mathcal{H}_{\delta_2}(E)$, so the limit on the right-hand side of (3.5) exists. The condition (3.2) is included to insure that $\mathcal{H}_\delta(\emptyset) = \mathcal{H}(\emptyset) = 0$, though some may feel this is redundant since the empty cover covers the empty set, and the empty sum is zero.

We leave it to the reader to verify the following easy result.

Lemma 3.1.2

(i) *For any $\delta > 0$, $\mathcal{H}_\delta$ is a measure, and if all elements of $\mathcal{F}$ are Borel sets, then every $E \subset \mathbb{R}^N$ is contained in a Borel set B with $\mathcal{H}_\delta(B) = \mathcal{H}_\delta(E)$.*

(ii) *$\mathcal{H}$ is a measure, and if all elements of $\mathcal{F}$ are Borel sets, then $\mathcal{H}$ is a Borel regular measure.*

We turn immediately to the applications of Carathéodory's construction that are of greatest importance.

Notation 3.1.3 *We will use Υ_k to denote the volume of the unit ball in $\mathbb{R}^k$. It is known (see Proposition A.2.6) that*
$$\Upsilon_k = 2\frac{\pi^{k/2}}{k\Gamma(k/2)}.$$

Let $\mathcal{F}$ be the collection of open Euclidean balls in $\mathbb{R}^N$. For $B = B(x,r) \in \mathcal{F}$ and $k = 1, 2, \ldots$, let $\sigma_k(B) = \Upsilon_k r^k$. Since Υ_k is the volume of the unit ball in $\mathbb{R}^k$, we see that $\sigma_N(B)$ is actually the Euclidean volume of B. One advantage of this last formula is that the right-hand side is defined for any positive real number α (instead of k) and allows us to define Υ_α for positive real α's. We also define σ_α for any positive real number α by

$$\sigma_\alpha(B) = 2\frac{\pi^{\alpha/2} r^\alpha}{\alpha \Gamma(\alpha/2)}. \tag{3.6}$$

The result of Carathéodory's construction from this $\mathcal{F}$ and this σ_α will be denoted by $\mathcal{S}^\alpha$. This measure is the α-**dimensional spherical measure** on $\mathbb{R}^N$. Let us consider spherical measure by way of several examples.

Example 3.1.4 Let $E \subset \mathbb{R}$ be a segment of Euclidean length ℓ. Then
$$\mathcal{S}^\alpha(E) = \begin{cases} +\infty & \text{if } \alpha < 1 \\ \ell & \text{if } \alpha = 1 \\ 0 & \text{if } \alpha > 1. \end{cases}$$

3.1. THE CARATHÉODORY CONSTRUCTION

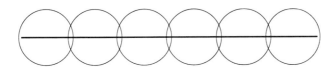

Figure 3.1: **Covering a Segment by Euclidean Balls**

To verify this assertion, we first make the geometrically obvious (and not difficult to check rigorously [Hurewicz and Wallman, 1]) statement that the most efficient way to cover E with Euclidean balls of radius less than δ is to lay them side by side with arbitrarily small overlap; see Figure 3.1. Taking the balls to be of radius about δ, we see that there are about $\ell/(2\delta)$ of these balls.

Now if $\alpha < 1$ then we see that the sum in (3.4) is about

$$(\ell/2\delta) \cdot \Upsilon_\alpha \delta^\alpha = (\Upsilon_\alpha \ell/2) \cdot \delta^{\alpha-1}. \qquad (3.7)$$

As $\delta \to 0^+$, we find that this expression converges to $+\infty$.

If $\alpha > 1$ then the calculation is the same, but now the expression in (3.7) converges to 0.

Finally, in case $\alpha = 1$ then (3.7) converges to $\Upsilon_1 \cdot \ell/2 = \ell$.

Example 3.1.5 Let us investigate how the last example changes if the segment E is now considered as a subset of $\mathbb{R}^N$. The beauty of this theory is that the picture (Figure 3.1) and the calculations from the last example are exactly the same. The result is

$$\mathcal{S}^\alpha(E) = \begin{cases} +\infty & \text{if } \alpha < 1 \\ \ell & \text{if } \alpha = 1 \\ 0 & \text{if } \alpha > 1. \end{cases}$$

Example 3.1.6 If $E \subset \mathbb{R}^N$ is *any* set and if $\alpha > N$ then $\mathcal{S}^\alpha(E) = 0$. To see this, first assume that E is a closed cube of side length ℓ. For any positive integer k, we can subdivide the closed cube E into k^N closed subcubes of side length ℓ/k that intersect only on their boundaries. Each subcube can be covered by a closed ball of diameter $\sqrt{N}\ell/k$ with center at the center of the subcube; see Figure 3.2. So for δ slightly larger than $\sqrt{N}\ell/k$ the sum in (3.4) becomes

$$k^N \cdot \Upsilon_\alpha \cdot \left(2^{-1}\sqrt{N}\ell/k\right)^\alpha. \qquad (3.8)$$

Since $\alpha > N$, it is easy to see that this expression tends to 0 as $k \to \infty$, thus as $\delta \to 0^+$.

Since $\mathcal{S}^\alpha$ is subadditive, the full result follows.

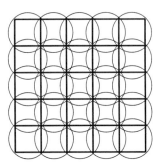

Figure 3.2: **Covering a Cube by Euclidean Balls**

Example 3.1.7 Let $E \subset \mathbb{R}^N$ be a compact set with interior. If $\alpha < N$ then $S^\alpha(E) = +\infty$. To see this, first assume that E is the closure of a cube. We proceed as in the last example, subdividing the cube into subcubes of side length ℓ/k. For δ less than or equal to ℓ/k, the side length of the subcubes, each ball can contain the center of no more than one subcube, so the sum in (3.4) is at least

$$k^N \cdot \Upsilon_\alpha \cdot \left(2^{-1}\sqrt{N}\ell/k\right)^\alpha. \tag{3.9}$$

Since $\alpha < N$, the expression (3.9) tends to $+\infty$ as $\delta \to 0^+$.

Finally, if E is any compact set with interior then E certainly contains the closure of a cube. Since S^α is subadditive, we are done with the general case.

Example 3.1.8 If $Q \subset \mathbb{R}^N$ is the cube

$$Q = \{(x_1, \ldots, x_N) \in \mathbb{R}^N : 0 \leq x_j \leq 1\}$$

then we have

$$S^\alpha(Q) = \begin{cases} +\infty & \text{if } \alpha < N \\ 1 & \text{if } \alpha = N \\ 0 & \text{if } \alpha > N. \end{cases}$$

The first and last cases are covered by the preceding two examples. Thus we need only treat the case $\alpha = N$.

In order to keep the length of this calculation under control, we shall invoke some facts from standard measure theory (see Federer [4] for details). Recall from Definition 1.3.20 that $\mathcal{L}^N$ denotes the usual Lebesgue measure in $|RR^N$. Now fix a $\delta > 0$. Also fix a small $\tau > 0$. By the Vitali covering lemma (see Section 3.5), we see that there are finitely many pairwise disjoint open balls B_j, all of radius less than δ, that lie in Q and are such that

$$\left|\mathcal{L}^N(Q) - \sum_j \mathcal{L}^N(B_j)\right| < \tau.$$

3.1. THE CARATHÉODORY CONSTRUCTION

The set $T = Q \setminus (\cup_j B_j)$ is still compact and has Lebesgue measure less than τ.

By the outer regularity of Lebesgue measure, we may find an open set $W \supseteq T$ that has measure not exceeding 2τ. Next, we may cover T with finitely many balls C_k, of radii less than δ, and lying in W. By the covering theorem of Lebesgue (see Section 3.5 or Hurewicz and Wallman [1]), we may assume (by passing to a subcovering) that no point of T is covered more than $N+1$ times.

In summary, the balls B_j, C_k are finite in number, have radii less than δ, and cover Q. Thus we have an upper bound for $\mathcal{S}_\delta^N(Q)$ given by

$$\sum_j \Upsilon_N [\tfrac{1}{2}\mathrm{diam}(B_j)]^N + \sum_k \Upsilon_N [\tfrac{1}{2}\mathrm{diam}(C_k)]^N$$
$$= \sum_j \mathcal{L}^N(B_j) + \sum_k \mathcal{L}^N(C_k)$$
$$\leq \mathcal{L}^N(Q) + (N+1) \cdot \mathcal{L}^N(T)$$
$$< \mathcal{L}^N(Q) + (N+1) \cdot \tau.$$

Since τ was arbitrary, we find that $\mathcal{S}_\delta^N(Q) \leq \mathcal{L}^N(Q) = 1$.

On the other hand, if the balls B_j have radii less than δ and cover Q then

$$\sum_j \Upsilon_N [\tfrac{1}{2}\mathrm{diam}(B_j)]^N = \sum_j \mathcal{L}^N(B_j) \geq \mathcal{L}^N(Q) = 1.$$

Thus $\mathcal{S}_\delta^N(Q) \geq 1$.

Combining our two estimates yields that $\mathcal{S}_\delta^N(Q) = 1$. Letting $\delta \to 0^+$ yields that $\mathcal{S}^N(Q) = 1$.

Hausdorff Dimension

Another measure closely related to spherical measure is the Hausdorff measure. The α-**dimensional Hausdorff measure**, denoted by $\mathcal{H}^\alpha$, is the result of Carathéodory's construction from the family of all bounded nonempty subsets of $\mathbb{R}^N$ and the gauge

$$\sigma_\alpha(A) = 2 \frac{\pi^{\alpha/2}}{\alpha \Gamma(\alpha/2)} \left(\frac{\mathrm{diam}(A)}{2} \right)^\alpha. \tag{3.10}$$

Since this gauge is an extension of σ_α to a larger family of sets, it is appropriate to use the same notation. It also follows immediately, from the fact that Hausdorff measure is defined by extending the gauge to a larger family of sets, that $\mathcal{H}^\alpha \leq \mathcal{S}^\alpha$.

Any non-empty bounded set A is contained in an open ball with radius larger than, but arbitrarily close to, the diameter of A. It follows from this observation that $\mathcal{S}^\alpha \leq 2^\alpha \mathcal{H}^\alpha$. In particular, we know that *spherical measure and Hausdorff measure have the same sets of measure zero.*

Calculations of the sort made in Example 3.1.8 can be used to show that if E is a polyhedron, or more generally the closure of a domain with C^1 boundary, then the $\mathcal{S}^N$ measure of E, the $\mathcal{H}^N$ measure of E, and the Lebesgue measure of E are all equal. See Federer [4].

Definition 3.1.9 *Let $E \subset \mathbb{R}^N$ be a compact set. Let*

$$\begin{aligned}\mathcal{D}(E) &= \inf\{\alpha > 0 : \mathcal{H}^\alpha(E) = 0\} \\ &= \inf\{\alpha > 0 : \mathcal{S}^\alpha(E) = 0\}.\end{aligned}$$

We call $\mathcal{D}(E)$ the **Hausdorff dimension** *of the set E.*

Remark 3.1.10 If $E \subset \mathbb{R}^N$ is compact and $\mathcal{D}(E) > 0$, then we also have

$$\begin{aligned}\mathcal{D}(E) &= \sup\{\alpha > 0 : \mathcal{H}^\alpha(E) = \infty\} \\ &= \sup\{\alpha > 0 : \mathcal{S}^\alpha(E) = \infty\}.\end{aligned}$$

In an informal treatment of the Cantor ternary set, we usually add up the lengths of the segments in the complement and find that the sum is 1. So, naively, the Cantor set has length zero. The measures we have introduced allow us to make that calculation more rigorous. Now we may ask the more precise question: What is the Hausdorff dimension of the Cantor set?

An elementary calculation, which we leave to the reader, determines that the Hausdorff dimension of the standard Cantor set is $\ln 2/\ln 3$. By varying the lengths of the intervals removed, and the rate at which they are removed, one can create Cantor sets of any Hausdorff dimension strictly between 0 and 1 (see Falconer [1], Section 1.5).

Although variants of the ideas presented here have reappeared in the mathematical literature in modern guise, and have been dubbed "fractal dimension", it should be noted that the original idea is due to F. Hausdorff and is at least eighty years old.

> **Exercise:** If $M \subset \mathbb{R}^N$ is a C^1, compact hypersurface, in particular if it is the boundary of a bounded domain, then the Hausdorff dimension of M is precisely $N - 1$. [**Hint:** First, the problem is local so work on a coordinate patch in the boundary. Also, the problem is invariant under diffeomorphisms so map the coordinate patch to a piece of $\mathbb{R}^{N-1}$. But in the examples above we have already (implicitly) determined the Hausdorff dimension of $\mathbb{R}^{N-1}$.]

Other Measures

The general scheme of Carathéodory's construction can be usefully applied for many choices of the family of subsets $\mathcal{F}$ and of the gauge σ.

3.1. THE CARATHÉODORY CONSTRUCTION

Definition 3.1.11 *The following measures are defined on $\mathbb{R}^N$ by applying Carathéodory's construction:*

(i) *The α-dimensional Hausdorff measure, $\mathcal{H}^\alpha$, $\alpha \in \mathbb{R} \cap \{t : t \geq 0\}$, results when*

$$\mathcal{F} \text{ is the family of all subsets of } \mathbb{R}^N \qquad (3.11)$$

and $\sigma = \sigma_\alpha$ where

$$\sigma_\alpha(A) = 2\frac{\pi^{\alpha/2}}{\alpha \Gamma(\alpha/2)} \left(\frac{\operatorname{diam}(A)}{2}\right)^\alpha. \qquad (3.12)$$

(ii) *The α-dimensional spherical measure, $\mathcal{S}^\alpha$, $\alpha \in \mathbb{R} \cap \{t : t \geq 0\}$, results when*

$$\mathcal{F} \text{ is the family of all open balls in } \mathbb{R}^N \qquad (3.13)$$

and $\sigma = \sigma_\alpha$ is as in (i).

(iii) *The m-dimensional integral geometric measure with exponent t, $\mathcal{I}_t^m$, $m \in \{0, 1, 2, \ldots\}$, $t \in \{r : 1 \leq r \leq \infty\}$, results when*

$$\mathcal{F} \text{ is the family of all Borel subsets of } \mathbb{R}^N \qquad (3.14)$$

and $\sigma = \sigma_{m,t}$ where

$$\sigma_{m,t}(A) = c_{m,t} \left(\int_{\substack{\text{orthogonal} \\ \text{projections} \\ \text{from } \mathbb{R}^N \text{ to } \mathbb{R}^m}} |\mathcal{L}^m[p(A)]|^t \, d\mu p \right)^{1/t} \quad \text{if } t < \infty, \quad (3.15)$$

with μ the invariant measure on the space of orthogonal projections from $\mathbb{R}^N$ to $\mathbb{R}^m$ [Federer 4, §2.7.16(6)] and with $c_{m,t}$ an appropriate normalizing constant, and

$$\sigma_{m,\infty}(A) \text{ is the essential supremum of } \mathcal{L}^m[p(A)], \qquad (3.16)$$

with $\mathcal{L}^m[p(A)]$ considered as a function on the space of orthogonal projections from $\mathbb{R}^N$ to $\mathbb{R}^m$ with the invariant measure.

(iv) *The m-dimensional Gross measure, $\mathcal{G}^m$, $m \in \{0, 1, 2, \ldots\}$, results when*

$$\mathcal{F} \text{ is the family of all Borel subsets of } \mathbb{R}^N \qquad (3.17)$$

and $\sigma = \sigma_{G,m}$ where

$$\sigma_G(A) = \sup\left\{\mathcal{L}^m[p(A)] : p \text{ orthogonally projects } \mathbb{R}^N \text{ to } \mathbb{R}^m \right\}. \qquad (3.18)$$

(v) *The m-dimensional convex integral geometric measure with exponent t, $\mathcal{Q}_t^m$, $m \in \{0, 1, 2, \ldots\}$, $t \in \{r : 1 \leq r \leq \infty\}$, results when*

$$\mathcal{F} \text{ is the family of all open convex subsets of } \mathbb{R}^N \tag{3.19}$$

and $\sigma = \sigma_{m,t}$ is as in (iii).

(vi) *The m-dimensional convex integral geometric measure with exponent 1, $\mathcal{Q}_1^m$, is also known as the m-dimensional **Gillespie measure**.*

(vii) *The m-dimensional convex integral geometric measure with exponent ∞, $\mathcal{Q}_\infty^m$, is also known as the m-dimensional **Carathéodory measure**.*

(viii) *The m-dimensional **symmetrized Hausdorff measure**, $\mathcal{T}^m$, $m \in \{0, 1, 2, \ldots\}$, results when*

$$\mathcal{F} \text{ is the family of all subsets of } \mathbb{R}^N \tag{3.20}$$

and $\sigma = \sigma_{S,m}$ where

$$\begin{aligned}\sigma_{S,m}(A) = 2^{-m} \Upsilon_m \sup\{ &\, m\text{-}dimensional \text{ area of } m\text{-}dimensional \\ &\, parallelepipeds \text{ formed from vectors } p_i - q_i, \\ &\, i = 1, 2, \ldots m, \text{ with } p_i, q_i \in A\} \end{aligned} \tag{3.21}$$

Remark 3.1.12 While all the measures defined in Definition 3.1.11 give the same value on any open piece of a smooth surface and all coincide with Lebesgue measure if $N = m$, they are not all equal on arbitrary sets. The interested reader should see the references cited in Federer [4], §2.10.6.

Definition 3.1.13 *For measures μ and ν on $\mathbb{R}^N$, we will write*

(i)
$$\mu \cong \nu$$

if and only if there exist non-negative real constants c_1 and c_2 such that
$$c_1 \mu(A) \leq \nu(A) \leq c_2 \mu(A)$$
holds for all $A \subset \mathbb{R}^N$,

(ii)
$$\mu \simeq \nu$$

if and only if
$$\mu(A) = 0 \iff \nu(A) = 0$$
for all $A \subset \mathbb{R}^N$.

3.1. THE CARATHÉODORY CONSTRUCTION

We collect the relevant facts in the next theorem. Proofs and examples to substantiate these claims can be found in Federer [4], §2.10.6 and the references cited therein.

Theorem 3.1.14

(i) For $m \in \{0, 1, 2, \ldots\}$, $1 \leq t < \infty$,
$$\mathcal{S}^m \geq \mathcal{H}^m \geq \mathcal{T}^m \geq \mathcal{Q}_\infty^m \geq \mathcal{G}^m \geq \mathcal{I}_\infty^m \geq \mathcal{I}_t^m.$$

(ii) For $m \in \{0, 1, 2, \ldots\}$,
$$\mathcal{Q}_1^m \geq \mathcal{Q}_\infty^m.$$

(iii) For $m \in \{0, 1, 2, \ldots\}$, $s, t \in \{r : 1 \leq r \leq \infty\}$,
$$\mathcal{S}^m \cong \mathcal{H}^m \cong \mathcal{T}^m \cong \mathcal{Q}_t^m \cong \mathcal{Q}_s^m.$$

(iv) For $m \in 1, 2, \ldots$, the measures defined in Definition 3.1.11 form the following three equivalence classes under the relation $\simeq$

$$\{\mathcal{S}^m, \mathcal{H}^m, \mathcal{T}^m\} \cup \{\mathcal{Q}_t^m : 1 \leq t \leq \infty\} \qquad (3.22)$$
$$\{\mathcal{G}^m\} \qquad (3.23)$$
$$\{\mathcal{I}_t^m : 1 \leq t \leq \infty\}. \qquad (3.24)$$

In general, the interaction of mappings and measures can be complicated, but for the two most important measures in Definition 3.1.11 certain facts follow easily by considering the effects of the mapping on the gauge.

Definition 3.1.15 *A function $f : X \to Y$, with X and Y subsets of Euclidean spaces, is a* **Lipschitz function**[1] *if there exists $C < \infty$ such that*

$$|f(x_1) - f(x_2)| \leq C|x_1 - x_2| \quad \text{holds for all} \quad x_1, x_2 \in X. \qquad (3.25)$$

The **Lipschitz constant** *of the Lipschitz function f is the smallest number C such that (3.25) holds. The Lipschitz constant of f is denoted by $\mathrm{Lip}(f)$.*

Theorem 3.1.16 *Suppose $\alpha \in \mathbb{R} \cap \{t : t \geq 0\}$. If $f : \mathbb{R}^N \to \mathbb{R}^M$ satisfies*

$$|f(x) - f(y)| \leq L|x - y|,$$

i.e. if f is a Lipschitz function with Lipschitz constant not exceeding L, then for any $A \subset \mathbb{R}^N$

$$\mathcal{H}^\alpha[f(A)] \leq L^\alpha \mathcal{H}^\alpha(A), \quad \mathcal{S}^\alpha[f(A)] \leq L^\alpha \mathcal{S}^\alpha(A). \qquad (3.26)$$

---
[1] Note that our Lipschitz functions are a special case of the Lipschitz or Hölder functions of harmonic analysis (see Krantz [1]).

Proof: For Hausdorff measure, note that the diameter of the f image of any set A is bounded by L times the diameter of A. For spherical measure, note that the f image of the sphere centered at P with radius r is contained in the sphere centered at $f(P)$ with radius Lr. ∎

The following special case of the theorem is so important that we state it separately.

Corollary 3.1.17 *Suppose $\alpha \in \mathbb{R} \cap \{t : t \geq 0\}$. If $\Pi : \mathbb{R}^N \to \mathbb{R}^M$ is an orthogonal projection or orthogonal injection, then for any $A \subset \mathbb{R}^N$*

$$\mathcal{H}^\alpha[\Pi(A)] \leq \mathcal{H}^\alpha(A), \quad \mathcal{S}^\alpha[\Pi(A)] \leq \mathcal{S}^\alpha(A). \tag{3.27}$$

3.2 Rectifiability

Definition 3.2.1 *By an **arc** is meant a homeomorphic image of some closed interval $[a,b] \subset \mathbb{R}$. If Γ is an arc in $\mathbb{R}^N$, say given as the homeomorphic image of $[a,b]$ under $\gamma : [a,b] \to \mathbb{R}^N$, then the **length** of Γ, denoted $\mathrm{len}(\Gamma)$, is given by*

$$\mathrm{len}(\Gamma) = \sup\left\{ \sum_{i=1}^m |\gamma(t_{i-1}) - \gamma(t_i)| : a = t_0 < t_1 < t_2 < \ldots < t_m = b \right\}. \tag{3.28}$$

*If the length of an arc is finite, then the arc is said to be **rectifiable**.*

Remark 3.2.2 Of course, the length of an arc may be infinite. Early in the twentieth century, W. F. Osgood was motivated by considering the boundary behavior of conformal mappings to construct an example of an arc with positive area in the plane (see Osgood [1]). In fact, examples show that an arc in $\mathbb{R}^N$ may have positive N-dimensional Lebesgue measure. (See Parks and Schori [1] for a construction based on the heuristic principle that given a Cantor set one can always run an arc through every point of it.) Note that because an arc is required to be the *homeomorphic* image of an interval, examples of space-filling curves (see Section 3.4) tell us nothing concrete about how badly behaved an arc may be.

Lemma 3.2.3 *If Γ is a rectifiable arc in $\mathbb{R}^N$, then Γ can be parametrized by arc length, meaning that with $\ell = \mathrm{len}(\Gamma)$, there exists a homeomorphism $\gamma_0 : [0, \ell] \to \mathbb{R}^N$ such that for each $0 < s \leq \ell$ the image of $[0, s]$ has length s.*

Proof: Let $\gamma : [a,b] \to \mathbb{R}^N$ be the given parametrization of Γ. Define $\gamma_0(s) = \gamma(t)$ where $t \in [a,b]$ is that unique number such that

$$\mathrm{len}\Big(\gamma([a,t])\Big) = s. \qquad \blacksquare$$

3.2. RECTIFIABILITY

Theorem 3.2.4 *If Γ is an arc, then*

$$\text{len}(\Gamma) = \mathcal{H}^1(\Gamma). \tag{3.29}$$

Proof: By the fact that orthogonal projection onto a line in $\mathbb{R}^N$ does not increase 1-dimensional Hausdorff measure, we can see that $\mathcal{H}^1(\Gamma) \geq \text{len}(\Gamma)$. The reverse inequality holds because the arc-length parametrization of a rectifiable curve is a map from $\mathbb{R}$ to $\mathbb{R}^N$ with Lipschitz constant 1. ∎

The preceding results show that the natural definition of length arising by considering inscribed polygons is well-behaved and agrees with the 1-dimensional Hausdorff measure. The situation in higher dimensions is far more complicated as the following example due to H. A. Schwarz shows (see Gelbaum and Olmstead [1] for more detail).

Example 3.2.5 The right circular cylinder of radius 1 and height 1, thus with area 2π, can be approximated by a family of $4mn$, $m, n \in \mathbb{N}$, congruent inscribed triangles determined by $(2m+1)n$ points

$$p_{jk} = \begin{cases} \left(\cos\frac{2j\pi}{n}, \sin\frac{2j\pi}{n}, \frac{k}{2m}\right) & k \text{ even}, \\ \left(\cos\frac{(2j+1)\pi}{n}, \sin\frac{(2j+1)\pi}{n}, \frac{k}{2m}\right) & k \text{ odd}. \end{cases}$$

lying equally spaced on $2m+1$ equally spaced cross-sectional circles. The area of this polyhedral approximation is

$$2\pi \frac{\sin\frac{\pi}{n}}{\frac{\pi}{n}} \sqrt{1 + 4m^2\left(1 - \cos\frac{\pi}{n}\right)}. \tag{3.30}$$

While the diameter of the triangles goes to zero whenever m and n approach infinity, the limit of the expression in (3.30) can approach any value in the interval $[2\pi, \infty]$ depending on *how* m and n grow.

The example of Schwarz shows that the area of a surface cannot be computed by taking the supremum of the areas of all possible polygonal approximations to the surface. The example does indicate that the limit inferior of the areas of approximating polyhedra *may* define the appropriate value for the area. Following Federer [2], we give the definition of Lebesgue area that is the appropriate extension of the notion of the length of an arc.

Definition 3.2.6

(i) *A set of $m+1$ points $p_0, p_1, \ldots, p_m$ in $\mathbb{R}^N$ is said to be in* **general position** *if, for any set of $m+1$ real numbers $\lambda_0, \lambda_1, \ldots, \lambda_m$, we have*

$$0 = \sum_{i=0}^{m} \lambda_i \quad \text{and} \quad 0 = \sum_{i=0}^{m} \lambda_i p_i \quad \text{imply} \quad 0 = \lambda_0 = \lambda_1 = \ldots = \lambda_m.$$

Equivalently, the points are in general position if and only if $p_1 - p_0, p_2 - p_0, \ldots, p_m - p_0$ are linearly independent.

(ii) *Given* $m+1$ *points* $p_0, p_1, \ldots, p_m$ *in* $\mathbb{R}^N$ *in general position, the m-***simplex** *determined by those points is denoted* $\lceil p_0, p_1, \ldots, p_m \rceil$ *and is defined by setting*

$$\lceil p_0, p_1, \ldots, p_m \rceil = \left\{ \sum_{i=0}^{m} \lambda_i p_i \ : \ 1 = \sum_{i=0}^{m} \lambda_i, \ \lambda_i \geq 0 \right\}.$$

The points $p_0, p_1, \ldots, p_m$ *are said to be the* **vertices** *of the simplex. The* $m+1$ *numbers* $\lambda_0, \lambda_1, \ldots, \lambda_m$ *are called the* **barycentric coordinates** *of the point* $\sum_{i=0}^{m} \lambda_i p_i$.

(iii) *For a given m-simplex* $\lceil p_0, p_1, \ldots, p_m \rceil$, *and for each integer* $0 \leq k < m$, *we call each k-simplex,* $\lceil p_{i_0}, p_{i_1}, \ldots, p_{i_k} \rceil$, *a* **face** *of the simplex* $\lceil p_0, p_1, \ldots, p_m \rceil$.

(iv) *A* **simplicial complex** *in* $\mathbb{R}^N$ *is a set* C *of simplices such that* (a) *every face of any simplex in* C *is also in* C, (b) *the intersection of two simplices in* C *is either empty or a face of both of them, and* (c) *each compact subset of* $\mathbb{R}^N$ *intersects only finitely many simplices in* C.

Definition 3.2.7

(i) *A* **finite triangulation** *of a topological space* X *in a Euclidean space* $\mathbb{R}^m$ *is a pair* (K, τ) *where* K *us a finite simplicial complex in* $\mathbb{R}^m$ *and* τ *is a homeomorphism mapping* X *onto the union of all elements of* K. *In this case,* X *is said to be* **finitely triangulable**

(ii) *A function* f *from the topological space* X *to* $\mathbb{R}^N$ *is* **quasi-linear** *if there exists a finite triangulation* (K, τ) *in some Euclidean space* $\mathbb{R}^m$ *and there exists a linear function* $L : \mathbb{R}^m \to \mathbb{R}^N$ *such that* $f = L \circ \tau$.

(iii) *If* X *is finitely triangulable and* $f : X \to \mathbb{R}^N$ *is continuous, then the* **k-dimensional Lebesgue area** *of* f, *denoted* $\mathrm{LA}_k(f)$, *is the infimum of all the real numbers* t *such that:*

For each $\epsilon > 0$ *there exists a quasi-linear function* $g : X \to \mathbb{R}^N$ *with*

$$|g(x) - f(x)| \leq \epsilon \text{ for all } x \in X$$

and

$$\int_{\mathbb{R}^N} N(g, y) \, d\mathcal{H}^k y < t,$$

where $N(g, y)$ *is the cardinality of* $g^{-1}(y)$, *i.e.*

$$N(g, y) = \mathrm{card}\, \{x \in X : g(x) = y\}.$$

3.2. RECTIFIABILITY

Strictly speaking, the k-dimensional Lebesgue area is a number associated with a *function* rather than a set, but the terminology is often abused by referring to the k-dimensional Lebesgue area of the set that is the image of the function. For smooth surfaces, the k-dimensional Lebesgue area agrees with the standard value of the k-dimensional area computed using multi-variable calculus, but unfortunately, even this more sophisticated definition of area comes to grief as is shown by the next example. The point of this example is that a set in $\mathbb{R}^3$ with finite Lebesgue area may also have positive 3-dimensional Lebesgue measure.

Example 3.2.8 (Besicovitch [2]) Let A be the closed unit cube in $\mathbb{R}^3$. We can subdivide A into 8 congruent subcubes and inside each of these we can construct another closed concentric cube with sides parallel to the sides of A, but having volume $\frac{3}{4} \cdot \frac{1}{8}$. These slightly smaller disjoint closed cubes will be denoted by $A_1, A_2, \ldots, A_8$.

Now considering A_i, $i = 1, 2, \ldots, 8$, we can similarly subdivide A_i into 8 congruent subcubes and inside each of these we can construct another closed concentric cube with sides parallel to the sides of A, but having volume $\frac{5}{8} \cdot \frac{1}{8^2}$. These cubes will be denoted by $A_{i,1}, A_{i,2}, \ldots, A_{i,8}$.

Supposing $A_{i_1, i_2, \ldots, i_k}$ with volume $\frac{1+2^k}{2^{k+1}} \cdot \frac{1}{8^k}$ to have been constructed, we subdivide it into 8 congruent subcubes and inside each of these construct another closed concentric cube with sides parallel to the sides of A, but having volume $\frac{1+2^{k+1}}{2^{k+2}} \cdot \frac{1}{8^{k+1}}$. We denote those cubes by $A_{i_1, i_2, \ldots, i_k, j}$, for $j = 1, 2, \ldots, 8$.

Setting
$$D_k = \bigcup_{i_1, i_2, \ldots, i_k} A_{i_1, i_2, \ldots, i_k},$$
we see that the volume of D_k is $\frac{1+2^k}{2^{k+1}}$ and $D_1 \supset D_2 \supset \cdots \supset D_k \supset \cdots$, so we may set
$$D = \bigcap_k D_k$$
and conclude that D is a compact set with Lebesgue measure $\frac{1}{2}$.

We will construct a continuous function f from the unit square in $\mathbb{R}^2$ into the unit cube A such that D is in the image, but the Lebesgue area of f is finite.

Choose a closed square B on some side of A and in the relative interior of B choose 8 pairwise disjoint closed squares $C_1, C_2, \ldots, C_8$. On each A_i, $i = 1, 2, \ldots, 8$, choose a closed square on some side of A_i and denote it by B_i.

Clearly, it is possible to construct non-intersecting polyhedral surfaces $P_1, P_2, \ldots, P_8$ such that each P_i is homeomorphic to a right circular cylinder of finite height and has as its boundary arc the edges of C_i and the edges of B_i. Further we can require that, except for the edges of C_i, P_i is interior to A and, except for the edges of B_i, P_i is exterior to D_1. Note that the area of each P_i can be made arbitrarily small by choosing small squares, so that the cylinder is thin. Indeed, the cylinders at each stage can be chosen to have small enough area that the total area of the union of all the cylinders is bounded above by any positive number given in advance.

To construct the continuous map, let B' be the unit square in $\mathbb{R}^2$. Let $C_1', C_2', \ldots, C_8'$ be disjoint squares interior to B', and for $i = 1, 2, \ldots, 8$, let B_i' be a concentric square to C_i' that is interior to C_i'. One can construct a quasi-linear map (see Definition 3.2.7(ii)) from $B' \setminus \cup_i C_i'$ onto $B \setminus \cup_i C_i$, from $C_i' \setminus B_i'$ onto P_i, and from B_i' onto B_i.

The construction is extended by repeating the above with A, B, and B' replaced by A_i, B_i, and B_i'.

The mapping extends to the entire unit square by continuity, and its image must contain all of D. But also the construction of the mapping may be terminated as a quasi-linear mapping at any stage by mapping each $B'_{i_1,i_2,\ldots,i_n}$ onto $B_{i_1,i_2,\ldots,i_n}$. Thus the Lebesgue area of the image is the sum of the areas of the cylinders, hence is finite.

Note that despite the fact that f has finite Lebesgue area, the image of f is a set with 3-dimensional Lebesgue measure $\frac{1}{2}$.

The upshot of the investigation of various phenomena in higher dimensions is to put the measure theoretic approach of Carathéodory in a commanding position. The various measures defined in Section 3.1 differ from each other on certain sets and agree with each other on more well-behaved sets. To describe the various classes of sets requires some definitions. The proofs of the main results are extremely technical, so we will state such results without proof and refer the interested reader to Federer [4].

Definition 3.2.9 *Suppose $S \subset \mathbb{R}^N$, ϕ is a measure on $\mathbb{R}^N$, and m is a positive integer.*

(i) *S is said to be **m-rectifiable** if S is the image of some bounded subset of $\mathbb{R}^m$ under a Lipschitz function.*

(ii) *S is said to be **countably m-rectifiable** if it is a countable union of m-rectifiable sets or equivalently the image of some subset of $\mathbb{R}^m$ under a locally Lipschitz function.*

(iii) *S is said to be **countably (ϕ, m)-rectifiable** if there is some countably m-rectifiable set that contains ϕ-almost all of S.*

(iv) *S is said to be **(ϕ, m)-rectifiable** if it is countably (ϕ, m)-rectifiable and $\phi(S) < \infty$.*

(v) *S is said to be **purely (ϕ, m)-unrectifiable** if S has no m-rectifiable subset A with $\phi(A) > 0$.*

All the m-dimensional measures of Section 3.1 agree on the m-rectifiable sets; indeed, a measure on Euclidean space that does not assign the same

3.2. RECTIFIABILITY

value to an m-rectifiable set as does $\mathcal{H}^m$ cannot reasonably be called an m-dimensional measure.

The next result shows us that the general agreement on the appropriate value for the measure of an m-dimensional set extends beyond the m-rectifiable sets.

Theorem 3.2.10 *Let m be a positive integer and suppose $1 \leq t \leq \infty$. If A is an $(\mathcal{H}^m, m)$-rectifiable subset of $\mathbb{R}^N$, then*

$$\mathcal{S}^m(A) = \mathcal{H}^m(A) = \mathcal{T}^m(A) = \mathcal{G}^m(A) = \mathcal{Q}_t^m(A) = \mathcal{I}_t^m(A). \tag{3.31}$$

As a consequence of Theorem 3.2.10, the class of $(\mathcal{H}^m, m)$-rectifiable subsets of $\mathbb{R}^N$ has a special place in the study of geometry and analysis in Euclidean space. The next theorem shows how close to C^1 such sets are.

Theorem 3.2.11 *Let m be a positive integer. A subset of $\mathbb{R}^N$ is countably $(\mathcal{H}^m, m)$-rectifiable if and only if, except for a set of $\mathcal{H}^m$ measure zero, it is contained in a countable union of C^1 submanifolds of $\mathbb{R}^N$.*

One of the deepest theorems in the subject is the Structure Theorem for Sets of Finite Hausdorff Measure. Originally proved in Besicovitch [1] for 1-dimensional sets in $\mathbb{R}^2$, it was extended to general m and N in Federer [1].

Theorem 3.2.12 (Structure Theorem) *Let m be a positive integer and suppose $1 \leq t \leq \infty$. If A is a subset of $\mathbb{R}^N$ with finite m-dimensional Hausdorff measure, then there exists a countably m-rectifiable set $R \subset \mathbb{R}^N$ such that*

(i) *$A \setminus R$ is purely $(\mathcal{H}^m, m)$-unrectifiable,*

(ii) *$\mathcal{I}_t^m(A \setminus R) = 0$.*

No easy proof of the Structure Theorem is known. The existing proofs involve careful study of the behavior of

$$A \cap X(a, r, V, s)$$

where A is the given set of finite Hausdorff measure, $a \in \mathbb{R}^N$, $0 < r < \infty$, V is an m-dimensional linear subspace of $\mathbb{R}^N$, $0 < s < \infty$, and

$$X(a, r, V, s) = \mathbb{B}(a, r) \cap \{x : \operatorname{dist}(x - a, V) < s|x - a|\}.$$

Another important tool used in the study of lower dimensional sets is the notion of the density of a set:

Definition 3.2.13 *Suppose $A \subset \mathbb{R}^N$.*

(i) *The α-**dimensional lower density** of A at x is denoted by $\Phi_*^\alpha(A, x)$ and is defined by setting*

$$\Phi_*^\alpha(A, x) = \liminf_{r \downarrow 0} \frac{\mathcal{H}^\alpha\left(A \cap \overline{\mathbb{B}}(x, r)\right)}{\Upsilon_\alpha r^\alpha}. \tag{3.32}$$

(ii) *The α-dimensional upper density of A at x is denoted by $\Phi^{*\alpha}(A,x)$ and is defined by setting*

$$\Phi^{*\alpha}(A,x) = \limsup_{r \downarrow 0} \frac{\mathcal{H}^\alpha\left(A \cap \overline{\mathbb{B}}(x,r)\right)}{\Upsilon_\alpha r^\alpha}. \tag{3.33}$$

(iii) *In case $\Phi_*^\alpha(A,x) = \Phi^{*\alpha}(A,x)$ we set*

$$\Phi^\alpha(A,x) = \Phi_*^\alpha(A,x) = \Phi^{*\alpha}(A,x). \tag{3.34}$$

We call $\Phi^\alpha(A,x)$ the α-dimensional density of A at x.

Theorem 3.2.14 *Suppose $0 \leq \alpha$. For any $\mathcal{H}^\alpha$ measurable set $A \subset \mathbb{R}^N$ with $0 < \mathcal{H}^\alpha(A) < \infty$,*

$$2^{-\alpha} \leq \Phi^{*\alpha}(A,x) \tag{3.35}$$

holds for $\mathcal{H}^\alpha$ almost every $x \in A$.

Proof: Arguing by contradiction, suppose A is $\mathcal{H}^\alpha$ measurable and

$$\{x : \Phi^{*\alpha}(A,x) < 2^{-\alpha}\}$$

has positive $\mathcal{H}^\alpha$ measure. We conclude that there exists $0 < t < 2^{-\alpha}$, $0 < r_0 < \infty$, and a $\mathcal{H}^\alpha$ measurable set $B \subset A$ with $0 < \mathcal{H}^\alpha(B) < \infty$ such that

$$\mathcal{H}^\alpha\left(B \cap \overline{\mathbb{B}}(x,r)\right) \leq t\Upsilon_\alpha r^\alpha \tag{3.36}$$

holds for $x \in B$ and $0 < r < r_0$.

Let $1 < \lambda$ be arbitrary. By the definition of Hausdorff measure, there is a covering of B by countably many sets S_i of diameter less than r_0 such that

$$\lambda \mathcal{H}^\alpha(B) > \sum_{i=1}^\infty 2^{-\alpha} \Upsilon_\alpha (\mathrm{diam}(S_i))^\alpha. \tag{3.37}$$

We may assume that each set S_i contains some point x_i of B. Let us also set $r_i = \mathrm{diam}(S_i)$. Then, by (3.36) and (3.37), we have

$$\begin{aligned}
\lambda \mathcal{H}^\alpha(B) &> \sum_{i=1}^\infty 2^{-\alpha} \Upsilon_\alpha (r_i)^\alpha \\
&\geq \sum_{i=1}^\infty 2^{-\alpha} t^{-1} \mathcal{H}^\alpha\left(B \cap \overline{\mathbb{B}}(x_i, r_i)\right) \\
&\geq 2^{-\alpha} t^{-1} \mathcal{H}^\alpha(B).
\end{aligned}$$

Since $0 < \mathcal{H}^\alpha(B) < \infty$, we conclude that $\lambda > 2^{-\alpha} t^{-1}$ or $t\lambda > 2^{-\alpha}$. But since the choice of $1 < \lambda$ was arbitrary, we have

$$t \geq 2^{-\alpha},$$

contradicting the choice of t. ∎

The following amazing theorem holds for the densities:

3.2. RECTIFIABILITY

Theorem 3.2.15 (Marstrand) *Suppose $A \subset \mathbb{R}^N$ is $\mathcal{H}^\alpha$ measurable and $0 < \mathcal{H}^\alpha(A) < \infty$. If*

$$0 < \Phi^\alpha(A, x) < \infty \text{ for all } x \in A, \tag{3.38}$$

then α is an integer.

The complete proof of this theorem is very hard. Following Falconer [1], we present an easy proof of part of the result.

Proof that $\alpha \notin (0,1)$: Arguing by contradiction, suppose $0 < \alpha < 1$.
For each $1 < \eta$, we consider an annular density

$$\widetilde{\Phi}^\alpha(A, x, \eta) = \lim_{r \downarrow 0} \frac{\mathcal{H}^\alpha\left(A \cap \overline{\mathbb{B}}(x, \eta r) \setminus \mathbb{B}(x, r)\right)}{\Upsilon_\alpha r^\alpha}.$$

Because the density exists at every point of A, we see that a simple difference of limits gives us

$$\widetilde{\Phi}^\alpha(A, x, \eta) = (\eta^\alpha - 1)\Phi^\alpha(A, x)$$

for each $x \in A$.

Now, by Theorem 3.2.14, we can find $r_0 > 0$ such that

$$X = \left\{x \in A : \frac{1}{2} \leq \frac{\mathcal{H}^\alpha\left(A \cap \overline{\mathbb{B}}(x, r)\right)}{\Upsilon_\alpha r^\alpha} \quad \forall\, 0 < r < r_0\right\}$$

has positive $\mathcal{H}^\alpha$ measure.

Let x_0 be an accumulation point of X. For points of $x \in X \setminus \{x_0\}$ set

$$r(x) = 2|x - x_0|/(1 + \eta).$$

Then $\overline{\mathbb{B}}(x, (\eta-1)r(x)/2) \subset \overline{\mathbb{B}}(x, \eta r(x)) \setminus \mathbb{B}(x, r(x))$, so, if $r(x)$ is sufficiently small,

$$\frac{\Upsilon_\alpha (\eta - 1)^\alpha r(x)^\alpha}{2^{1+\alpha}} \leq \mathcal{H}^\alpha\left(A \cap \overline{\mathbb{B}}(x_0, \eta r(x)) \setminus \mathbb{B}(x_0, r(x))\right)$$

holds. As $x \to x_0$ we find

$$\frac{(\eta - 1)^\alpha}{2^{1+\alpha}} \leq \widetilde{\Phi}^\alpha(A, x_0, \eta) = (\eta^\alpha - 1)\Phi^\alpha(A, x_0).$$

Thus

$$\frac{(\eta-1)^\alpha}{\eta^\alpha - 1} < 2^{1+\alpha}\Phi^\alpha(A, x_0) \tag{3.39}$$

holds for every $1 < \eta$. The left-hand side of (3.39) has an infinite limit as $\eta \downarrow 1$, but the right-hand side is finite by hypothesis. This contradiction proves the special case of the theorem. ∎

3.3 Minkowski Content

Definition 3.3.1 *Suppose $A \subset \mathbb{R}^N$ and $0 \leq K \leq N$. The K-**dimensional upper Minkowski content** of A, denoted by $\mathcal{M}^{*K}(A)$, is defined by*

$$\mathcal{M}^{*K}(A) = \limsup_{r \downarrow 0} \frac{\mathcal{L}^N \{x : \text{dist}(x,A) < r\}}{\Upsilon_{N-K} r^{N-K}}.$$

Similarly, the K-dimensional lower Minkowski content of A is denoted by $\mathcal{M}_*^K(A)$ and defined by

$$\mathcal{M}_*^K(A) = \liminf_{r \downarrow 0} \frac{\mathcal{L}^N \{x : \text{dist}(x,A) < r\}}{\Upsilon_{N-K} r^{N-K}}.$$

*In case the K-dimensional upper Minkowski content and the K-dimensional lower Minkowski content of A are equal, then their common value is called the K-dimensional **Minkowski content** of A and is denoted by $\mathcal{M}^K(A)$.*

We leave it to the reader to verify the following easy result.

Proposition 3.3.2 *Fix $0 \leq K \leq N$.*

(i) *If the sets A and B are separated by a positive distance, then*

$$\begin{aligned}\mathcal{M}^{*K}(A \cup B) &= \mathcal{M}^{*K}(A) + \mathcal{M}^{*K}(B), \\ \mathcal{M}_*^K(A \cup B) &= \mathcal{M}_*^K(A) + \mathcal{M}_*^K(B).\end{aligned}$$

(ii) *The K-dimensional upper Minkowski content is a finitely sub-additive set function.*

Unfortunately, the finite sub-additivity described in the preceding Proposition does not extend to countable additivity (consider $K > 0$ and a countable dense set of points). Thus, the Minkowski content is not a measure, and in general the relationship between Minkowski content and the commonly used measures of K-dimensional size such as the Hausdorff measure is not very good. Clearly, the Minkowski content of a set and its closure is the same, but that is not the fundamental difficulty, for there are known to exist compact sets in $\mathbb{R}^3$ with finite 2-dimensional Hausdorff measure and infinite Minkowski content (see Federer [4], §3.2.40). On the other hand, the next proposition gives us a general result bounding the Hausdorff measure by a multiple of the lower Minkowski content.

Proposition 3.3.3 *For $0 \leq K \leq N$ and any set $A \subset \mathbb{R}^N$, we have*

$$\mathcal{H}^K(A) \leq 3^K \frac{\Upsilon_{N-K} \Upsilon_K}{\Upsilon_N} \mathcal{M}_*^K(A). \qquad (3.40)$$

3.3. MINKOWSKI CONTENT

Proof: We may assume that $\mathcal{M}_*^K(A) < \infty$. Let $\epsilon > 0$ be arbitrary. Then there are arbitrarily small positive numbers r such that

$$\frac{\mathcal{L}^N\{x : \text{dist}(x, A) < r\}}{\Upsilon_{N-K} r^{N-K}} \leq \epsilon + \mathcal{M}_*^K(A). \tag{3.41}$$

Fixing such an r, let $\mathbb{B}(a_i, r)$, $i = 1, 2, \ldots, n$, be a maximal pairwise disjoint family $\mathcal{F}$ of open balls of radius r centered in A—maximal in the sense that there is no open ball of radius r centered in A that is disjoint from the balls in $\mathcal{F}$. By (3.41), we have

$$n \leq \frac{\mathcal{L}^N\{x : \text{dist}(x, A) < r\}}{\Upsilon_N r^N} \leq \frac{\Upsilon_{N-K}}{\Upsilon_N} r^{-K} [\epsilon + \mathcal{M}_*^K(A)]. \tag{3.42}$$

We also have

$$A \subset \bigcup_{i=1}^n \mathbb{B}(a_i, 3r), \tag{3.43}$$

else there would exist another open ball of radius r centered in A and disjoint from all the balls in the maximal family. It follows from (3.42) and (3.43) that the $6r$ approximation to the K-dimensional Hausdorff measure is bounded by

$$n \, \Upsilon_K (3r)^K \leq 3^K \frac{\Upsilon_{N-K} \Upsilon_K}{\Upsilon_N} [\epsilon + \mathcal{M}_*^K(A)]. \tag{3.44}$$

First letting $r \downarrow 0$ and then using the fact that $\epsilon > 0$ was arbitrary, we see that the result follows. ∎

As mentioned above, there is no bound for the K-dimensional Minkowski content of a set by a multiple of its K-dimensional Hausdorff measure. The best result available is the following.

Theorem 3.3.4 (Federer [4] §3.2.39) *Suppose $1 \leq K \leq N$ is an integer. If A is closed and $A \subset f(\mathbb{R}^K)$ for some Lipschitz function $f : \mathbb{R}^K \to \mathbb{R}^N$, then*

$$\mathcal{M}^K(A) = \mathcal{H}^K(A).$$

The proof of this result is fairly technical. It depends on a key lemma, Federer [4], §3.2.38, which bounds the upper K-dimensional Minkowski content of the Lipschitz image of a compact subset of $\mathbb{R}^K$ by the integral of the K-dimensional Jacobian (the K-dimensional Jacobian will be defined in Section 3.8). The difference in complexity between dealing with the image of a Lipschitz map and the image of a C^1 map is very significant, and it is perhaps more instructive to consider instead the proof of the following simplified form of that key lemma.

Proposition 3.3.5 *Suppose $f : \mathbb{R}^K \to \mathbb{R}^N$ is C^1 and A is compact with*

$$A \subset \{x : |Df(x)| \leq \rho\},$$

then

$$\mathcal{M}^*K[f(A)] \leq \rho^K \mathcal{L}^K(A).$$

Proof: For $0 < t$, set

$$\sigma(t) = \sup\{|Df(x) - Df(a)| : a \in A, \ |x - a| \leq t\}.$$

Since f is C^1, we see that σ is monotone increasing and continuous with $\sigma(t) \downarrow 0$ as $t \downarrow 0$. Also we have

$$|f(x) - f(a)| \leq |x - a|\,\sigma(|x - a|), \tag{3.45}$$

whenever $a \in A$.

Fix a number $0 < \lambda < 1$. For any given sufficiently small positive r we can find a positive h such that

$$r = \sqrt{K}\,h\,\sigma\left(\sqrt{K}\,h\right)^{\lambda}. \tag{3.46}$$

Clearly, we have $h \downarrow 0$ as $r \downarrow 0$.

With r given and h chosen as in (3.46), divide $\mathbb{R}^K$ into cubes of side length h. Thus we may cover A by a family of closed cubes $I_1, I_2, \ldots, I_n$, all of side length h which intersect only on their faces, such that

$$\bigcup_{j=1}^{n} I_j \subset \{x : \text{dist}(x, A) \leq \sqrt{K}\,h\}.$$

We may and shall suppose each cube contains a point of A. We easily estimate the number, n, of cubes by

$$n \leq h^{-K} \mathcal{L}^1 \{x : \text{dist}(x, A) \leq \sqrt{K}\,h\}. \tag{3.47}$$

Now suppose $a \in A \cap I_j$, and consider the image of I_j under the affine map, g, sending x to $f(a) + \langle x - a, Df(a) \rangle$. We see that $g(I_j)$ is a K-dimensional parallelepiped with side lengths not exceeding ρh. Further, we have

$$f(I_j) \subset \{x : \text{dist}(x, g(I_j)) \leq \sqrt{K}\,h\,\sigma(\sqrt{K}\,h)\},$$

and

$$\{x : \text{dist}(x, f(I_j)) < r\} \subset \{x : \text{dist}(x, g(I_j)) < r + \sqrt{K}\,h\,\sigma(\sqrt{K}\,h)\}.$$

For a K-dimensional parallelepiped, P, in $\mathbb{R}^N$ with side lengths bounded by s it is elementary to estimate

$$\mathcal{L}^N\{x : \text{dist}(x, P) < t\} \leq \sum_{i=0}^{K-1} C_i\, s^i\, t^{N-i} + \Upsilon_{N-K}\, s^K\, t^{N-K}, \tag{3.48}$$

3.3. MINKOWSKI CONTENT

where the C_i are constants. Applying (3.48) to $g(I_j)$ with s replaced by ρh and t replaced by $r + \sqrt{K}\, h\, \sigma(\sqrt{K}\, h)$, we see that

$$\mathcal{L}^N\{x : \operatorname{dist}(x, g(I_j)) < r + \sqrt{K}\, h\, \sigma(\sqrt{K}\, h)\} =$$
$$\sum_{i=0}^{K-1} C_i\, (\rho h)^i\, (r + \sqrt{K}\, h\, \sigma(\sqrt{K}\, h))^{N-i}$$
$$+ \Upsilon_{N-K}\, (\rho h)^K\, (r + \sqrt{K}\, h\, \sigma(\sqrt{K}\, h))^{N-K}. \qquad (3.49)$$

Thus we have

$$\mathcal{L}^N\{x : \operatorname{dist}(x, f(I_j)) < r\} \leq$$
$$n\left(\sum_{i=0}^{K-1} C_i\, (\rho h)^i\, (r + \sqrt{K}\, h\, \sigma(\sqrt{K}\, h))^{N-i}\right.$$
$$\left. + \Upsilon_{N-K}\, (\rho h)^K\, (r + \sqrt{K}\, h\, \sigma(\sqrt{K}\, h))^{N-K}\right). \qquad (3.50)$$

Dividing (3.49) by $\Upsilon_{N-K}\, r^{N-K}$ and using (3.46) and (3.47), we estimate

$$\frac{\mathcal{L}^N\{x : \operatorname{dist}(x, f(I_j)) < r\}}{\Upsilon_{N-1}\, r^{N-1}} \leq \mathcal{L}^1\{x : \operatorname{dist}(x, A) \leq h\} \times$$
$$\left(\sum_{i=0}^{K-1} \frac{C_i\, \rho^i\, h^i\, r^{K-i}\, (r + \sqrt{K}\, h\, \sigma(\sqrt{K}\, h))^{N-i}}{\Upsilon_{N-K}\, h^K\, r^{N-i}}\right.$$
$$\left. + \frac{\Upsilon_{N-K}\, \rho^K\, h^K\, (r + \sqrt{K}\, h\, \sigma(\sqrt{K}\, h))^{N-K}}{\Upsilon_{N-K}\, h^K\, r^{N-K}}\right). \qquad (3.51)$$

By (3.46), we have

$$\frac{(r + \sqrt{K}\, h\, \sigma(\sqrt{K}\, h))^{N-i}}{r^{N-i}} \to 1 \text{ as } r \downarrow 0, \qquad (3.52)$$

for $i = 0, 1, \ldots, K$. But (3.46) also implies

$$\frac{h^i\, r^{K-i}}{h^K} \to 0 \text{ as } r \downarrow 0, \qquad (3.53)$$

for $i = 0, 1, \ldots, K-1$. The result now follows from (3.51), (3.52), and (3.53) since

$$\mathcal{L}^K\{x : \operatorname{dist}(x, A) \leq \sqrt{K}h\} \downarrow \mathcal{L}^K(A) \text{ as } r \downarrow 0. \qquad \blacksquare$$

It is interesting to observe that the Minkowski contents are well-behaved relative to cartesian products.

Theorem 3.3.6 *If $A \subset \mathbb{R}^M$ and $B \subset \mathbb{R}^N$, then*

$$2^{-[(M+N)-(K+L)]/2} \frac{\Upsilon_{M-K} \Upsilon_{N-L}}{\Upsilon_{(M+N)-(K+L)}} \mathcal{M}_*^K(A) \cdot \mathcal{M}_*^L(B) \leq$$
$$\mathcal{M}_*^{(K+L)}(A \times B) \leq \mathcal{M}^{*(K+L)}(A \times B) \leq$$
$$\frac{\Upsilon_{M-K} \Upsilon_{N-L}}{\Upsilon_{(M+N)-(K+L)}} \mathcal{M}^{*K}(A) \cdot \mathcal{M}^{*L}(B).$$

Proof: This follows from Fubini's Theorem since a point (x,y) within distance r of $A \times B$ must have x within distance r of X and have y within distance r of y. Similarly, if x is within distance r of some point of A and y is within distance r of some point of B, then (x,y) is within distance $\sqrt{2}\,r$ of a point of $A \times B$. ∎

The **upper Minkowski dimension** of a set A can be defined to be the infimum of the numbers K such that $M^{*K}(A) = 0$, or equivalently the supremum of the numbers K such that $M^{*K}(A) > 0$. A similar definition is made for the **lower Minkowski dimension**. Unfortunately, these dimensions do not have the fundamental property that the dimension of a countable union of sets is the supremum of the dimensions of those sets. This is another reflection of the fact that the Minkowski contents are not measures.

We can make an additional definition to create a countably sub-additive set function based on the upper Minkowski content, and thus we will obtain a measure in the usual sense.

Definition 3.3.7 *Suppose $A \subset \mathbb{R}^N$ and $0 \leq K \leq N$. The K-dimensional upper Minkowski measure of A, denoted $\overline{\mathcal{M}}^K(A)$, is defined by setting*

$$\overline{\mathcal{M}}^K(A) = \inf \left\{ \sum_{B \in F} \mathcal{M}^{*K}(B) : F \subset 2^{\mathbb{R}^N}, \ \mathrm{card}(F) \leq \aleph_0, \ A \subset \cup F \right\}. \tag{3.54}$$

Similarly, the K-dimensional lower Minkowski measure of A is denoted by $\underline{\mathcal{M}}^K(A)$ and defined by

$$\underline{\mathcal{M}}^K(A) = \inf \left\{ \sum_{B \in F} \mathcal{M}_*^K(B) : F \subset 2^{\mathbb{R}^N}, \ \mathrm{card}(F) \leq \aleph_0, \ A \subset \cup F \right\}. \tag{3.55}$$

We leave it to the reader to verify the following easy result.

Lemma 3.3.8 *For each $0 \leq K \leq N$, $\overline{\mathcal{M}}^K$ and $\underline{\mathcal{M}}^K$ are countably sub-additive.*

Remark 3.3.9 The preceding lemma shows us that $\overline{\mathcal{M}}^K$ and $\underline{\mathcal{M}}^K$ are outer measures. The issue of which sets are measurable is settled by the next proposition.

3.3. MINKOWSKI CONTENT

Proposition 3.3.10 *All open sets of $\mathbb{R}^N$ are $\overline{\mathcal{M}}^K$ and $\underline{\mathcal{M}}^K$ measurable, and both $\overline{\mathcal{M}}^K$ and $\underline{\mathcal{M}}^K$ are Borel regular.*

Proof: This is a consequence of Carathéodory's criterion, Federer [4], §2.3.2(9), and Proposition 3.3.2. To see this, suppose that A and B are separated by a positive distance and let $\epsilon > 0$ be arbitrary. Let F be a countable cover of $A \cup B$ such that

$$\epsilon + \overline{\mathcal{M}}(A \cup B) \geq \sum_{C \in F} \mathcal{M}^{*K}(C).$$

Then

$$\begin{aligned}
\sum_{C \in F} \mathcal{M}^{*K}(C) &\geq \sum_{C \in F} \mathcal{M}^{*K}(C \cap (A \cup B)) \\
&= \sum_{C \in F} \mathcal{M}^{*K}((C \cap A) \cup (C \cap B)) \\
&= \sum_{C \in F} \mathcal{M}^{*K}(C \cap A) + \sum_{C \in F} \mathcal{M}^{*K}(C \cap B) \\
&\geq \overline{\mathcal{M}}^K(A) + \overline{\mathcal{M}}(B).
\end{aligned}$$

Since $\epsilon > 0$ was arbitrary, we have

$$\overline{\mathcal{M}}^K(A \cup B) \geq \overline{\mathcal{M}}^K(A) + \overline{\mathcal{M}}^K(B),$$

as required.

The argument for $\underline{\mathcal{M}}^K$ is similar.

To see that the measure $\overline{\mathcal{M}}^K$ (respectively, $\underline{\mathcal{M}}^K$) is Borel regular, simply use the fact that the upper (respectively, lower) Minkowski content of a set and its closure are equal. ∎

To relate these measures to each other and to other quantities we have the following result.

Proposition 3.3.11 *Fix $0 \leq K \leq N$. Then the following hold.*

(i) $\underline{\mathcal{M}}^K \leq \overline{\mathcal{M}}^K \leq \mathcal{M}^{*K}$.

(ii) $\underline{\mathcal{M}}^K \leq \mathcal{M}_*^K$.

(iii) *If K is an integer, then the K dimensional lower Minkowski measure is bounded below by the K dimensional Gross measure (see Definition 3.1.11), that is, $\mathcal{G}^K \leq \underline{\mathcal{M}}^K$.*

(iv) *If A is contained in the image of a compact subset of $\mathbb{R}^K$ under a Lipschitz function, then*

$$\overline{\mathcal{M}}^K(A) = \underline{\mathcal{M}}^K(A) = \mathcal{M}^K(A) = \mathcal{H}^K(A).$$

Proof:
(i) and **(ii)** Easy exercises.
(iii) Recall that the K dimensional Gross measure results from Carathéodory's construction using as the gauge the supremum of the Lebesgue measure of the orthogonal projection onto $\mathbb{R}^K$. Since Fubini's Theorem tells us that

$$\mathcal{M}^K_*(S) \geq \mathcal{L}^K[\Pi(S)]$$

holds for any orthogonal projection $\Pi : \mathbb{R}^M \to \mathbb{R}^K$, it is easy to see that

$$\mathcal{M}^K_*(S) \geq \mathcal{G}^K(S)$$

must hold for any $S \subset \mathbb{R}^N$. Then it follows that for any countable cover F of a set A we have

$$\mathcal{G}^K(A) \leq \sum_{B \in F} \mathcal{G}^K(B) \leq \sum_{B \in F} \mathcal{M}^K_*(B),$$

and the inequality $\mathcal{G}^K(A) \leq \underline{\mathcal{M}}^K(A)$ follows by taking the infimum over all such countable covers of A.
(iv) By Federer [4], §3.2.26 and §3.2.39, we know that $\mathcal{G}^K(A) = \mathcal{H}^K(A) = \mathcal{M}^K(A)$. Thus we have

$$\mathcal{H}^K(A) = \mathcal{G}^K(A) \leq \underline{\mathcal{M}}^K(A) \leq \overline{\mathcal{M}}^K(A) \leq \mathcal{M}^{*K}(A) = \mathcal{H}^K(A). \quad \blacksquare$$

Corollary 3.3.12 *In $\mathbb{R}^N$, the measures $\overline{\mathcal{M}}^N$, $\underline{\mathcal{M}}^N$, and $\mathcal{L}^N$ agree.*

Proof: All three are translation-invariant measures that assign the same measure to the unit ball. $\quad \blacksquare$

Remark 3.3.13 Note that it is a consequence of the corollary that a set in $\mathbb{R}^N$ with Lebesgue measure zero can be covered by countably many sets whose upper Minkowski contents add to less than any given positive number.

Using the measures $\overline{\mathcal{M}}^K$ and $\underline{\mathcal{M}}^K$, we define two more notions of dimension. To agree with the literature, we call them *packing dimensions*.

Definition 3.3.14 *Suppose $A \subset \mathbb{R}^N$. The* **upper packing dimension** *of A, denoted by $\overline{\dim}_p A$, is defined by*

$$\overline{\dim}_p A = \inf \left\{ K : \overline{\mathcal{M}}^K(A) = 0 \right\} = \sup \left\{ K : \overline{\mathcal{M}}^K(A) > 0 \right\}. \quad (3.56)$$

Similarly, the **lower packing dimension** *of A is denoted $\underline{\dim}_p A$ and is defined by*

$$\underline{\dim}_p A = \inf \left\{ K : \underline{\mathcal{M}}^K(A) = 0 \right\} = \sup \left\{ K : \underline{\mathcal{M}}^K(A) > 0 \right\}. \quad (3.57)$$

The interested reader should see Mattila [1] for much more information on these matters.

3.4 A Space-Filling Curve

The first example of a space-filling curve was presented in Peano [1]. The existence of such a curve was startling, and to this day remains surprising to those first learning of it. Typical constructions are quite geometric as in Gelbaum and Olmstead [1], but Peano's original construction was not. Here we give another interesting non-geometric construction.

For motivation we define a map, $g : [0,1) \times [0,1) \to [0,1)$ as follows: Given $(x,y) \in [0,1) \times [0,1)$ express x and y as non-terminating decimals which do not have an infinite sequence of 9's. Thus we have

$$x = 0.c_1 c_2 \ldots,$$
$$y = 0.d_1 d_2 \ldots,$$

and for any integer n there exist n' and n'' such that $n' > n$, $n'' > n$, $c_{n'} \neq 9$ and $d_{n''} \neq 9$. Then $g(x,y)$ will be the real number with decimal expansion

$$0.c_1 d_1 c_2 d_2 \ldots$$

Lemma 3.4.1 *The map $g : [0,1) \times [0,1) \to [0,1)$ is one-to-one.*

Proof: This is obvious, since the decimal expansion of $g(x,y)$ cannot have an infinite sequence of 9's. ∎

One would like to show that the inverse of g is continuous where it is defined, but unfortunately that is just not true: To see this, let the inverse of g be denoted by f and consider the two numbers s and t

$$s = 0.7 \, \{2n \text{ 9's}\} \, 00 \ldots$$
$$t = 0.8 \, \{2n \text{ 0's}\} \, 00 \ldots,$$

the difference between which is 10^{-2n-1}. Both s and t are in the image of g, and we have

$$f(s) = (\ 0.7\ \{n\ 9\text{'s}\}\ 00\ldots,\ 0.\ \{n\ 9\text{'s}\}\ 00\ldots\ ),$$
$$f(t) = (\ 0.800\ldots,\ 0.00\ldots\ )$$

the distance between which always exceeds 0.9. ∎

The lack of continuity in f above is caused by the arbitrarily long sequences of 9's that may exist in the decimal expansions of elements of the image of g. To avoid arbitrarily long sequences of 9's, we will introduce an alternative representation of real numbers. Simply put, the real number in $[0, 1)$ is to be expressed in decimal form without an infinite sequence of 9's, and then any finite sequence of k copies of the digit 9 is to be replaced by "9 {the binary expression for k} 9". We introduce the notation

$$\star d_1 d_2 d_3 \ldots$$

for numbers in $[0, 1)$ expressed in this way. Note that the sequence of digits $d_1, d_2, d_3, \ldots$ cannot be arbitrarily given; the 9's must fall into pairs separated by binary expressions for positive integers. To illustrate, we have

$$0.9799799977\ldots = \star 919791097911977\ldots$$

Using this new notation, we can now define $G : [0, 1) \times [0, 1) \to [0, 1)$ by writing

$$x = \star c_1 c_2 \ldots,$$
$$y = \star d_1 d_2 \ldots,$$

and setting

$$G(x, y) = 0.c_1 d_1 c_2 d_2 \ldots.$$

Note that the preceding expression for $G(x, y)$ must be an ordinary decimal expansion, since we cannot guarantee the proper pairing of 9's. As before we have

Lemma 3.4.2 *The map $G : [0, 1) \times [0, 1) \to [0, 1)$ is one-to-one.*

We denote the image of G by $\mathcal{X}$, and let $F : \mathcal{X} \to [0, 1) \times [0, 1)$ be the inverse of G.

Lemma 3.4.3 *The maximal length for any sequence of consecutive 9's in the decimal expansion of any number in $\mathcal{X}$ is 2.*

Proof: This is clear since neither of the expressions

$$x = \star c_1 c_2 \ldots,$$
$$y = \star d_1 d_2 \ldots,$$

can have a pair of consecutive 9's. ∎

Lemma 3.4.4 *If two numbers in $\mathcal{X}$ differ by less than 10^{-n}, then they must agree in the first $n-3$ decimal places.*

Proof: This is clear because there are at worst a pair of 9's occurring in the nth and $(n-1)^{st}$ decimal places. ∎

Lemma 3.4.5 *F is uniformly continuous on $\mathcal{X}$.*

Proof: The fact that numbers in $\mathcal{X}$ differing by less than 10^{-n} must have the same first $n-3$ digits insures that the inverse images must agree in each coordinate up to the first k digits where k is the greatest integer less than $(n-3)/2$. ∎

Theorem 3.4.6 *F extends to a continuous function mapping from $[0,1]$ onto $[0,1] \times [0,1]$*

Proof: Since F is uniformly continuous, it extends continuously to the closure of $\mathcal{X}$. The further continuous extension to $[0,1]$ follows from Tietze's Extension Theorem (see for example Dugundji [1]).

Since F is continuous on $[0,1]$, the image of F must be compact. Also F is an extension of G^{-1}, so F maps onto $[0,1) \times [0,1)$. Thus we have

$$[0,1] \times [0,1] = \overline{[0,1) \times [0,1)} \subset F([0,1]).$$ ∎

3.5 Covering Lemmas

Preliminary Remarks

The technique of covering lemmas has become an entire area of mathematical analysis (see, for example, de Guzman [1]). It is intimately connected with problems of differentiation of integrals, with certain maximal operators (such as the Hardy-Littlewood maximal operator), with the boundedness of multiplier operators in harmonic analysis, and concomitantly with questions of summation of Fourier series.

Our intention in this section is to give an overview of the geometric aspects of covering lemmas of various kinds—not with an aim for completeness, but rather to give the reader a taste of some of the basic covering lemmas and how they can be used. Along the way, we shall provide a variety of references for ancillary reading.

The four basic types of covering lemmas on which we shall concentrate are the Wiener covering lemma, the Besicovitch covering lemma, the Lebesgue covering lemma, and the Vitali covering lemma. The Wiener and Vitali covering lemmas are of interest because of their interaction with measure theory. The Besicovitch and Lebesgue lemmas have a strong intuitive geometric appeal. They are also useful because they are independent of measure theory, and depend instead on critical geometric artifacts of space.

Suppose that a compact subset $K \subset \mathbb{R}^N$ is covered by a family of open sets. Our first reaction to this hypothesis is that, because K is compact, we may pass to a finite subcovering. However, in the theory of covering lemmas, we wish to find a subcovering that is "thin" in some sense. For instance, we could have a finite covering $\{U_\alpha\}$ of $K \subset \mathbb{R}^2$ by open sets with the property that some points in K are covered 5 times (that is, by five of the U_α) and other points in K are covered 10^{10} times. (The maximum number of times that any point is covered is called the **valence** of the covering.) This last seems somehow wasteful. One would like to think that, if a compact set in $\mathbb{R}^2$ is covered by open sets, then it could be done fairly efficiently—so that the valence is fairly small.

Figure 3.3 shows a covering of the unit square in the plane that has valence 10. (Note also that no element of the covering in Figure 3.3 can be discarded without leaving some points in the unit square uncovered.) Clearly, with simple modifications the valence could be increased to equal any large positive integer. Thus, interpreted *ipso facto*, the quest posed in the last paragraph is not possible. But if we are willing to pass to a *refinement* of the given open cover, or to restrict to covers by certain fairly regular types of open sets, then in fact the sort of "efficient covering lemma" that we seek is possible. The first such lemma that we shall discuss is the so-called Wiener covering lemma.

The Wiener Covering Lemma

We begin with a formal statement of the lemma:

Lemma 3.5.1 (Wiener) *Let K be a compact subset of $\mathbb{R}^N$. Let $\mathcal{B} = \{B_\alpha\}_{\alpha \in \mathcal{A}}$ be a covering of K by open Euclidean balls. Then there is a subcollection $\{B_{\alpha_j}\}_{j=1}^m$ of the balls such that*

(i) *the balls $\{B_{\alpha_j}\}_{j=1}^m$ are pairwise disjoint,*

(ii) *if we write $B_{\alpha_j} = \mathbb{B}(P_{\alpha_j}, r_{\alpha_j})$, then the dilated balls $3B_{\alpha_j} \equiv \mathbb{B}(P_{\alpha_j}, 3r_{\alpha_j})$ cover K.*

Proof: Since K is compact, we may as well suppose at the outset that the collection $\{B_\alpha\}$ is finite. We write $\mathcal{B} = \{B_\alpha\}_{\alpha=1}^M$.

Now select B_{α_1} to have radius as large as possible (chosen from among the M given balls)—if there are several balls with this same largest radius, then just pick one of them arbitrarily. Note that there is no problem in selecting B_{α_1} since we are working with only finitely many balls.

3.5. COVERING LEMMAS

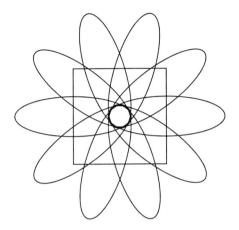

Figure 3.3: **Covering with Valence 10**

Having selected B_{α_1}, we now select B_{α_2} subject to the two conditions that (1) it be disjoint from B_{α_1} and that (2) it have radius as large as possible. Again, if more than one ball satisfies these two conditions, then just pick one of them arbitrarily.

Continue in the preceding fashion. If $B_{\alpha_1}, B_{\alpha_2}, \ldots B_{\alpha_p}$ have been selected, then we select $B_{\alpha_{p+1}}$ to satisfy the two conditions that (1) it be disjoint from $B_{\alpha_1}, B_{\alpha_2}, \ldots, B_{j_p}$ and that (2) it have radius as large as possible.

Clearly this process must stop, because our universe of balls is just the finite collection $\{B_\alpha\}_{\alpha=1}^M$. Let $\mathcal{B}' \equiv \{B_{\alpha_j}\}_{j=1}^m$ be the subcollection of balls that results from our selection process. We claim that the subcollection $\mathcal{B}'$ satisfies the conclusions of the lemma.

First, by design, the subcollection $\mathcal{B}'$ is pairwise disjoint. We claim now that the dilated balls $\{3B_{\alpha_j}\}$ cover K. Of course it will suffice for us to prove that these dilated balls cover each of the original balls $\{B_\alpha\}_{\alpha=1}^M$. Let $B_{\tilde{\alpha}}$ be one of these latter balls. If it is equal to one of the selected balls B_{α_j}, then of course it is covered by the dilated ball $3B_{\alpha_j}$. If instead it is *not* one of the selected balls, then let $B_{\alpha_q} = \mathbb{B}(P_{\alpha_q}, r_{\alpha_q})$ be the *first* selected ball that intersects $B_{\tilde{\alpha}} = \mathbb{B}(P_{\tilde{\alpha}}, r_{\tilde{\alpha}})$. Then the radius r_{α_q} of B_{α_q} must be at least as great as the radius $r_{\tilde{\alpha}}$ of $B_{\tilde{\alpha}}$, otherwise we would have selected $B_{\tilde{\alpha}}$ *instead* of B_{α_q} at the qth step. It follows then from the triangle inequality that $3\mathbb{B}(P_{\alpha_q}, r_{\alpha_q}) \equiv \mathbb{B}(P_{\alpha_q}, 3r_{\alpha_q})$ covers $\mathbb{B}(P_{\tilde{\alpha}}, r_{\tilde{\alpha}})$. That is what we wished to show. ∎

One of the most important (sublinear) operators in classical analysis is the Hardy-Littlewood maximal operator. We shall apply the Wiener covering lemma to prove the boundedness, in a certain sense, of the Hardy-

Littlewood operator. We shall then apply this boundedness to prove the Lebesgue differentiation theorem.

Definition 3.5.2 Let f be a locally integrable function on $\mathbb{R}^N$. We define the **Hardy-Littlewood maximal function** Mf by setting

$$Mf(x) = \sup_{0 < r < \infty} \frac{1}{\mathcal{L}^N(\mathbb{B}(x,r))} \int_{\mathbb{B}(x,r)} |f(t)|\, dt,$$

for each $x \in \mathbb{R}^N$.

It is obvious that the operator M satisfies

$$\|Mf\|_{L^\infty} \leq \|f\|_{L^\infty}.$$

What is more useful is the subtle bound, called a **weak type (1,1) bound**, that we derive in the next lemma.

Lemma 3.5.3 *There is a constant $C > 0$—depending on the dimension N, but not on the particular function f—such that the operator M satisfies*

$$\mathcal{L}^N\{x \in \mathbb{R}^N : Mf(x) > \lambda\} \leq \frac{C}{\lambda} \|f\|_{L^1(\mathbb{R}^N)}$$

for any scalar $\lambda > 0$. The constant C is independent of λ and of f; indeed, $C = 3^N$ will do.

Proof: Fix $f \in L^1(\mathbb{R}^N)$. Fix $\lambda > 0$. Let

$$S_\lambda = \{x \in \mathbb{R}^N : Mf(x) > \lambda\}.$$

We wish to estimate the measure of S_λ. By the inner regularity of Lebesgue measure (Proposition 1.3.7), it suffices to measure any compact subset K of S_λ. If K is such a compact set, and if $k \in K$, then there is an $r_k > 0$ such that

$$\frac{1}{\mathcal{L}^N(\mathbb{B}(k, r_k))} \int_{\mathbb{B}(k, r_k)} |f(t)|\, dt > \lambda.$$

In other words,

$$\mathcal{L}^N(\mathbb{B}(k, r_k)) < \frac{1}{\lambda} \int_{\mathbb{B}(k, r_k)} |f(t)|\, dt.$$

Obviously the balls $\{\mathbb{B}(k, r_k)\}_{k \in K}$ cover K. Thus they satisfy the hypotheses of Lemma 3.5.1. So we may select a subcollection $\{\mathbb{B}(k_j, r_{k_j})\}_{j=1}^m$ consisting of pairwise disjoint balls such that $\{\mathbb{B}(k_j, 3r_{k_j})\}_{j=1}^m$ covers K.

3.5. COVERING LEMMAS

Now we calculate that

$$\begin{aligned}
\mathcal{L}^N(K) &\leq \mathcal{L}^N\left(\bigcup_{j=1}^{m} \mathbb{B}(k_j, 3r_{k_j})\right) \\
&\leq \sum_{j=1}^{m} \mathcal{L}^N\left(\mathbb{B}(k_j, 3r_{k_j})\right) \\
&= \sum_{j=1}^{m} 3^N \mathcal{L}^N\left(\mathbb{B}(k_j, r_{k_j})\right) \\
&\leq 3^N \sum_{j=1}^{m} \frac{1}{\lambda} \int_{\mathbb{B}(k_j, r_{k_j})} |f(t)|\, dt.
\end{aligned}$$

But the balls $\{\mathbb{B}(k_j, r_{k_j})\}$ are pairwise disjoint. Thus the last line does not exceed

$$\frac{3^N}{\lambda} \int_{\mathbb{R}^N} |f(t)|\, dt \equiv \frac{C}{\lambda} \|f\|_{L^1(\mathbb{R}^N)}.$$

That is the estimate that we wished to prove. ∎

Just to familiarize the reader with a typical, if elementary, application of maximal function estimates, we now prove a version of the Lebesgue differentiation theorem.

Proposition 3.5.4 *Let f be a locally integrable function on $\mathbb{R}^N$. Then, for almost every $x \in \mathbb{R}^N$, it holds that*

$$\lim_{r \to 0^+} \frac{1}{\mathcal{L}^N(\mathbb{B}(x,r))} \int_{\mathbb{B}(x,r)} f(t)\, dt$$

exists and equals $f(x)$.

Remark 3.5.5 Perhaps a few comments are in order before we proceed with the proof. First, it is common in measure theory books to formulate a stronger version of Lebesgue's result. It is this:

> Let f be a locally integrable function on $\mathbb{R}^N$. For almost every $x \in \mathbb{R}^N$ we let
>
> $$f(x) \equiv \lim_{r \to 0^+} \frac{1}{\mathcal{L}^N(\mathbb{B}(x,r))} \int_{\mathbb{B}(x,r)} f(t)\, dt, \qquad (3.58)$$
>
> which limit is guaranteed to exist by Lebesgue's theorem. Then, for almost every x, we have that
>
> $$\lim_{r \to 0^+} \frac{1}{\mathcal{L}^N(\mathbb{B}(x,r))} \int_{\mathbb{B}(x,r)} |f(t) - f(x)|\, dt = 0. \qquad (3.59)$$

It is easy to see that statement (3.59) is stronger, on a formal level, than the statement in the proposition (logically speaking, it turns out that the two are equivalent in the sense that either one implies the other).

Notice also that the stronger statement (3.59) contains implicitly an interesting fact. Strictly speaking, an L^p function is an *equivalence class* of functions—it is only defined up to a set of measure zero. In other words, any given representative of this equivalence class may differ from any other representative on a set of measure zero. However equation (3.58) gives a way to pick a canonical representative from the equivalence class: no matter what the choice of f from the equivalence class, (3.58) gives the same definition of $f(x)$. The set of Lebesgue points—those points at which (3.59) holds—is independent of the representative f with which we work. And the resulting function $f(x)$, defined by (3.58), does not depend on which representative is used to calculate the limit.

Proof of the Proposition: Let $U \subset \mathbb{R}^N$ be an open set, and let V be an open set that contains $\overline{U}$. By Urysohn's Lemma 1.1.3, we may let ϕ be a cutoff function that is identically equal to 1 on U and that is supported in V. By multiplying f by ϕ, we may concentrate attention on points in U and may thereby assume, without loss of generality, that $f \in L^1(\mathbb{R}^N)$. We may further assume that f is real-valued.

Let $\epsilon > 0$. Choose a continuous function g with compact support such that $\|f - g\|_{L^1} < \epsilon^2$. Set

$$T_\epsilon = \left\{ x \in \mathbb{R}^N : \left| \limsup_{r \to 0^+} \frac{1}{\mathcal{L}^N(\mathbb{B}(x,r))} \int_{\mathbb{B}(x,r)} f(t)\, dt \right. \right.$$
$$\left. \left. - \liminf_{r \to 0^+} \frac{1}{\mathcal{L}^N(\mathbb{B}(x,r))} \int_{\mathbb{B}(x,r)} f(t)\, dt \right| > \epsilon \right\}.$$

We wish to estimate the measure of T_ϵ.

Now

$$\mathcal{L}^N(T_\epsilon)$$
$$\leq \mathcal{L}^N\left(\left\{ x \in \mathbb{R}^N : \limsup_{r \to 0^+} \frac{1}{\mathcal{L}^N(\mathbb{B}(x,r))} \int_{\mathbb{B}(x,r)} |f(t) - g(t)|\, dt > \epsilon/3 \right\} \right)$$
$$+ \mathcal{L}^N\left(\left\{ x \in \mathbb{R}^N : \left| \limsup_{r \to 0^+} \frac{1}{\mathcal{L}^N(\mathbb{B}(x,r))} \int_{\mathbb{B}(x,r)} g(t)\, dt \right. \right. \right.$$
$$\left. \left. \left. - \liminf_{r \to 0^+} \frac{1}{\mathcal{L}^N(\mathbb{B}(x,r))} \int_{\mathbb{B}(x,r)} g(t)\, dt \right| > \epsilon/3 \right\} \right)$$
$$+ \mathcal{L}^N\left(\left\{ x \in \mathbb{R}^N : \liminf_{r \to 0^+} \frac{1}{\mathcal{L}^N(\mathbb{B}(x,r))} \int_{\mathbb{B}(x,r)} |g(t) - f(t)|\, dt > \epsilon/3 \right\} \right)$$
$$\equiv A_1 + A_2 + A_3.$$

It is plain that $A_2 = 0$ since g is continuous. We estimate A_1 and A_3 in

3.5. COVERING LEMMAS

just the same way, so we shall concentrate on the former. Now

$$\begin{aligned}
A_1 &\leq \mathcal{L}^N\left(\left\{x \in \mathbb{R}^N : \sup_r \frac{1}{\mathcal{L}^N(\mathbb{B}(x,r))} \int_{\mathbb{B}(x,r)} |f(t) - g(t)|\, dt > \epsilon/3\right\}\right) \\
&= \mathcal{L}^N\left(\left\{x \in \mathbb{R}^N : M(f-g) > \epsilon/3\right\}\right) \\
&\leq \frac{C}{\epsilon/3} \|f - g\|_{L^1} \\
&< \frac{C}{\epsilon/3} \cdot \epsilon^2 \\
&= C \cdot \epsilon.
\end{aligned}$$

It follows from this last estimate that $A_1 = 0$ (if instead $A_1 = \lambda > 0$ then take $\epsilon = \lambda/(2C)$ to derive a contradiction). By similar reasoning, $A_3 = 0$. Thus, for ϵ sufficiently small, $\mathcal{L}^N(T_\epsilon) = 0$. This gives the result that the desired limit exists. That the limit agrees with the original function (representative) f almost everywhere follows from a similar argument that we leave to the reader. ∎

Corollary 3.5.6 *If $A \subset \mathbb{R}^N$ is Lebesgue measurable, then, for almost every $x \in \mathbb{R}^N$, it holds that*

$$\chi_A(x) = \lim_{r \to 0^+} \frac{\mathcal{L}^N(A \cap \mathbb{B}(x,r))}{\mathcal{L}^N(\mathbb{B}(x,r))}.$$

Proof: Set $f = \chi_A$. Then

$$\int_{\mathbb{B}(x,r)} f(t)\, dt = \mathcal{L}^N(A \cap \mathbb{B}(x,r))$$

and the corollary follows from Proposition 3.5.4. ∎

Definition 3.5.7 *A function $f : \mathbb{R}^N \to \mathbb{R}$ is said to be **approximately continuous** if, for almost every $x_0 \in \mathbb{R}^N$ and for each $\epsilon > 0$, the set $\{x : |f(x) - f(x_0)| > \epsilon\}$ has density 0 at x_0, that is,*

$$0 = \lim_{r \to 0^+} \frac{\mathcal{L}^N(\{x : |f(x) - f(x_0)| > \epsilon\} \cap \mathbb{B}(x_0, r))}{\mathcal{L}^N(\mathbb{B}(x_0, r))}.$$

Corollary 3.5.8 *If a function $f : \mathbb{R}^N \to \mathbb{R}$ is Lebesgue measurable, then it is approximately continuous.*

Proof: Suppose that f is Lebesgue measurable. Let $q_1, q_2, \ldots$ be an enumeration of the rational numbers. For each positive integer i, let E_i be the set of points $x \notin \{z : f(z) < q_i\}$ for which

$$0 < \limsup_{r \to 0^+} \frac{\mathcal{L}^N(\{z : f(z) < q_i\} \cap \mathbb{B}(x,r))}{\mathcal{L}^N(\mathbb{B}(x,r))}$$

and let E^i be the set of points $x \notin \{z : q_i < f(z)\}$ for which

$$0 < \limsup_{r \to 0^+} \frac{\mathcal{L}^N(\{z : q_i < f(z)\} \cap \mathbb{B}(x,r))}{\mathcal{L}^N(\mathbb{B}(x,r))}.$$

By Corollary 3.5.6 and the Lebesgue measurability of f, we know that $\mathcal{L}^N(E_i) = 0$ and $\mathcal{L}^N(E^i) = 0$. Thus we see that

$$E = \bigcup_{i=1}^{\infty} (E_i \cup E^i)$$

is also a set of Lebesgue measure zero.

Consider any point $x_0 \notin E$ and any $\epsilon > 0$. There exist rational numbers q_i and q_j such that

$$f(x_0) - \epsilon < q_i < f(x_0) < q_j < f(x_0) + \epsilon.$$

We have $\{x : |f(x) - f(x_0)| > \epsilon\} \subset \{z : f(z) < q_i\} \cup \{z : q_j < f(z)\}$. By the definition of E_i and E^j we have

$$0 = \lim_{r \to 0^+} \frac{\mathcal{L}^N(\{z : f(z) < q_i\} \cap \mathbb{B}(x_0,r))}{\mathcal{L}^N(\mathbb{B}(x_0,r))}$$

and

$$0 = \lim_{r \to 0^+} \frac{\mathcal{L}^N(\{z : q_j < f(z)\} \cap \mathbb{B}(x_0,r))}{\mathcal{L}^N(\mathbb{B}(x_0,r))}.$$

It follows that

$$0 = \lim_{r \to 0^+} \frac{\mathcal{L}^N(\{x : |f(x) - f(x_0)| > \epsilon\} \cap \mathbb{B}(x_0,r))}{\mathcal{L}^N(\mathbb{B}(x_0,r))}.$$

Since $x_0 \notin E$ and $\epsilon > 0$ were arbitrary, we conclude that f is approximately continuous. ∎

Remark 3.5.9 The converses of Corollaries 3.5.6 and 3.5.8 are also true, but there is more technical difficulty involved when one deals with a set A or a function f that may not be measurable. The interested reader should see Federer [4], §2.9.12 and §2.9.13.

The Besicovitch Covering Theorem

We next turn to Besicovitch's covering theorem because its proof is similar to that of Wiener's theorem. It is a purely geometric lemma, as it speaks to how balls may be packed in space.

3.5. COVERING LEMMAS

Theorem 3.5.10 (Besicovitch) *Let N be a positive integer. There is a constant $K = K(N)$ with the following property. Let $\mathcal{B} = \{B_j\}_{j=1}^{M}$ be any finite collection of open balls in $\mathbb{R}^N$ with the property that no ball contains the center of any other. Then we may write*

$$\mathcal{B} = \mathcal{B}_1 \cup \cdots \cup \mathcal{B}_K$$

so that each $\mathcal{B}_j$, $j = 1, \ldots, K$, is a collection of pairwise disjoint *balls.*

In general it is not known what the best possible K is for any given dimension N. Interesting progress on this problem has been made in Sullivan [1]. Certainly our proof will give little indication of the best K.

We shall see that the heart of this theorem is the following lemma about balls. We shall give two different proofs of this lemma. One, contrary to our avowed philosophy in the present subsection, will in fact depend on measure—or at least on the notion of volume. The second proof will rely instead on trigonometry.

Lemma 3.5.11 *There is a constant $K = K(N)$, depending only on the dimension of our space $\mathbb{R}^N$, with the following property: Let $B_0 = \mathbb{B}(x_0, r_0)$ be a ball of fixed radius. Let $B_1 = \mathbb{B}(x_1, r_1), B_2 = \mathbb{B}(x_2, r_2), \ldots, B_p = \mathbb{B}(x_p, r_p)$ be balls such that*

(i) *Each B_j has non-empty intersection with B_0, $j = 1, \ldots, p$;*

(ii) *The radii $r_j \geq r_0$ for all $j = 1, \ldots, p$;*

(iii) *No ball B_j contains the center of any other B_k for $k \neq j$, $j, k = 0, \ldots, p$.*

Then $p \leq K$.

Here is what the lemma says in simple terms: fix the ball B_0. Then at most K pairwise disjoint balls of (at least) the same size can touch B_0. Note here that being 'pairwise disjoint' and 'intersecting but not containing the center of any other ball' are essentially equivalent: if the balls intersect, but none contains the center of another, then shrinking each ball by a factor of one half makes the balls pairwise disjoint; if the balls are already pairwise disjoint, have equal radii, and are close together, then doubling their size arranges for them to intersect without any ball containing the center of another.

First Proof of the Lemma: The purpose of providing this particular proof, even though it relies on the concept of volume, is that it is quick and intuitive. The second proof is less intuitive, but it introduces the important idea of 'directionally limited.'

Since any ball B_j, $j = 1, 2, \ldots, p$, for which $r_j > r_0$ can be replaced by the ball of radius r_0 internally tangent to B_j at the point of intersection

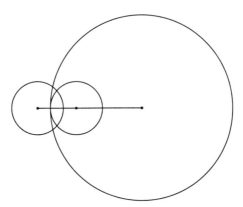

Figure 3.4: **Replacing a Larger Ball**

of the segment connecting the center of B_j and B_0 while maintaining the validity of all three conditions (i)–(iii) (see Figure 3.4), it will suffice to prove the lemma with all balls having radius r_0.

We now assume that all balls have the same radius. With the balls as given, the smaller balls $\mathbb{B}(x_1, r_0/2), \mathbb{B}(x_2, r_0/2), \ldots, \mathbb{B}(x_p, r_0/2)$ are pairwise disjoint and are all contained in $\mathbb{B}(x_0, 3r_0)$. We calculate that

$$\begin{aligned} p &= \frac{\mathcal{L}^N(\cup_{j=1}^p B_j)}{(r_0/2)^N \Upsilon_N} \\ &\leq \frac{\mathcal{L}^N(\mathbb{B}(x_0, 3r_0))}{(r_0/2)^N \Upsilon_N} \\ &= 6^N. \end{aligned}$$

As a result of this calculation, we see that $K(N)$ exists and does not exceed 6^N. ∎

Second Proof of the Lemma: For this argument, see Krantz and Parsons [1]. In fact we shall prove the following more technical statement

> Let the universe be the two dimensional plane, $\mathbb{R}^2$, and let $\Sigma = \{re^{i\theta} : 0 \leq r < \infty, 0 \leq \theta \leq \pi/6\}$. Set $S = \{z \in \Sigma : |z| \geq 3\}$. If $a, b \in S$ and if each of the balls $\mathbb{B}(a, r), \mathbb{B}(b, s)$ intersects $\mathbb{B}(0, 1)$, then

$$|a - b| < \max(r, s). \tag{3.60}$$

3.5. COVERING LEMMAS

A moment's thought reveals that this yields the desired sparseness condition in dimension two. The N-dimensional result is obtained by slicing with two dimensional planes.

To prove (3.60), we first note the inequalities

$$(\alpha - 1)^2 - (2 - \sqrt{3})\alpha^2 \geq 0 \quad \text{if } \alpha \geq 3; \tag{3.61}$$

$$(\beta - 1)^2 - (\alpha^2 - \sqrt{3}\alpha\beta + \beta^2) \geq 0 \quad \text{if } \beta \geq \alpha \geq 3. \tag{3.62}$$

The first of these is proved by noting that the derivative of the left side of (3.61), in the variable α, is positive when $\alpha \geq 3$; and the inequality is satisfied when $\alpha = 3$. So the result follows from the fundamental theorem of calculus.

Similarly, the derivative of the left side of (3.62), in the variable β, is positive when $\beta \geq \alpha \geq 3$, and the case $\beta = \alpha = 3$ is just inequality (3.61), which has already been established.

With these inequalities in hand, we introduce polar coordinates in the plane, writing $a = \alpha e^{i\theta}$ and $b = \beta e^{i\phi}$. We assume without loss of generality that $\alpha \leq \beta$. The hypothesis that $\mathbb{B}(b, s) \cap \mathbb{B}(0, 1) \neq \emptyset$ entails $s > \beta - 1$; thus it suffices to show that

$$|a - b|^2 \leq (\beta - 1)^2. \tag{3.63}$$

The law of cosines tells us that

$$|a - b|^2 = a^2 - 2\alpha\beta \cos(\phi - \theta) + \beta^2. \tag{3.64}$$

Since $\cos(\phi - \theta) \geq \cos \pi/6 = \sqrt{3}/2$, it follows that the right side of (3.64) does not exceed $a^2 - \sqrt{3}ab + b^2$. The inequality (3.63) now follows from (3.62). ∎

H. Federer's concept of a **directionally limited metric space**—see Federer [4]—formalizes the geometry that goes into the proof of our last lemma. More precisely, it generalizes to abstract contexts the notion that a cone in a given direction can contain only a certain number of points with distance $\eta > 0$ from the vertex and distance η from each other. The interested reader is advised to study that primary source.

Now we can present the proof of Besicovitch's covering lemma:

Proof of Theorem 3.5.10: Begin as in the proof of the Wiener covering lemma. Select B_1^1 to be a ball of maximum radius. Then select B_2^1 to be a ball of maximum radius that is disjoint from B_1^1. Continue until this selection procedure is no longer possible (remember that there are only finitely many balls in total). Set $\mathcal{B}_1 = \{B_j^1\}$.

Now work with the remaining balls. Let B_1^2 be the ball with greatest radius. Then select B_2^2 to be the remaining ball with greatest radius, disjoint from B_1^2. Continue in this fashion until no further selection is possible. Set $\mathcal{B}_2 = \{B_j^2\}$.

Working with the remaining balls, we now produce the family $\mathcal{B}_3$, and so forth. Clearly, since in total there are only finitely many balls, this procedure must stop. We will have produced finitely many—say p—nonempty families of pairwise disjoint balls, $\mathcal{B}_1, \ldots, \mathcal{B}_p$. It remains to say how large p can be.

Suppose that $p > K(N) + 1$, where $K(N)$ is as in the lemma. Let B_1^p be the first ball in the family $\mathcal{B}_p$. That ball must have intersected a ball in each of the preceding families; by our selection procedure, each of those balls must have been at least as large in radius as B_1^p. Thus B_1^p is an open ball with at least $K(N) + 1$ "neighbors" as in the lemma. But the lemma says that a ball can only have $K(N)$ neighbors. That is a contradiction.

We conclude that $p \leq K(N) + 1$. That proves the theorem. ∎

Using the Besicovitch covering theorem, we now can extend the earlier results on the maximal function and differentiation to Radon measures other than the Lebesgue measure.

Definition 3.5.12 Let μ be a Radon measure on $\mathbb{R}^N$ and let f be a locally μ-integrable function on $\mathbb{R}^N$. We define the **maximal function with respect to** μ, denoted $M_\mu f$, by setting $M_\mu f(x) = 0$ if there exists $0 < r < \infty$ with $\mu(\mathbb{B}(x,r)) = 0$, and

$$M_\mu f(x) = \sup_{0 < r < \infty} \frac{1}{\mu(\mathbb{B}(x,r))} \int_{\mathbb{B}(x,r)} |f(t)|\, d\mu t$$

otherwise.

Lemma 3.5.13 *There is a constant $C > 0$—depending on the dimension N and the Radon measure μ, but not on the particular function f—such that the operator M satisfies*

$$\mu\{x \in \mathbb{R}^N : M_\mu f(x) > \lambda\} \leq \frac{C}{\lambda} \|f\|_{L^1_\mu(\mathbb{R}^N)}$$

for any scalar $\lambda > 0$. The constant C is independent of λ and of f.

Proof: The same proof may be used for Lemma 3.5.13 as was used for Lemma 3.5.3, except that the Besicovitch covering theorem replaces the Wiener covering lemma and a little more care must be exercised in constructing the initial cover of K so as to arrange that no ball contain the center of any other ball. ∎

As a corollary, we get the following more general differentiation theorem:

Proposition 3.5.14 *Let μ be a Radon measure on $\mathbb{R}^N$ and let f be a locally μ-integrable function on $\mathbb{R}^N$. Then for μ-almost every $x \in \mathbb{R}^N$ it holds that*

$$\lim_{r \to 0^+} \frac{1}{\mu(\mathbb{B}(x,r))} \int_{\mathbb{B}(x,r)} f(t)\, d\mu t$$

Lebesgue's Covering Theorem

The covering theorems presented thus far can be adapted to a variety of situations (besides Euclidean space); notable among these is the setting of spaces of homogeneous type (see Christ [1], Coifman and Weiss [1], [2], Krantz [5]). The next theorem is an integral part of dimension theory (for which see Hurewicz and Wallman [1]), which is usually studied on a separable metric space. The theorem has important geometric content, and is of wide utility.

Note that if $\mathcal{U} = \{U_\alpha\}_{\alpha \in A}$ is a covering of a set S by open sets then a **refinement** $\mathcal{V} = \{V_\beta\}_{\beta \in B}$ of $\mathcal{U}$ is an open covering of S with the property that each V_β lies in some U_α.

Theorem 3.5.15 *Let $S \subset \mathbb{R}^N$ be any set. Let $\mathcal{U} = \{U_\alpha\}_{\alpha \in A}$ be a covering of S by open sets. Then there is a refinement $\mathcal{V} = \{V_\beta\}_{\beta \in B}$ of $\mathcal{U}$ that has valence at most $N + 1$. That is, each point of S is an element of at most $N + 1$ of the V_β.*

The proof will proceed in several steps. We will introduce a few simple concepts from dimension theory.

First, the empty set is declared to have dimension -1. It is the only set with dimension -1. A point P in a set S is said to have dimension $\leq N$ if P has a neighborhood basis $\{U_j\}$ with the property that $\partial U_j \cap S$ has dimension $\leq (N-1)$. A set is said to have dimension N if it has dimension $\leq N$ but not $\leq (N-1)$.

A moment's thought reveals that the set of rational numbers, lying in $\mathbb{R}$, has dimension 0. So does the set of irrational numbers. For the first of these, simply note that if P is rational then a ball around P with irrational radius will have boundary possessing intersection with $\mathbb{Q}$ that is empty (hence has dimension -1). It will be useful for us to note that if $1 \leq m \leq N$ then the set T_m of points $(x_1, x_2, \ldots, x_N)$ in $\mathbb{R}^N$ with $x_1, x_2, \ldots, x_m$ rational has dimension 0. Thus we see that

$$\mathbb{R}^N = (T_1 \cup T_2 \cup \cdots \cup T_N) \cup S,$$

where S is the set of points in $\mathbb{R}^N$ with all irrational coordinates. And each of these sets has dimension zero. Thus we have decomposed $\mathbb{R}^N$ into $(N+1)$ sets of dimension 0.

While it is certainly true that a discrete set is of dimension 0, the preceding examples show that the converse is not true. Nevertheless, sets of dimension 0 share certain properties with discrete sets. The one of greatest interest for us is contained in the following lemmas. Before beginning, we note that a set $C \subset \mathbb{R}^N$ is said to **separate** two disjoint subsets A_1, A_2 of $\mathbb{R}^N$ if $\mathbb{R}^N \setminus C = U_1 \cup U_2$ with U_1, U_2 being relatively open in $\mathbb{R}^N \setminus C$, disjoint, and $A_j \subset U_j$ for $j = 1, 2$. As an exercise, the reader may wish to verify that if a set $S \subset \mathbb{R}^N$ is zero dimensional then any two disjoint relatively closed sets in S may be separated by some set $C \subset S$ (or see statement (E), page 15, in Hurewicz and Wallman [1]).

Lemma 3.5.16 *If E_1, E_2 are closed, disjoint subsets of $\mathbb{R}^N$ and if B is a zero dimensional subset of $\mathbb{R}^N$, then there is a closed set C in $\mathbb{R}^N$ such that (i) C separates E_1 and E_2 and (ii) $B \cap C = \emptyset$.*

Proof: There certainly exist open sets W_1, W_2 such that $E_j \subset W_j, j = 1, 2$, and $\overline{W}_1 \cap \overline{W}_2 = \emptyset$. [This is just the property of **normality** for $\mathbb{R}^N$.] Now $\overline{W}_1 \cap B$ and $\overline{W}_2 \cap B$ are disjoint and closed in B. Since B has dimension 0, these two sets can be separated in B (by the remark preceding the statement of this lemma). We conclude that there are disjoint sets U'_1, U'_2 satisfying

$$B = U'_1 \cup U'_2$$

with $\overline{W}_j \cap B \subset U'_j$, $j = 1, 2$, and U'_1, U'_2 both closed and open in B.
Now

$$(U'_1 \cap \overline{W}_2) \cup (U'_2 \cap \overline{W}_1) = \emptyset, \tag{3.65}$$

$$(U'_1 \cap \overline{U'}_2) \cup (U'_2 \cap \overline{U'}_1) = \emptyset, \tag{3.66}$$

and therefore

$$(U'_1 \cap \overline{E_2}) \cup (U'_2 \cap \overline{E_1}) = \emptyset. \tag{3.67}$$

But W_1, W_2 are open, so that (3.65) implies that $(\overline{U'}_1 \cap W_2) \cup (\overline{U'}_2 \cap W_1) = \emptyset$. As a result, by the first sentence of the proof, we conclude that $(\overline{U'}_1 \cap E_2) \cup (\overline{U'}_2 \cap E_1) = \emptyset$. Equations (3.65)–(3.67) allow us now to conclude, since $\overline{E_1} \cap \overline{E_2} = \emptyset$, that neither of the disjoint sets $E_1 \cup U'_1$ and $E_2 \cup U'_2$ contains a cluster point of the other. Since every subspace of $\mathbb{R}^N$ is normal, we find that there is an open set $\mathcal{O}$ such that

$$E_1 \cup U'_1 \subset \mathcal{O}$$

and

$$\overline{\mathcal{O}} \cap (E_2 \cup U'_2) = \emptyset.$$

In conclusion, the boundary $C = \overline{\mathcal{O}} \setminus \mathcal{O}$ separates E_1 and E_2 and is disjoint from $U'_1 \cup U'_2 = B$. ∎

Lemma 3.5.17 *Let S be a zero dimensional subset of $\mathbb{R}^N$. Let $\mathcal{U} = \{U_1, U_2\}$ be open subsets of $\mathbb{R}^N$ that cover S. Then there is a refinement $\mathcal{V} = \{V_1, V_2\}$ of $\mathcal{U}$ that still covers S and such that $V_1 \cap V_2 = \emptyset$.*

3.5. COVERING LEMMAS

Proof: It is convenient to replace the universe $\mathbb{R}^N$ with the set $X = U_1 \cup U_2$ and we do so without further comment.

Thus $D_1 = X \setminus U_2$ and $D_2 = X \setminus U_1$ are closed, disjoint sets. Since S has dimension 0, we may apply the preceding lemma to get a closed set C that is disjoint from S and separates D_1 from D_2. Thus there exist sets W_1, W_2 which are open, disjoint, and satisfy $W_j \supset D_j$, $j = 1, 2$, and $X \setminus C = W_1 \cup W_2$.

As a consequence, $W_j \subset U_j$, $j = 1, 2$. Also $S \cap C = \emptyset$ and $W_1 \cup W_2 \supset S$. This is the desired conclusion. ∎

Lemma 3.5.18 *Let S be a zero dimensional subset of $\mathbb{R}^N$. Let $\mathcal{U} = \{U_j\}_{j=1}^p$ be open sets that cover S. Then there exists a refinement $\mathcal{V} = \{V_j\}_{j=1}^p$ of $\mathcal{U}$ that still covers S and such that $V_j \cap V_k = \emptyset$ when $j \neq k$.*

Proof: We proceed by induction on p. Note that if $p = 1$ then the statement is trivial.

Now assume that the result has been established for $p = k - 1$; we use that hypothesis to prove the result for $p = k$. Thus we are given a covering $\mathcal{U} = \{U_j\}_{j=1}^k$ of a set $S \subset \mathbb{R}^N$. Let $U'_{k-1} \equiv U_{k-1} \cup U_k$. Then the collection $\mathcal{U}'$ given by

$$\mathcal{U}' = \left\{ U_1, U_2, \ldots, U_{k-2}, U'_{k-1} \right\}$$

is still an open covering of S; of course it has $(k-1)$ elements. The inductive hypothesis applies to the covering $\mathcal{U}'$ of S, and we obtain a refinement

$$\mathcal{V}' = \{V_1, V_2, \ldots, V_{k-2}, V'_{k-1}\}.$$

of $\mathcal{U}'$ that still covers S and is such that the elements of $\mathcal{V}'$ are pairwise disjoint.

Observe that the set $T \equiv S \cap V'_{k-1}$ has dimension not exceeding 0 (since S does). Notice also that U_{k-1} and U_k cover T. By the last lemma, there exist open sets V_{k-1}, V_k that refine the cover U_{k-1}, U_k of T and which are disjoint. But then

$$V_1, V_2, \ldots, V_{k-2}, V_{k-1}, V_k$$

form an open covering of S with all the desired properties. ∎

The proof of the Lebesgue covering theorem is now easy:

Proof of Theorem 3.5.15: Begin by writing

$$S = S_1 \cup S_2 \cup \cdots \cup S_{N+1},$$

where each S_j has dimension at most 0. Since our open covering certainly covers each S_j, we may apply the last lemma to obtain $(N+1)$ refinements

$\mathcal{V}^1, \ldots, \mathcal{V}^{N+1}$ of $\mathcal{U}$ such that

$$\mathcal{V}^i = \{V_1^i, \ldots, V_{p(i)}^i\}; \tag{3.68}$$

$$\mathcal{V}^i \text{ covers } S_i; \tag{3.69}$$

$$V_s^i \cap V_t^i = \emptyset \quad \text{if } s \neq t. \tag{3.70}$$

Now let $\mathcal{V}$ be the covering consisting of all the open sets V_j^i for all possible values of i and j. This covering certainly covers S, for it covers each of the S_i. We claim that $\mathcal{V}$ has valence at most $N+1$. For note that any selection of $N+2$ elements of $\mathcal{V}$ must contain two elements from a single $\mathcal{V}^i$ which are therefore disjoint. So it is impossible for $N+2$ elements of $\mathcal{V}$ to have a point in common. ∎

It turns out that the property of any open covering having a refinement of valence at most $N+1$ characterizes sets of dimension N. Since this is not a study of dimension theory, it would take us far afield to provide all the details of this assertion. We refer the interested reader to Hurewicz and Wallman [1].

The Vitali Covering Theorem

The Vitali covering theorem is probably the most broadly used covering theorem in all of measure theory. It has the dual advantage of being both intuitively appealing and rather profound. Indeed it says something very important about the structure of sets in Euclidean space.

We derive our treatment of Vitali's ideas from Evans and Gariepy [1]; the interested reader may consult that source for further details. We begin with a modification of the Wiener covering theorem:

Proposition 3.5.19 *Let $\mathcal{U}$ be a collection of closed balls in $\mathbb{R}^N$ such that*

$$\sup\{\operatorname{diam} B : B \in \mathcal{U}\} = M < \infty.$$

Then there is a countable, disjoint subfamily $\mathcal{V} \subset \mathcal{U}$ such that

$$\bigcup_{B \in \mathcal{U}} B \subset \bigcup_{B \in \mathcal{V}} 5B.$$

[Here, as usual the notation $5B$ denotes the ball with the same center as B but 5 times the radius.]

The proof of this proposition is left to the reader.

Now the principal theorem of Vitali type that we wish to consider is this:

3.5. COVERING LEMMAS

Theorem 3.5.20 (Vitali) *Let $U \subset \mathbb{R}^N$ be an open set. Let $\delta > 0$. Then there exists a countable collection $\mathcal{W} = \{\overline{B}_j\}_{j=1}^{\infty}$ of disjoint closed balls, each lying in U, such that*

(i) $\operatorname{diam} \overline{B}_j \leq \delta$ *for each j;*

(ii) $\mathcal{L}^N \left[U \setminus \bigcup_{j=1}^{\infty} \overline{B}_j \right] = 0.$

The gist of the theorem is that one may "fill up" an arbitrary open set U with small balls, to the extent that the portion not covered has measure zero. Of course it is too much to hope that the *entire* open set can be exhausted by balls. For example, if U is the interior of a unit square in the plane, then it is easy to see that there is no way to fill U completely with countably many discs of radius not exceeding $1/4$. In fact, one can see in this example that the set of points not covered will be uncountable and nowhere dense.

Proof: Assume for convenience that U has finite Lebesgue measure. Some remarks about the general case will be given at the end. Fix a number λ such that $1 - 1/5^N < \lambda < 1$, where N is the dimension.

We will construct the required balls by an iterative procedure. Each step of our process will exhaust a proportion $(1 - \theta)$ of that portion of U that is not already covered by balls.

For the first step, set $U_1 = U$ and let

$$\mathcal{G}_1 = \{\overline{B} : \overline{B} \text{ is a closed ball lying in } U_1 \text{ and having diameter less than } \delta\}.$$

By the preceding proposition, there is a countable, disjoint family $\mathcal{H}_1 \subset \mathcal{G}_1$ with the property that

$$U_1 \subset \bigcup_{\overline{B} \in \mathcal{H}_1} 5\overline{B}.$$

As a result, we may calculate that

$$\begin{aligned} \mathcal{L}^N(U_1) &\leq \sum_{\overline{B} \in \mathcal{H}_1} \mathcal{L}^N(5\overline{B}) \\ &= 5^N \sum_{\overline{B} \in \mathcal{H}_1} \mathcal{L}^N(\overline{B}) \\ &= 5^N \mathcal{L}^N \left(\bigcup_{\overline{B} \in \mathcal{H}_1} \overline{B} \right). \end{aligned}$$

We conclude that

$$\mathcal{L}^N \left(\bigcup_{\overline{B} \in \mathcal{H}_1} \overline{B} \right) \geq \frac{1}{5^N} \mathcal{L}^N(U_1).$$

Since each element of $\mathcal{H}_1$ is a subset of U_1, we see that

$$\mathcal{L}^N\left(U_1 \setminus \bigcup_{\overline{B} \in \mathcal{H}_1} \overline{B}\right) \leq \left(1 - \frac{1}{5^N}\right)\mathcal{L}^N(U_1).$$

Of course the collection of balls $\mathcal{H}_1$ is countable; thus there is a subcollection $\overline{B}_1, \overline{B}_2, \ldots, \overline{B}_{K_1}$ satisfying

$$\mathcal{L}^N\left(U_1 \setminus \bigcup_{j=1}^{K_1} \overline{B}_j\right) \leq \lambda m(U_1).$$

For the second step, set

$$U_2 = U_1 \setminus \bigcup_{j=1}^{K_1} \overline{B}_j$$

and let

$\mathcal{G}_2 = \{\overline{B} : \overline{B}$ is a closed ball lying in U_2 and having diameter less than $\delta\}$.

Repeating the arguments from Step 1, we find a collection $\overline{B}_{K_1+1}, \overline{B}_{K_1+2}, \ldots, \overline{B}_{K_2}$ of disjoint elements of $\mathcal{G}_\epsilon$ such that

$$\mathcal{L}^N\left(U \setminus \bigcup_{j=1}^{K_2}\overline{B}_j\right) = \mathcal{L}^N\left(U_2 \setminus \bigcup_{j=M_1+1}^{M_2} \overline{B}_j\right)$$
$$\leq \lambda \mathcal{L}^N(U_2)$$
$$\leq \lambda^2 \mathcal{L}^N(U).$$

We continue this process countably many times. The result is a pairwise disjoint collection of closed balls $\overline{B}_j$, all lying in U, all having diameter less than δ, such that

$$\mathcal{L}^N\left(U \setminus \bigcup_{j=1}^{K_p}\overline{B}_j\right) \leq \lambda^p \mathcal{L}^N(U),$$

for $p = 1, 2, \ldots$. Since $0 < \lambda < 1$, our conclusion follows provided that $\mathcal{L}^N(U) < \infty$.

In case $\mathcal{L}^N(U) = \infty$, write $U = (\cup_\ell U_\ell) \cup T$, where each U_ℓ is open with finite measure and T has measure zero. ∎

3.6 Functions of Bounded Variation

Functions of bounded variation from $\mathbb{R}$ to $\mathbb{R}$ are well understood (see Federer [4], §§2.5.16–2.5.18). The situation for functions from $\mathbb{R}^N$ to $\mathbb{R}$ is

3.6. FUNCTIONS OF BOUNDED VARIATION

less settled and there would seem to be opportunities for future clarification. Our main references for this section are Evans and Gariepy [1] and Ziemer [1].

The following example motivates the definition of functions of bounded variation.

Example 3.6.1 Suppose $f \in C^2(\Omega)$, where Ω is an open subset of $\mathbb{R}^N$. For any $g = (g_1, g_2, \ldots, g_N) \in C_c^1(\Omega; \mathbb{R}^N)$, with $|g| \leq 1$ for all $x \in \Omega$, we can use integration by parts to see that

$$\int_\Omega f \operatorname{div}(g)\, dx = -\int_\Omega (\operatorname{grad} f) \cdot g\, dx \leq \int_\Omega |\operatorname{grad} f|\, dx.$$

Now, set $\Omega' = \Omega \cap \{x : \operatorname{grad} f(x) \neq 0\}$. If we let K be an arbitrary compact subset of Ω', then we can choose a non-negative C^∞ function ϕ with $0 \leq \phi \leq 1$, $\operatorname{supp} f \subset \Omega'$, and with $\phi(x) = 1$ for $x \in K$. Setting

$$g = -\phi \operatorname{grad} f |\operatorname{grad} f|^{-1},$$

we have $g \in C_0^1(\Omega; \mathbb{R}^N)$, with $|g(x)| \leq 1$ for all $x \in \Omega$, and in this case, we have

$$\int_\Omega f \operatorname{div}(g)\, dx = \int_\Omega \phi |\operatorname{grad} f|\, dx \geq \int_K |\operatorname{grad} f|\, dx.$$

Since $K \subset \Omega'$ was an arbitrary compact set, we conclude that

$$\int_\Omega |\operatorname{grad} f|\, dx =$$
$$\sup\left\{\int_\Omega f \operatorname{div}(g)\, dx : g \in C_c^1(\Omega; \mathbb{R}^N),\ |g(x)| \leq 1\ \forall x \in \Omega\right\}. \quad (3.71)$$

The reader should note that with more effort one can see that (3.71) also holds for $f \in C^1(\Omega)$.

Definition 3.6.2 Let $\Omega \subset \mathbb{R}^N$ be open. A function $f \in L^1(\Omega)$ is said to be of **bounded variation** on Ω if

$$\int_\Omega |Df| :=$$
$$\sup\left\{\int_\Omega f \operatorname{div}(g)\, dx : g \in C_c^1(\Omega, \mathbb{R}^N),\ |g(x)| \leq 1\ \forall x \in \Omega\right\} < \infty. \quad (3.72)$$

The space of functions of bounded variation on Ω will be denoted $BV(\Omega)$ and will be normed by

$$\|f\|_{BV} = \|f\|_{L^1} + \int_\Omega |Df|. \quad (3.73)$$

Basic, but important, properties of BV functions follow easily from the definition.

Proposition 3.6.3 *If $f \in BV(\Omega)$ and $g \in C_c^1(\Omega; \mathbb{R}^N)$, then*

$$\left| \int_\Omega f \, \mathbf{div}(g) \, dx \right| \leq \sup_\Omega |g| \int_\Omega |DF|. \tag{3.74}$$

Proof: Immediate from the definition. ∎

Proposition 3.6.4 *If $f \in BV(\Omega)$ and $u \in C^1(\Omega)$ with $|u|$ and $|\mathbf{grad}\, u|$ both bounded, then $uf \in BV(\Omega)$ with*

$$\int_\Omega |D(uf)| \leq \sup_\Omega |u| \int_\Omega |Df| + \sup_\Omega |\mathbf{grad}\, u| \int_\Omega |f| \, dx. \tag{3.75}$$

Proof: Use the identity

$$u \, \mathbf{div}(g) = \mathbf{div}(ug) - (\mathbf{grad}\, u \cdot g)$$

and (3.72). ∎

Theorem 3.6.5 (Semicontinuity) *Suppose $\Omega \subset \mathbb{R}^N$ is open. If $f_1, f_2, \ldots$ is a sequence in $BV(\Omega)$ that converges in $L^1_{\mathrm{loc}}(\Omega)$ to a function f, then*

$$\int_\Omega |Df| \leq \liminf_{i \to \infty} \int_\Omega |Df_i|. \tag{3.76}$$

Proof: Let $g \in C_c^1(\Omega; \mathbb{R}^N)$ with $|g| \leq 1$ be arbitrary. Then for any i we have

$$\int_\Omega f \, \mathbf{div}(g) \, dx = \lim_{i \to \infty} \int_\Omega f_i \, \mathbf{div}(g) \, dx \leq \liminf_{i \to \infty} |Df_i|.$$

The result follows by taking the supremum over all such g. ∎

Corollary 3.6.6 (Completeness) *The space of functions of bounded variation with the norm given in (3.73) is a Banach space.*

Proof: The reader may verify that (3.73) does in fact define a norm. We will verify completeness.

Suppose $f_1, f_2, \ldots$ is a Cauchy sequence in the norm $\|\cdot\|_{BV}$. A fortiori, $f_1, f_2, \ldots$ is a Cauchy sequence in the L^1 norm, and thus converges in the L^1 norm to a function $f \in L^1(\Omega)$.

We know that $f_i - f_j$ converges to $f_i - f$ in $L^1(\Omega)$ as $j \to \infty$, so by the Semicontinuity Theorem 3.6.5 we have

$$\limsup_{i \to \infty} \int_\Omega |D(f_i - f)| \leq \limsup_{i \to \infty} \left(\liminf_{j \to \infty} \int_\Omega |D(f_i - f_j)| \right)$$

$$\leq \limsup_{i,j \to \infty} \int_\Omega |D(f_i - f_j)| = 0,$$

3.6. FUNCTIONS OF BOUNDED VARIATION

from which the result follows. ∎

The next theorem shows that BV functions can also be characterized as those functions whose partial derivatives are measures.

Theorem 3.6.7 *Let $\Omega \subset \mathbb{R}^N$ be open. If $f \in BV(\Omega)$ and $1 \leq i \leq N$, then there exists a signed Borel measure μ_i of finite total variation on Ω such that*

$$-\int_\Omega f \frac{\partial \phi}{\partial x_i} \, dx = \int_\Omega \phi \, d\mu_i \qquad (3.77)$$

for each $\phi \in C_c^\infty(\Omega)$.

Conversely, if there are signed Borel measures $\mu_1, \mu_2, \ldots, \mu_N$ such that (3.77) holds for each $1 \leq i \leq N$ and $\phi \in C_c^\infty(\Omega)$, then $f \in BV(\Omega)$.

Remark 3.6.8 Note that the left-hand side of (3.77) represents the action on the test function ϕ of the partial derivative of f in the sense of distributions.

Proof: Suppose that $f \in BV(\Omega)$. The left-hand side of (3.77) defines a linear operator L_i on $C_c^\infty(\Omega)$. By (3.74), we see that L is bounded in the supremum norm. The Riesz representation theorem (see Rudin [2], Theorem 6.19) then insures the existence of a signed measure such that

$$L_i(\phi) = \int_\Omega \phi \, d\mu_i, \qquad (3.78)$$

for $\phi \in C_c^\infty(\Omega)$. (Note that since L_i is bounded in the supremum norm, it extends to $C_0(\Omega)$.)

To see the converse, simply observe that for $g \in C_c^1(\Omega; \mathbb{R}^N)$ with $|g| \leq 1$, we have

$$\int_\Omega f \, \mathbf{div}(g) \, dx = -\sum_{i=1}^N \int_\Omega g_i \, d\mu_i \leq \sum_{i=1}^N |\mu_i|(\Omega). \qquad \blacksquare$$

Definition 3.6.9 *If $f \in BV(\Omega)$, then Df is used to denote the derivative in the sense of distributions.*

Corollary 3.6.10 *If $f \in BV(\Omega)$, then there exists a Radon measure μ on Ω and a μ-measurable function σ, with $|\sigma(x)| = 1$ for μ almost every $x \in \Omega$, such that*

$$Df = \sigma \mu.$$

An important point to make in reference to Theorem 3.6.7 is that the measures representing the partial derivatives of a function of bounded variation typically have a non-trivial singular part with respect to Lebesgue measure. Whenever the singular part with respect to Lebesgue measure is non-zero, then the function in question cannot be in any Sobolev space (Sobolev spaces will be treated in some detail in the next chapter). The next example illustrates this phenomenon.

Example 3.6.11 Set
$$f = \chi_{\overline{\mathbb{B}}(0,1)},$$
the characteristic function of the closed unit ball in $\mathbb{R}^N$. By the divergence theorem[2], for $g \in C_c^1(\mathbb{R}^N, \mathbb{R}^N)$ we have
$$\int_{\overline{\mathbb{B}}(0,1)} \mathbf{div}(g)\, dx = \int_{\mathbb{S}(0,1)} g \cdot \nu \, d\mathcal{H}^{N-1},$$
where ν is the outward unit normal to the unit sphere. Comparing this equation with (3.77), we recognize that
$$Df = -\nu \mathcal{H}^{N-1} \llcorner \mathbb{S}(0,1),$$
the measure $\mathcal{H}^{N-1} \llcorner \mathbb{S}(0,1)$ is singular with respect to Lebesgue measure.

The closure of the space of C^∞ functions in the BV norm is the Sobolev space $H^{1,1}$ of functions whose partial derivatives in the sense of distributions are represented by L^1 functions. Thus in general it is impossible to approximate a BV function in the BV norm by a smooth function. On the other hand, the next theorem gives us some positive information on smooth approximations to BV functions.

Theorem 3.6.12 *Let $f \in BV(\Omega)$. Then there exists a sequence of functions $f_1, f_2, \ldots$ in $C^\infty(\Omega)$ such that*

(i) $f_i \to f$ in $L^1(\Omega)$,

(ii) $\int_\Omega |Df_i| \to \int_\Omega |Df|$,

(iii) $Df_i \rightharpoonup Df$

Proof: Let ψ_σ be a family of mollifiers as in Section 1.3. While it is not essential, it is convenient to suppose that ψ_σ satisfies the symmetry (evenness) condition
$$\psi_\sigma(-x) = \psi_\sigma(x). \tag{3.79}$$
In any case, one can always arrange for (3.79) to hold by replacing $\psi_\sigma(x)$ by $\frac{1}{2}(\psi_\sigma(x) + \psi_\sigma(-x))$. Technical difficulties are minimized if we assume that $\Omega = \mathbb{R}^N$, so we shall first make that assumption.

---

[2]We use the terms Gauss-Green theorem and divergence theorem interchangeably. Both are multidimensional versions of the Fundamental Theorem of Calculus (see Section 5.5)

3.6. FUNCTIONS OF BOUNDED VARIATION

Fix $f \in BV(\mathbb{R}^N)$. By Theorem 1.3.31, we know that $f * \psi_\sigma \to f$ in $L^1(\mathbb{R}^N)$, proving (i). Suppose $g \in C_c^\infty(\mathbb{R}^N; \mathbb{R}^N)$. Then using Fubini's Theorem and the theorem on differentiating under the integral sign, we compute

$$\int_{\mathbb{R}^N} (f * \psi_\sigma)(x) \operatorname{div}(g)(x) \, dx$$

$$= \int_{\mathbb{R}^N} \left(\int_{\mathbb{R}^N} f(y) \, \psi_\sigma(x-y) \, dy \right) \operatorname{div}(g)(x) \, dx$$

$$= \int_{\mathbb{R}^N} f(y) \left(\int_{\mathbb{R}^N} \psi_\sigma(x-y) \operatorname{div}(g)(x) \, dx \right) dy$$

$$= \int_{\mathbb{R}^N} f(y) \left(\int_{\mathbb{R}^N} \psi_\sigma(-x) \operatorname{div}(g)(y-x) \, dx \right) dy$$

$$= \int_{\mathbb{R}^N} f(y) \operatorname{div} \left(\int_{\mathbb{R}^N} \psi_\sigma(-x) \, g(y-x) \, dx \right) dy$$

$$= \int_{\mathbb{R}^N} f(y) \operatorname{div}(g * \psi_\sigma)(y) \, dy;$$

conclusion (iii) now follows, because Theorem 1.3.31 tells us that $g * \psi_\sigma$ converges to g in the supremum norm as $\sigma \to 0$. Also note that $|(g * \psi_\sigma)(x)| \leq 1$ holds for all x, showing that

$$\int_{\mathbb{R}^N} |D(f * \psi_\sigma)| \leq \int_{\mathbb{R}^N} |Df|.$$

But of course Theorem 3.6.5 implies that

$$\int_{\mathbb{R}^N} |Df| \leq \liminf_{\sigma \to 0} \int_{\mathbb{R}^N} |D(f * \psi_\sigma)|,$$

so (ii) also follows.

To handle the general case, that $f \in BV(\Omega)$ when $\Omega \neq \mathbb{R}^N$, we must show how to construct a function $u \in C^\infty(\Omega)$ with

$$\int_\Omega |f - u| \, dx < \epsilon \quad \text{and} \quad \int_\Omega |Du| < \int_\Omega |Df| + \epsilon,$$

where $\epsilon > 0$ is given.

To this end, choose an open $\Omega_0 \subset\subset \Omega$ with

$$\int_{\Omega \setminus \Omega_0} |Df| \leq \frac{\epsilon}{3}$$

and construct a sequence of open subsets of Ω with $\Omega_0 \subset\subset \Omega_1 \subset\subset \ldots \subset \Omega$ and $\cup_{i=0}^\infty \Omega_i = \Omega$.

Next, introduce a partition of unity ϕ_i, $i = 0, 1, \ldots$, such that

$$\phi_0 = 1 \text{ on } \Omega_0, \quad \operatorname{supp}(\phi_0) \subset \Omega_1,$$

and in general, supp $(\phi_i) \subset \Omega_{i+1} \setminus \overline{\Omega}_{i-1}, i \geq 1$. Of course, we have $\sum_{i=0}^{\infty} \phi_i \equiv 1$ on Ω.

For each $i = 0, 1, \ldots$, we choose $\sigma_i > 0$ so that with

$$u_i = (f\phi_i) * \psi_{\sigma_i}, \quad v_i = (f \operatorname{\mathbf{grad}} \phi_i) * \psi_{\sigma_i},$$

the following conditions hold:

$$\operatorname{supp}(u_0) \subset \Omega_1, \tag{3.80}$$

$$\operatorname{supp}(u_i) \subset \Omega_{i+1} \setminus \overline{\Omega}_{i-1}, \; i \geq 1, \tag{3.81}$$

$$\int_\Omega |u_i - f\phi_i|\, dx < 2^{-(i+1)} \frac{\epsilon}{4}, \tag{3.82}$$

$$\int_\Omega |v_i - f\operatorname{\mathbf{grad}}\phi_i|\, dx < 2^{-(i+1)} \frac{\epsilon}{4}. \tag{3.83}$$

Set

$$u = \sum_{i=0}^{\infty} u_i,$$

and notice that $u \in C^\infty(\Omega)$ because each point of Ω is only in the support of finitely many u_i.

The estimate of the L^1 norm of the difference between f and u is simply

$$\int_\Omega |u - f|\, dx \leq \sum_{i=0}^{\infty} \int_\Omega |(f\phi_i) * \psi_{\sigma_i} - f\phi_i|\, dx < \epsilon.$$

The more interesting estimate bounds the quantity $\int_\Omega |Du|$. Using the fact that

$$0 = \operatorname{\mathbf{grad}}(\sum_{i=0}^{\infty} \phi_i) = \sum_{i=0}^{\infty} \operatorname{\mathbf{grad}} \phi_i,$$

we compute

$$\begin{aligned} Du &= \sum_{i=0}^{\infty} (f \operatorname{\mathbf{grad}} \phi_i) * \psi_{\sigma_i} + \sum_{i=0}^{\infty} (\phi_i\, Df) * \psi_{\sigma_i} \\ &= \sum_{i=0}^{\infty} [(f \operatorname{\mathbf{grad}} \phi_i) * \psi_{\sigma_i} - f \operatorname{\mathbf{grad}}\phi_i] + \sum_{i=0}^{\infty} (\phi_i\, Df) * \psi_{\sigma_i}. \end{aligned}$$

Arguing with Fubini's Theorem as in the case $\Omega = \mathbb{R}^N$, we see that

$$\int_\Omega |(\phi_i\, Df) * \psi_{\sigma_i}| \leq \int_{\Omega_{i+1} \setminus \overline{\Omega}_{i-1}} |Df|;$$

since each $x \in \Omega$ belongs to at most three of the sets $\Omega_{i+1} \setminus \overline{\Omega}_{i-1}$, we are able to estimate

$$\begin{aligned} \int_\Omega |Du| &\leq \int_{\Omega_0} |Df| + \tfrac{1}{4}\epsilon + \sum_{i=1}^{\infty} \int_{\Omega_{i+1}\setminus\overline{\Omega}_{i-1}} \\ &\leq \int_\Omega |Df| + \epsilon. \end{aligned}$$

∎

The final result we will present in this section concerns compactness in the space of functions of bounded variation. The compactness result is a consequence of the smooth approximation theorem just proved and Rellich's theorem which we will state without proof. A general treatment of Rellich's Theorem can be found in Ziemer [1], Section 2.5. Decreasing the regularity of $\partial\Omega$ increases the difficulty of proving the result. The relevant version of the theorem with nearly the weakest possible regularity assumption for $\partial\Omega$ is the following:

Theorem 3.6.13 (Rellich) *Suppose $\Omega \subset \mathbb{R}^N$ is a bounded open set the boundary of which can locally be represented as the graph of a Lipschitz function. If $f_1, f_2, \ldots$ is a sequence of functions in $L^1(\Omega)$ whose first partial derivatives in the sense of distributions are also in $L^1(\Omega)$, and if there is $M < \infty$ such that*

$$\int_\Omega |f_i|\,dx + \int_\Omega |Df|\,dx \leq M \text{ for } i = 1, 2, \ldots,$$

then a subsequence f_{i_j} converges to a function $f \in L^1(\Omega)$.

Corollary 3.6.14 (Compactness) *If $\Omega \subset \mathbb{R}^N$ is a bounded open set the boundary of which can locally be represented as the graph of a Lipschitz function, then any set of functions bounded in the BV norm has a subsequence that converges in $L^1(\Omega)$ to a function in $BV(\Omega)$.*

Proof: By the Approximation Theorem 3.6.12 we can choose functions $h_1, h_2, \ldots$ in $C^\infty(\Omega)$ with

$$\int_\Omega |f_i - h_i|\,dx \leq \frac{1}{i} \quad \text{and} \quad \int_\Omega |Dh_i| \leq M + 1.$$

Then Rellich's Theorem allows us to extract a subsequence h_{i_j} that converges in the L^1 norm to $f \in L^1(\Omega)$. A fortiori we know that f_{i_j} converges to f in the L^1 norm and we have $f \in BV(\Omega)$ by the Semicontinuity Theorem 3.6.5. ∎

3.7 Domains with Finite Perimeter

Definition 3.7.1
(i) *If S is a Borel set and $\Omega \subset \mathbb{R}^N$ is open, then the **perimeter of S in Ω** is denoted by $P(S, \Omega)$ and is defined by*

$$\begin{aligned} P(S, \Omega) &= \int_\Omega |D\chi_S| \\ &= \sup\left\{ \int_S \operatorname{div}(g)\,d\mathcal{L}^N : g \in C_c^1(\Omega; \mathbb{R}^N),\ |g| \leq 1 \right\}, \end{aligned} \quad (3.84)$$

where χ_S denotes the characteristic function of S.

(ii) *We say S is of* **locally finite perimeter** *if*

$$P(S, \Omega) < \infty \tag{3.85}$$

holds for every bounded open set Ω. Sets of locally finite perimeter are also called **Caccioppoli sets**.

(iii) *If S is of locally finite perimeter, then there is a positive Radon measure μ and a μ-measurable $\mathbb{R}^N$-valued function σ, with $|\sigma(x)| = 1$ for μ-almost every x, such that $D\chi_S = \sigma\mu$ (see Corollary 3.6.10). It is customary to use the notation $\|\partial S\|$ for the Radon measure μ and to write ν_S for the function σ, so that*

$$D\chi_S = \nu_S \|\partial S\|. \tag{3.86}$$

(iv) *We write $P(S) = P(S, \mathbb{R}^N)$, and we say S is of* **finite perimeter** *if $P(S) < \infty$.*

Example 3.7.2 *If S is a C^1 domain, then S is of locally finite perimeter.* To see this, suppose ρ is a C^1 defining function for S so that

$$S = \{x : \rho(x) < 0\}$$

and $\nabla \rho(x) \neq 0$ for $x \in \partial S$. Let $\phi : \mathbb{R} \to \mathbb{R}$ be any C^∞ function with $\phi'(t) \leq 0$ for all t and

$$\phi(t) = \begin{cases} 1 & \text{if } t \leq -1, \\ 0 & \text{if } 0 \leq t. \end{cases}$$

Fixing a bounded open set Ω and a function $g \in C_c^1(\Omega; \mathbb{R}^N)$ with $|g| \leq 1$, we use integration by parts to estimate

$$\begin{aligned}
\int_S \mathbf{div}(g)\, d\mathcal{L}^N &= \lim_{t \downarrow 0} \int_{\mathbb{R}^N} \phi(\rho/t)\, \mathbf{div}(g)\, d\mathcal{L}^N \\
&= -\lim_{t \downarrow 0} \int_{\mathbb{R}^N} \nabla\left[\phi(\rho/t)\right] \cdot g\, d\mathcal{L}^N \\
&= -\lim_{t \downarrow 0} t^{-1} \int_{\mathbb{R}^N} \phi'(\rho/t) (\nabla \rho) \cdot g\, d\mathcal{L}^N \\
&\leq -(\inf \phi')\lim_{t \downarrow 0} t^{-1} \int_{\Omega \cap \{x : -t \leq \rho(x) \leq 0\}} |\nabla \rho|\, d\mathcal{L}^N.
\end{aligned}$$

The last integral is finite because Ω is bounded and $\nabla \rho$ is non-vanishing on ∂S.

3.7. DOMAINS WITH FINITE PERIMETER

In fact, the term (inf ϕ') in the preceding argument can be made as close to -1 as we wish. Also, careful use of the change of variables formula for integration allows one to see that

$$\lim_{t \downarrow 0} t^{-1} \int_{\Omega \cap \{x : -t \leq \rho(x) \leq 0\}} |\nabla \rho| \, d\mathcal{L}^N = \mathcal{H}^{N-1}(\Omega \cap \partial S).$$

We conclude that, for a C^1 domain S,

$$P(S, \Omega) = \mathcal{H}^{N-1}(\Omega \cap \partial S) \tag{3.87}$$

holds for each bounded open set Ω.

The following lemma is an immediate consequence of the fact that the integral of the divergence of a function with compact support is zero.

Lemma 3.7.3 *Suppose S has locally finite perimeter and Ω is a bounded open set. Then*

$$\|\partial(\mathbb{R}^N \setminus S)\| = \|\partial S\|, \qquad \nu_{\mathbb{R}^N \setminus S} = -\nu_S$$

and

$$P(\Omega \setminus S, \Omega) = P(S, \Omega)$$

hold.

Proof: Note that, for $g \in C_c^1(\mathbb{R}^N; \mathbb{R}^N)$, we have

$$\int_S \mathbf{div}(g) \, \mathcal{L}^N = -\int_{\mathbb{R}^N \setminus S} \mathbf{div}(g) \, \mathcal{L}^N. \qquad \blacksquare$$

By Theorem 3.6.7, if $S \subset \mathbb{R}^N$ is a domain with finite perimeter, then the following formula holds for any $g \in C_c^1(\mathbb{R}^N; \mathbb{R}^N)$:

$$\int_S \mathbf{div}(g) \, d\mathcal{L}^N = -\int g \cdot \nu_S \, d\|\partial S\|. \tag{3.88}$$

Equation (3.88) can be considered to be an abstract form of the classical Gauss-Green formula (or synonymously the divergence theorem) which we will discuss in more detail in Section 5.5. One aim of the study of sets of finite perimeter is to improve our understanding of the right-hand side of (3.88). As a first step, the next lemma extends the applicability of (3.88).

Lemma 3.7.4 *Suppose $S \subset \mathbb{R}^N$ is of finite perimeter. If $g \in C_c^0(\mathbb{R}^N; \mathbb{R}^N)$ is differentiable $\mathcal{L}^N$-almost everywhere and*

$$\int_{\mathbb{R}^N} |\mathbf{div}(g)| \, d\mathcal{L}^N < \infty, \tag{3.89}$$

then (3.88) holds for g.

Proof: Let ψ_σ, $\sigma > 0$, be a family of mollifiers as in Section 1.3. Then (3.88) applies to $\psi_\sigma * g$, so that

$$\int_S \psi_\sigma * (\operatorname{div} g) \, d\mathcal{L}^N = \int_S \operatorname{div}(\psi_\sigma * g) \, d\mathcal{L}^N = -\int (\psi_\sigma * g) \cdot \nu_S \, d\|\partial S\| \quad (3.90)$$

holds. As $\sigma \downarrow 0$ the left-hand side of (3.90) converges to $\int_S \operatorname{div}(g) \, d\mathcal{L}^N$ and the right-hand side converges to $\int g \cdot \nu_S \, d\|\partial S\|$, both by Theorem 1.3.31. ∎

Remark 3.7.5 The preceding lemma admits the use of vector fields that are piecewise continuously differentiable and of those that have singular differentials on small sets in the Gauss-Green formula.

The next lemma contains a particular instance of the Gauss-Green formula that will be useful later in this section.

Lemma 3.7.6 *Suppose $S \subset \mathbb{R}^N$ is of locally finite perimeter and $g \in C_c^1(\mathbb{R}^N; \mathbb{R}^N)$. For $x \in \mathbb{R}^N$,*

$$\int_{S \cap \mathbb{B}(x,r)} \operatorname{div}(g) \, d\mathcal{L}^N = \int_{\mathbb{B}(x,r)} g \cdot \nu_S \, d\|\partial S\| + \int_{S \cap \mathbb{S}(x,r)} g \cdot \tfrac{y-x}{|y-x|} \, d\mathcal{H}^{N-1} y, \quad (3.91)$$

holds for $\mathcal{L}^1$-almost every $r > 0$.

Proof: Fix $r > 0$. For $\epsilon > 0$, define

$$g_\epsilon(y) = \begin{cases} g(y) & \text{if } |y - x| \leq r, \\ \tfrac{r + \epsilon - |y-x|}{\epsilon} g(y) & \text{if } r \leq |y - x| \leq r + \epsilon, \\ 0 & \text{if } r + \epsilon \leq |y - x|. \end{cases}$$

We compute

$$\operatorname{div} g_\epsilon(y) = \begin{cases} \operatorname{div} g(y) & \text{if } |y - x| \leq r, \\ \tfrac{r + \epsilon - |y-x|}{\epsilon} \operatorname{div} g(y) - \tfrac{y-x}{\epsilon |y-x|} \cdot g(y) & \text{if } r \leq |y - x| \leq r + \epsilon, \\ 0 & \text{if } r + \epsilon \leq |y - x|. \end{cases}$$

By Lemma 3.7.4 we have

$$\int_S \operatorname{div} g_\epsilon \, d\mathcal{L}^N = \int g_\epsilon \cdot \nu_S \, d\|\partial S\|.$$

Assume r is such that $S \cap \mathbb{S}(x, r)$ is $\mathcal{H}^{N-1}$-measurable, which is true for $\mathcal{L}^1$-almost every $r > 0$. Then the result follows by letting ϵ decrease to 0. ∎

Area Minimization and the Isoperimetric Inequality

Another reason for our interest in sets of finite perimeter is that they provide a context in which to study problems of area minimization.

3.7. DOMAINS WITH FINITE PERIMETER

Theorem 3.7.7 (Existence of Area Minimizers) *Let $U \subset \mathbb{R}^N$ be a bounded open set, and let B be a set of finite perimeter. Then there exists a set S of finite perimeter with $S \setminus U = B \setminus U$ such that*

$$P(S) \leq P(T) \text{ for any } T \text{ with } T \setminus U = B \setminus U.$$

Proof: Let $S_1, S_2, \ldots$ be a sequence of sets of finite perimeter with $S_i \setminus U = B \setminus U$, for $i = 1, 2, \ldots$, and

$$\lim_{i \to \infty} P(S_i) = \inf\{P(T) : T \setminus U = B \setminus U\}.$$

Let K be a compact set that contains U. Using the Compactness Theorem 3.6.14 to pass to a subsequence, but without changing notation, we may suppose that $\chi_{S_i \cap K}$ converges in $L^1(K)$ to a function f. Passing to another subsequence if necessary, we can suppose that $\chi_{S_i \cap K}$ also converges almost everywhere to f, thus f is the characteristic function of a set C. We set

$$S = C \cup (B \setminus K),$$

and the result follows from the Semicontinuity Theorem 3.6.5. ∎

One of the central facts of classical Euclidean geometry is the Isoperimetric Inequality. A version of this result can be easily proved in the present context of geometric analysis. Later, in Sections 7.2 and 7.3, we will approach the Isoperimetric Inequality using Steiner symmetrization

For this purpose we need a simple Sobolev type lemma. The general thrust of a "Sobolev Inequality" is to use an integral norm of a higher order derivative to bound an integral norm of a lower order derivative. Actually, this was first done by Poincaré on a bounded domain for the L^2 norm of the first derivative of a function and the L^2 norm of the function, so that particular type inequality is often called Poincaré's Inequality.

Lemma 3.7.8 *If $f \in BV(\mathbb{R}^N)$, then the following inequality is valid:*

$$\left(\int_{\mathbb{R}^N} |f|^{\frac{N}{N-1}} d\mathcal{L}^N \right)^{\frac{N-1}{N}} \leq \int_{\mathbb{R}^N} |\nabla f| d\mathcal{L}^N. \tag{3.92}$$

Remark 3.7.9 This result can be generalized to other exponents and applies to functions in the appropriate function space, namely Sobolev space (see Evans and Gariepy [1]). In the next chapter, we will give the formal definition of Sobolev space and present the main theorems concerning such spaces.

Proof: We will give the proof in the case $N = 3$. This case illustrates all the salient points without excessive notational complication.

By Theorem 3.6.12, it will suffice to prove the result when $f \in C_c^\infty(\mathbb{R}^N)$. For any $(x, y, z) \in \mathbb{R}^3$, we have

$$|f(x,y,z)| = \left| \int_{u=-\infty}^{x} \frac{\partial f}{\partial x}(u,y,z) \, du \right| \leq \int_{u \in \mathbb{R}} |\nabla f(u,y,z)| \, du,$$

and similarly,
$$|f(x,y,z)| \leq \int_{v\in\mathbb{R}} |\nabla f(x,v,z)|\, dv, \quad |f(x,y,z)| \leq \int_{w\in\mathbb{R}} |\nabla f(x,y,w)|\, dw.$$

Accordingly we can estimate
$$|f(x,y,z)|^{\frac{3}{2}} \leq \left(\int_{u\in\mathbb{R}} |\nabla f(u,y,z)|\, du\right)^{\frac{1}{2}} \times$$
$$\left(\int_{v\in\mathbb{R}} |\nabla f(x,v,z)|\, dv\right)^{\frac{1}{2}} \times \left(\int_{w\in\mathbb{R}} |\nabla f(x,y,w)|\, dw\right)^{\frac{1}{2}}.$$

Integrating with respect to x and using Hölder's inequality, we have
$$\int_{x\in\mathbb{R}} |f(x,y,z)|^{\frac{3}{2}}\, dx \leq \int_{x\in\mathbb{R}^N} \left[\left(\int_{u\in\mathbb{R}} |\nabla f(u,y,z)|\, du\right)^{\frac{1}{2}} \times\right.$$
$$\left.\left(\int_{v\in\mathbb{R}} |\nabla f(x,v,z)|\, dv\right)^{\frac{1}{2}} \times \left(\int_{w\in\mathbb{R}} |\nabla f(x,y,w)|\, dw\right)^{\frac{1}{2}}\right] dx$$
$$= \left(\int_{u\in\mathbb{R}} |\nabla f(u,y,z)|\, du\right)^{\frac{1}{2}} \times$$
$$\int_{x\in\mathbb{R}} \left(\int_{v\in\mathbb{R}} |\nabla f(x,v,z)|\, dv\right)^{\frac{1}{2}} \times \left(\int_{w\in\mathbb{R}} |\nabla f(x,y,w)|\, dw\right)^{\frac{1}{2}} dx$$
$$\leq \left(\int_{u\in\mathbb{R}} |\nabla f(u,y,z)|\, du\right)^{\frac{1}{2}} \times$$
$$\left(\int_{x\in\mathbb{R}}\int_{v\in\mathbb{R}} |\nabla f(x,v,z)|\, dv\, dx\right)^{\frac{1}{2}} \times \left(\int_{x\in\mathbb{R}}\int_{w\in\mathbb{R}} |\nabla f(x,y,w)|\, dw\, dx\right)^{\frac{1}{2}}.$$

Then we integrate with respect to y, again use Hölder's inequality, and estimate
$$\int_{y\in\mathbb{R}}\int_{x\in\mathbb{R}} |f(x,y,z)|^{\frac{3}{2}}\, dx\, dy \leq \left(\int_{x\in\mathbb{R}}\int_{v\in\mathbb{R}} |\nabla f(x,v,z)|\, dv\, dx\right)^{\frac{1}{2}} \times$$
$$\int_{y\in\mathbb{R}} \left(\int_{u\in\mathbb{R}} |\nabla f(u,y,z)|\, du\right)^{\frac{1}{2}} \times \left(\int_{x\in\mathbb{R}}\int_{w\in\mathbb{R}} |\nabla f(x,y,w)|\, dw\, dx\right)^{\frac{1}{2}} dy$$
$$\leq \left(\int_{x\in\mathbb{R}}\int_{v\in\mathbb{R}} |\nabla f(x,v,z)|\, dv\, dx\right)^{\frac{1}{2}} \times \left(\int_{y\in\mathbb{R}}\int_{u\in\mathbb{R}} |\nabla f(u,y,z)|\, du\, dy\right)^{\frac{1}{2}}$$
$$\times \left(\int_{y\in\mathbb{R}}\int_{x\in\mathbb{R}}\int_{w\in\mathbb{R}} |\nabla f(x,y,w)|\, dw\, dx\, dy\right)^{\frac{1}{2}}.$$

Finally, we integrate with respect to z, use Hölder's inequality, and estimate
$$\int_{z\in\mathbb{R}}\int_{y\in\mathbb{R}}\int_{x\in\mathbb{R}} |f(x,y,z)|^{\frac{3}{2}}\, dx\, dy\, dz \leq$$

3.7. DOMAINS WITH FINITE PERIMETER

$$\left(\int_{y \in \mathbb{R}} \int_{x \in \mathbb{R}} \int_{w \in \mathbb{R}} |\nabla f(x,y,w)| \, dw \, dx \, dy \right)^{\frac{1}{2}} \times$$

$$\int_{z \in \mathbb{R}} \left[\left(\int_{x \in \mathbb{R}} \int_{v \in \mathbb{R}} |\nabla f(x,v,z)| \, dv \, dx \right)^{\frac{1}{2}} \times \right.$$

$$\left. \left(\int_{y \in \mathbb{R}} \int_{u \in \mathbb{R}} |\nabla f(u,y,z)| \, du \, dy \right)^{\frac{1}{2}} \right] dz$$

$$\leq \left(\int_{y \in \mathbb{R}} \int_{x \in \mathbb{R}} \int_{w \in \mathbb{R}} |\nabla f(x,y,w)| \, dw \, dx \, dy \right)^{\frac{1}{2}} \times$$

$$\left(\int_{z \in \mathbb{R}} \int_{x \in \mathbb{R}} \int_{v \in \mathbb{R}} |\nabla f(x,v,z)| \, dv \, dx \, dz \right)^{\frac{1}{2}} \times$$

$$\left(\int_{z \in \mathbb{R}} \int_{y \in \mathbb{R}} \int_{u \in \mathbb{R}} |\nabla f(u,y,z)| \, du \, dy \, dz \right)^{\frac{1}{2}}$$

$$= \left(\int_{\mathbb{R}^3} |\nabla f| \, d\mathcal{L}^3 \right)^{\frac{3}{2}}.$$

∎

Corollary 3.7.10 (Isoperimetric Inequality) *If $S \subset \mathbb{R}^N$ is a set of finite perimeter and finite Lebesgue measure, then the following inequality is valid:*

$$\left[\mathcal{L}^N(S) \right]^{\frac{N-1}{N}} \leq P(S). \tag{3.93}$$

Proof: Apply Lemma 3.7.8 to χ_S. ∎

Remark 3.7.11 Recall that Remark 3.87 showed us that the perimeter of a C^1 domain in $\mathbb{R}^N$ agrees with the $(N-1)$-dimensional Hausdorff measure of the topological boundary of the domain. Thus, when $N = 2$, the Isoperimetric Inequality gives a lower bound for the length required to enclose a given area, and when $N = 3$, the Isoperimetric Inequality gives a lower bound for the area required to enclose a given volume. Similar interpretations apply in higher dimensions. The version of the Isoperimetric Inequality given in Corollary 3.7.10 is not sharp: Simply consider the unit ball in $\mathbb{R}^2$ for which the left-hand side of (3.93) equals $\pi^{1/2}$ and is *strictly less* than the right-hand side of (3.93) which equals 2π.

A fairly simple geometric argument, which we sketch, allows us to obtain the following inequality as a corollary of the above lemma. We will call this a **Poincaré Inequality**, because it applies to the unit cube (even though it is not for the L^2 norm).

Corollary 3.7.12 *Let Q denote the closed unit cube in $\mathbb{R}^N$ and suppose Ω is open with $Q \subset \Omega$. There exists a constant C, depending only on N, such that*

$$\left(\int_Q |f - f_Q|^{\frac{N}{N-1}} d\mathcal{L}^N \right)^{\frac{N-1}{N}} \leq C \int_Q |\nabla f| d\mathcal{L}^N, \tag{3.94}$$

holds for $f \in BV(\Omega)$, where

$$f_Q = \int_Q f \, d\mathcal{L}^N. \tag{3.95}$$

Proof: Again it follows from Theorem 3.6.12 that it suffices to prove the result under the stronger assumption that $f \in C_c^1(\Omega)$. We define a function $g : [-1,3]^N \to \mathbb{R}^N$ by reflecting through the faces of the unit cube. That is, $g(x_1, x_2, \ldots, x_N) = f(x_1', x_2', \ldots, x_N')$, where

$$x_i' = \begin{cases} -x_i & \text{if } -1 \leq x_i \leq 0 \\ x_i & \text{if } 0 \leq x_i \leq 1 \\ x_i - 1 & \text{if } 1 \leq x_i \leq 2 \end{cases}.$$

Since f is continuous, there is a point $x_0 \in Q$ such that $f_Q = f(x_0)$. Let $U_{x_0} : \mathbb{R}^N \to [-1,3]^N$ be any smooth transformation that is the identity on Q and at distance more than $1/2$ from Q maps all points to x_0. The result follows by applying Lemma 3.7.8 to the function $(g \circ U) - f_Q$ and using the change of variables formula for multiple integrals. This yields a constant $C(x_0)$ for which (3.94) will hold whenever $f_Q = f(x_0)$. Further supposing that the transformation U_{x_0} is defined in a way that depends continuously on x_0, we see the value of C needed to make (3.94) generally valid is the maximum of the values $C(x_0)$, as x_0 ranges over Q. ∎

Remark 3.7.13 We have used the unit cube because the notion of reflection through a face is easily understood. Poincaré's Inequality 3.7.12 is typically stated with Q a ball of radius r. An elementary change of variables shows that the specific choice of Q as a cube, ball, *etc.* only affects the value of the constant, not the basic validity of the result. One caveat is that if non-symmetrical figures (for example, ellipsoids) are used, then the ratio of the smallest dimension to the largest dimension will need to be bounded above and below.

Using our version of Poincaré's Inequality, we can prove a local form of the Isoperimetric Inequality which will be useful for analyzing the fine structure of sets of locally finite perimeter.

Corollary 3.7.14 (Local Isoperimetric Inequality) *If $S \subset \mathbb{R}^N$ is a set with locally finite perimeter, then for each $x \in \mathbb{R}^N$ and each $r > 0$ the*

3.7. DOMAINS WITH FINITE PERIMETER

following inequality is valid:

$$\inf\left\{\mathcal{L}^N(\mathbb{B}(x,r)\cap S),\ \mathcal{L}^N(\mathbb{B}(x,r)\setminus S)\right\} \leq C\left[\|\partial S\|\mathbb{B}(x,r)\right]^{N/(N-1)}. \quad (3.96)$$

Here C is a constant depending only on N.

Proof: Apply Corollary 3.7.12 to $\mathcal{X}_S$, but with Q replaced by the open ball of radius r about x (as we may by Remark 3.7.13). ∎

Structure Theory for Sets with Finite Perimeter

In an earlier example, we showed that a domain with a C^1 boundary is a set of locally finite perimeter. The main goal of our study of sets of locally finite perimeter is to show the extent to which the converse of this statement is true. In other words, we wish to show that a set of locally finite perimeter has a C^1 boundary in a measure-theoretic sense. The term "boundary" here cannot mean the topological boundary; point set topology is too crude a tool to isolate the relevant points. Instead we must use the "reduced boundary" as defined next.

Definition 3.7.15 *Suppose $S \subset \mathbb{R}^N$ is a set of locally finite perimeter. The* **reduced boundary** *of S, denoted by $\partial^* S$, is the set of $x \in \mathbb{R}^N$ such that*

(i) $\|\partial S\|(\mathbb{B}(x,r)) > 0$ *holds for $r > 0$,*

(ii) $\nu_S^*(x) = \lim_{r\downarrow 0} \dfrac{\int_{\mathbb{B}(x,r)} \nu_S\, d\|\partial S\|}{\|\partial S\|(\mathbb{B}(x,r))}$ *exists,*

(iii) $|\nu_S^*| = 1$.

Theorem 3.7.16 *If S is of locally finite perimeter, then the reduced boundary of S contains $\|\partial S\|$-almost all of S.*

Proof: This is a consequence of the Lebesgue differentiation theorem for general Radon measures on $\mathbb{R}^N$. ∎

At a boundary point of a smooth domain, the boundary will infinitesimally be an $N-1$ dimensional ball that separates a half-ball in the domain from a half-ball in the complement of the domain. Such behavior can be interpreted in terms of densities at the boundary point. In particular, each element of the boundary is also an element of the reduced boundary. In the next lemma, we obtain some density estimates that apply at each point of the reduced boundary. These estimates are a step toward showing the reduced boundary is much like the boundary of a smooth domain.

Lemma 3.7.17 *There exists a constant $0 < C = C(N) < \infty$ such that for any set S with locally finite perimeter the following inequalities hold for $x \in \partial^* S$:*

(i) $C^{-1} \leq \liminf_{r \downarrow 0} r^{-N} \mathcal{L}^N\bigl(\mathbb{B}(x,r) \cap S\bigr),$

(ii) $C^{-1} \leq \liminf_{r \downarrow 0} r^{-N} \mathcal{L}^N\bigl(\mathbb{B}(x,r) \setminus S\bigr),$

(iii) $C^{-1} \leq \liminf_{r \downarrow 0} r^{-(N-1)} \|\partial S\|\bigl(\mathbb{B}(x,r)\bigr)$

$\leq \limsup_{r \downarrow 0} r^{-(N-1)} \|\partial S\|\bigl(\mathbb{B}(x,r)\bigr) \leq C.$

Proof: For convenience of notation, let us suppose that $0 = x \in \partial^* S$ and that $\nu_S^*(0) = \mathbf{e}_N$.

We apply Lemma 3.7.6 with g chosen to be the constant vector $\mathbf{e}_N$ in some neighborhood of 0. Then, for all sufficiently small $r > 0$, the left-hand side of (3.91) vanishes. Thus we conclude that for $\mathcal{L}^1$-almost all sufficiently small $r > 0$

$$\mathbf{e}_N \cdot \int_{\mathbb{B}(0,r)} \nu_S \, d\|\partial S\| = \int_{\mathbb{B}(0,r)} \mathbf{e}_N \cdot \nu_S \, d\|\partial S\|$$

$$= -\int_{S \cap \mathbb{S}(0,r)} \mathbf{e}_N \cdot \tfrac{y}{|y|} \, d\mathcal{H}^{N-1} y,$$

$$\leq \mathcal{H}^{N-1}[S \cap \mathbb{S}(0,r)].$$

Dividing both ends of the preceding formula by $\|\partial S\|\bigl(\mathbb{B}(0,r)\bigr)$, we see that Definition 3.7.15(ii) implies

$$1 \leq \lim_{r \downarrow 0} \frac{\mathcal{H}^{N-1}[S \cap \mathbb{S}(0,r)]}{\|\partial S\|\bigl(\mathbb{B}(0,r)\bigr)}. \tag{3.97}$$

Notice that

$$\limsup_{r \downarrow 0} r^{-(N-1)} \|\partial S\|\bigl(\mathbb{B}(0,r)\bigr) \leq C$$

follows from (3.97), proving the second half of part (iii).

Note that Lemma 3.7.6 also implies that

$$P\bigl(S \cap \mathbb{B}(0,r)\bigr) \leq \|\partial S\|\bigl(\mathbb{B}(0,r)\bigr) + \mathcal{H}^{N-1}[S \cap \mathbb{S}(0,r)] \tag{3.98}$$

holds for $\mathcal{L}^1$-almost every $r > 0$. Combining the Isoperimetric Inequality of Corollary 3.7.10 and equations (3.97) and (3.98), we see that for each $1 < \lambda$ there exists $0 < r(\lambda)$ such that

$$\bigl[\mathcal{L}^N\bigl(S \cap \mathbb{B}(0,r)\bigr)\bigr]^{\frac{N-1}{N}} \leq (1+\lambda)\mathcal{H}^{N-1}[S \cap \mathbb{S}(0,r)] \tag{3.99}$$

3.7. DOMAINS WITH FINITE PERIMETER

holds for $\mathcal{L}^1$-almost every $0 < r < r(\lambda)$.

Equation (3.99) is actually a slightly disguised differential inequality: Notice that if we set
$$f(r) = \mathcal{L}^N(S \cap \mathbb{B}(0,r)),$$
then
$$f'(r) = \mathcal{H}^{N-1}[S \cap \mathbb{S}(0,r)].$$

Thus (3.99) tells us that (after we make a choice of λ and rename the constants)
$$C \le [f(r)]^{\frac{1}{N}-1} f'(r) = N \frac{d}{dr}[f(r)]^{\frac{1}{N}} \qquad (3.100)$$

holds for $\mathcal{L}^1$-almost all sufficiently small $r > 0$. Part (i) of the lemma now follows by integrating (3.100). As an additional bonus we also obtain the inequality
$$C \le \liminf_{r \downarrow 0} r^{-(N-1)} \mathcal{H}^{N-1}[S \cap \mathbb{S}(0,r)]. \qquad (3.101)$$

We obtain part (ii) of the lemma by noting that if 0 is in the reduced boundary of S, then it is also in the reduced boundary of $\mathbb{R}^N \setminus S$. Finally, the first half of part (iii) follows from parts (i) and (ii) and the Relative Isoperimetric Inequality. ∎

Example 3.7.18 The theorem tells us that $\mathcal{H}^{N-1}(\partial^* S) \le \|\partial S\|(\mathbb{R}^N)$, so for a set of finite perimeter we have
$$\mathcal{H}^{N-1}(\partial^* S) < \infty. \qquad (3.102)$$

Equation (3.102) is false if $\partial^* S$ is replaced by the topological boundary: For example if S is a union of balls centered at the points of a countable dense subset of $\mathbb{B}(0,1)$ and with radii chosen so that the total $\mathcal{L}^N$ measure of S is strictly less than that of $\overline{\mathbb{B}}(0,1)$ and the sum of the $\mathcal{H}^{N-1}$ measure of the surfaces of the balls is finite, then S will be of finite perimeter, but ∂S will contain $\overline{\mathbb{B}}(0,1) \setminus S$, and thus have positive $\mathcal{L}^N$ measure and infinite $\mathcal{H}^{N-1}$ measure.

Notation 3.7.19 *For a set $S \subset \mathbb{R}^N$ and for $r > 0$, we set*
$$rS = \{rx : x \in S\}.$$

An important tool in proving the Structure Theorem for sets of finite perimeter is blowing-up[3]. Blowing-up produces local information that relates the vector ν_S^* to the geometry of S. The easy part of blowing-up is in the following lemma.

---
[3]This is not the *blowing-up* of algebraic geometry (as discussed in Krantz and Parks [2], for example).

Lemma 3.7.20 *If S is of locally finite perimeter, $0 \in \partial^* S$, and r_i is a sequence with $r_i \to \infty$ as $i \to \infty$, then there exist a subsequence r_{i_j} and a set H of locally finite perimeter such that*

$$\chi_{r_{i_j} S} \to \chi_H \quad \text{in } L^1_{loc} \quad \text{as } j \to \infty \tag{3.103}$$

and such that $\nu_{r_{i_j} S} \|\partial(r_{i_j} S)\|$ converges weakly to $\nu_H \|\partial H\|$.

Proof: The existence of H as in (3.103) follows from the Compactness Theorem 3.6.14 and the density estimate in part (iii) of Lemma 3.7.17. The weak convergence is a consequence of the theorem of de la Vallée Poussin that states that *any sequence of uniformly bounded Radon measures has a subsequence that weakly converges to Radon measure* (a consequence of the Banach-Alaoglu Theorem, Rudin [3], 3.15, or see Appendix A of Giusti [1] for a direct proof). ∎

The geometric import of blowing-up is shown in the following theorem.

Theorem 3.7.21 (Blowing-Up) *If S is of locally finite perimeter, $0 \in \partial^* S$, and $\nu_S^*(0) = \mathbf{e}_N$, then*

$$\chi_{rS} \to \chi_{\mathbb{H}^-} \quad \text{in } L^1_{loc} \quad \text{as } r \to \infty, \tag{3.104}$$

$$\lim_{R \downarrow 0} \frac{\mathcal{L}^N[\mathbb{H}^+ \cap (\mathbb{B}(0,R) \cap S)]}{R^N} = 0, \tag{3.105}$$

$$\lim_{R \downarrow 0} \frac{\mathcal{L}^N[\mathbb{H}^- \cap (\mathbb{B}(0,R) \setminus S)]}{R^N} = 0, \tag{3.106}$$

$$\lim_{R \downarrow 0} \frac{\|\partial S\| \mathbb{B}(0,R)}{\omega_{N-1} R^{N-1}} = 1. \tag{3.107}$$

Here $\mathbb{H}^-$ is the lower half-space $\{x : x \cdot \mathbf{e}_N < 0\}$ and $\mathbb{H}^+$ is the upper half-space $\{x : x \cdot \mathbf{e}_N > 0\}$

Proof: We concentrate on (3.104). The other conclusions will then follow easily. It suffices to show that any sequence $r_i \to \infty$ has a subsequence r_{i_j} such that

$$\chi_{r_{i_j} S} \to \chi_{\mathbb{H}^-} \quad \text{in } L^1_{loc} \quad \text{as } j \to \infty. \tag{3.108}$$

By the lemma, we know there is a subsequence r_{i_j} such that

$$\chi_{r_{i_j} S} \to \chi_H \quad \text{in } L^1_{loc} \quad \text{as } j \to \infty \tag{3.109}$$

for some set H of locally finite perimeter and that additionally

$$\nu_{r_{i_j} S} \|\partial(r_{i_j} S)\|$$

converges weakly to $\nu_H \|\partial H\|$. We must show that $H = \mathbb{H}^-$.

3.7. DOMAINS WITH FINITE PERIMETER

First, we show that $\nu_H = \mathbf{e}_N$ for $\|\partial H\|$ almost every point. For $\phi \in C_c^1(\mathbb{R}^N; \mathbb{R}^N)$, we use the change of variables $r_{i_j} y = x$ and the notation $\psi(y) = \phi(r_{i_j} y)$ to see that

$$\begin{aligned}
\int \phi \cdot \nu_{r_{i_j} S} \, d\|\partial r_{i_j} S\| &= \int_{\mathbb{R}^N} \mathrm{div}\phi(x) \chi_{r_{i_j} S}(x) \, d\mathcal{L}^N x \\
&= r_{i_j}^{N-1} \int_{\mathbb{R}^N} \mathrm{div}\psi(y) \chi_S(y) \, d\mathcal{L}^N y \\
&= r_{i_j}^{N-1} \int \psi \cdot \nu_S \, d\|\partial S\|.
\end{aligned}$$

We conclude that

$$\|\partial r_{i_j} S\| \mathbb{B}(0,R) = r_{i_j}^{N-1} \|\partial S\| \mathbb{B}(0, r_{i_j}^{1-N} R) \tag{3.110}$$

and

$$\int_{\mathbb{B}(0,R)} \nu_{r_{i_j} S} \, \|\partial r_{i_j} S\| = r_{i_j}^{N-1} \int_{\mathbb{B}(0, r_{i_j}^{1-N} R)} \nu_S \, \|\partial S\|. \tag{3.111}$$

Because $0 \in \partial^* S$, (3.110) and (3.111) allow us to conclude that

$$\lim_{j \to \infty} \frac{\int_{\mathbb{B}(0,R)} \nu_{r_{i_j} S} \, d\|\partial r_{i_j} S\|}{\|\partial r_{i_j} S\| \mathbb{B}(0,R)} = \lim_{j \to \infty} \frac{\int_{\mathbb{B}(0, r_{i_j}^{1-N} R)} \nu_S \, d\|\partial S\|}{\|\partial S\| \mathbb{B}(0, r_{i_j}^{1-N} R)} = \mathbf{e}_N. \tag{3.112}$$

Considering $R > 0$ for which $\|\partial H\| \partial \mathbb{B}(0,R) = 0$, we see by the Semicontinuity Theorem 3.6.5, by (3.112), and by the weak convergence of $\nu_{r_{i_j} S} \|\partial (r_{i_j} S)\|$ to $\nu_H \|\partial H\|$ that

$$\begin{aligned}
\|\partial H\| \overline{\mathbb{B}}(0,R) &\leq \liminf_{j \to \infty} \|\partial r_{i_j} S\| \overline{\mathbb{B}}(0,R) \\
&= \lim_{j \to \infty} \int_{\overline{\mathbb{B}}(0,R)} \mathbf{e}_N \cdot \nu_{r_{i_j} S} \, d\|\partial r_{i_j} S\| \\
&= \int_{\overline{\mathbb{B}}(0,R)} \mathbf{e}_N \cdot \nu_H \, d\|\partial H\|.
\end{aligned}$$

Since $|\nu_H| = 1$ holds $\|\partial H\|$-almost everywhere, we must have

$$\nu_H = \mathbf{e}_N \tag{3.113}$$

$\|\partial H\|$-almost everywhere.

While it seems intuitively clear that (3.113) implies that H is a half-space with its bounding hyperplane orthogonal to $\mathbf{e}_N$, the point of this theorem is to prove that the vector ν_S^* really does have that desired geometric significance. Thus we argue as follows: Let ψ_σ be a family of mollifiers as in Section 1.3. For $\phi \in C_c^1(\mathbb{R}^N; \mathbb{R}^N)$, we have

$$-\int \mathbf{grad}(\psi_\sigma * \chi_H) \cdot \phi \, d\mathcal{L}^N = \int \psi_\sigma * \chi_H \, \mathrm{div}\phi \, d\mathcal{L}^N$$

$$= \int \chi_H \, \mathbf{div}(\psi_\sigma * \phi) \, d\mathcal{L}^N$$

$$= \int \mathbf{e}_N \cdot (\psi_\sigma * \phi) \, d\|\partial H\|$$

$$= \int \psi_\sigma * (\phi \cdot \mathbf{e}_N) \, d\|\partial H\|.$$

We conclude that the smooth function $\psi_\sigma * \chi_H$ is independent of the first $N-1$ coordinates and is a non-increasing function of the Nth coordinate. Since $\psi_\sigma * \chi_H$ converges $\mathcal{L}^N$-almost everywhere to χ_H we conclude that H is some translate of $\mathbb{H}^-$ in the $\mathbf{e}_N$ direction.

Finally, since $\chi_{r_{i_j} S}$ converges weakly to χ_H in L^1_{loc} we see that parts (i) and (ii) of Lemma 3.7.17 imply that indeed $H = \mathbb{H}^-$. ∎

The following lemma shows that if the results of blowing-up are sufficiently uniform then significant global information can be obtained. The proof of the lemma is highly technical, but it seems to be an unavoidable fact of geometric analysis that a highly technical construction lies at the heart of any argument going from local information to global information.

Lemma 3.7.22 *Suppose that*

(i) $K \subset \mathbb{R}^N$ *is compact,*

(ii) $\nu : K \to \mathbb{S}(0,1)$ *is continuous,*

(iii) $\lim_{r \downarrow 0} \left(\sup \left\{ \left| \nu(x) \cdot \frac{(x-y)}{|x-y|} \right| : 0 < |x-y| < r, \ x, y \in K \right\} \right) = 0.$

Then there exist countably many Lipschitz functions $f_i : \mathbb{R}^{N-1} \to \mathbb{R}$ and orthonormal bases $\{\mathbf{e}_{i1}, \mathbf{e}_{i2}, \ldots, \mathbf{e}_{iN}\}$, $i = 1, 2, \ldots$, such that

$$K \subset \bigcup_{i=1}^{\infty} G_i \tag{3.114}$$

where

$$G_i = \left\{ \left(\sum_{j=1}^{N-1} u_j \mathbf{e}_{ij}, f_i(u) \mathbf{e}_{iN} \right) : u \in \mathbb{R}^{N-1} \right\}. \tag{3.115}$$

Proof: First, we observe that, for each j, since K is compact, K is contained in finitely many open balls of radius 2^{-j} centered in K. Taking the union of all these finite families and replacing each open ball by its closure, we construct a countable set $\mathcal{B}$ of closed balls centered in K such that, for each $\epsilon > 0$,

$$\{ B \in \mathcal{B} : \mathrm{diam}(B) \geq \epsilon \} \text{ is finite} \tag{3.116}$$

3.7. DOMAINS WITH FINITE PERIMETER

and
$$K \subset \cup\{ B \in \mathcal{B} : \text{diam}(B) < \epsilon \}. \tag{3.117}$$

Next, we construct a countable family $\mathcal{V}$ of orthonormal bases of $\mathbb{R}^N$ such that the set consisting of the Nth vector from each basis in $\mathcal{V}$ forms a countable dense subset of $\mathbb{S}(0,1)$. Thirdly, we arrange $\mathcal{B} \times \mathcal{V} \times \mathbb{N}$ into a sequence
$$\left\{ \left(\overline{\mathbb{B}}(p_i, r_i), \{\mathbf{e}_{i1}, \mathbf{e}_{i2}, \ldots, \mathbf{e}_{iN}\}, M_i \right) \right\}_{i=1}^{\infty}. \tag{3.118}$$

For each i, let G_i be a maximal subset of $K \cap \overline{\mathbb{B}}(p_i, r_i)$ satisfying
$$|(x-y) \cdot \mathbf{e}_{iN}| \le M_i \left| (x-y) - [(x-y) \cdot \mathbf{e}_{iN}] \mathbf{e}_{iN} \right| \text{ for all } x, y \in G_i. \tag{3.119}$$

The function f_i is defined by setting
$$f_i(x \cdot \mathbf{e}_{i1}, x \cdot \mathbf{e}_{i2}, \ldots, x \cdot \mathbf{e}_{i(N-1)}) = x \cdot \mathbf{e}_{iN} \text{ for } x \in G_i. \tag{3.120}$$

The domain of f_i is
$$A_i = \{ (x \cdot \mathbf{e}_{i1}, x \cdot \mathbf{e}_{i2}, \ldots, x \cdot \mathbf{e}_{i(N-1)}) : x \in G_i \}. \tag{3.121}$$

We see that f_i is Lipschitz from the defining condition (3.119) for G_i, and that the Lipschitz constant of f_i does not exceed M_i. We can extend the domain of f_i to all of $\mathbb{R}^{N-1}$ by using the following formula (due to H. Whitney):
$$f_i(x) = \sup\{ f_i(z) - M_i \text{dist}(x, z) : z \in A_i \}. \tag{3.122}$$

It remains to show that (3.114) holds. To this end, we argue by contradiction and assume to the contrary that (3.114) does not hold. Suppose $x^* \in K$, but $x^* \notin G_i$, for $i = 1, 2, \ldots$.

By (3.116) and (3.117), there must be infinitely many balls in $\mathcal{B}$ that contain x^*. Set
$$\mathcal{B}' = \{ B \in \mathcal{B} : x^* \in B \}.$$

We now choose a sequence in $\mathcal{B}' \times \mathcal{V} \times \mathbb{N}$ with some particular properties: Let i_j be such that that $x^* \in \overline{\mathbb{B}}(p_{i_j}, r_{i_j})$ and
$$\lim_{j \to \infty} r_{i_j} = 0, \tag{3.123}$$
$$\lim_{j \to \infty} \mathbf{e}_{i_j N} = \nu(x^*), \tag{3.124}$$
$$\lim_{j \to \infty} M_{i_j} = \infty. \tag{3.125}$$

Now, for each j, since $x^* \notin G_{i_j}$, there must be
$$y_j \in G_{i_j} \subset K \cap \overline{\mathbb{B}}(p_{i_j}, r_{i_j}) \tag{3.126}$$

such that
$$|(x^* - y_j) \cdot \mathbf{e}_{i_j N}| > M_{i_j} \left| (x^* - y_j) - [(x^* - y_j) \cdot \mathbf{e}_{i_j N}] \mathbf{e}_{i_j N} \right|. \tag{3.127}$$

Equations (3.123)–(3.127) contradict part (iii) of the hypotheses. ∎

Remark 3.7.23 The functions f_i defined in the proof of Lemma 3.7.22 can be extended to be C^1 functions, instead of just Lipschitz functions, if we are willing to appeal to the Whitney Extension Theorem. While the Whitney Extension Theorem is not proved until Chapter 5, its proof is independent of the results of this section.

Measure theory allows the local information provided by blowing-up to be made sufficiently uniform that the preceding lemma can be applied. The result is the following theorem.

Theorem 3.7.24 (Structure Theorem) *If $S \subset \mathbb{R}^N$ has locally finite perimeter, then there exist countably many $(N-1)$-rectifiable sets, $G_1, G_2, \ldots$ and a set E with $\|\partial S\|(E) = 0$ such that*

$$\partial^* S = E \cup \bigcup_{i=1}^{\infty} G_i. \qquad (3.128)$$

Proof: Egoroff's Theorem (see Evans and Gariepy [1], Section 1.2, or Federer [4], 2.3.7) tells us that, for each $\epsilon > 0$, pointwise convergence of measurable functions implies uniform convergence off a set of measure ϵ. Applying Egoroff's Theorem countably many times, we see that there exist disjoint $\|\partial S\|$-measurable subsets of $\partial^* S$, covering $\|\partial S\|$-almost all of $\partial^* S$, on which the convergence in (3.105), (3.106), and (3.107) is uniform. Similarly, Lusin's Theorem (see Evans and Gariepy [1], Section 1.2, or Federer [4], 2.3.5) tells us that, for each $\epsilon > 0$, a measurable function is equal to a continuous function off a set of measure ϵ. Applying Lusin's Theorem countably many times on each of the preceding subsets, we see that each is $\|\partial S\|$-almost the union of disjoint compact subsets on which the normal vector ν_S is a continuous function. We wish to apply Lemma 3.7.22 to each of these compact subsets. To do so, we need to verify that hypothesis (iii) of Lemma 3.7.22 is satisfied, but the density estimates of Lemma 3.7.17 insures that (3.105) and (3.106) imply that hypothesis. ∎

Remark 3.7.25 As in Remark 3.7.23, the Whitney Extension Theorem allows one to conclude that the sets G_i can be required to be subsets of the images of C^1 functions.

The final result of this section gives a very general form of the Gauss-Green Theorem. This result was discovered independently by E. De Giorgi and H. Federer.

Theorem 3.7.26 (Generalized Gauss-Green Theorem) *If $S \subset \mathbb{R}^N$ has finite perimeter, then for each $\phi \in C_c^1(\mathbb{R}^N; \mathbb{R}^N)$ we have*

$$\int_S \operatorname{div} \phi \, d\mathcal{L}^N = \int_{\partial^* S} \phi \cdot \nu_S \, d\mathcal{H}^{N-1}. \qquad (3.129)$$

3.8. THE AREA FORMULA

Proof: We need to show that on $\partial^* S$ the measure $\|\partial S\|$ equals the $N-1$ dimensional Hausdorff measure. This follows from (3.107) and the Structure Theorem 3.7.24. ∎

Remark 3.7.27 If the set S in Theorem 3.7.26 is not of finite perimeter—or even locally finite perimeter—one can still use Equation 3.129 as the *definition* of the boundary. Of course, it is more interesting if both sides of the equation can be given by independent definitions. We direct the reader's attention to just such a non-trivial extension of the Gauss-Green Theorem that has been given in Harrison and Norton [1].

3.8 The Area Formula

The main result of this section is the following theorem.

Theorem 3.8.1 (The Area Formula) *If $f : \mathbb{R}^M \to \mathbb{R}^N$ is a Lipschitz function and $M \leq N$, then*

$$\int_A J_M f(x) \, d\mathcal{L}^M x = \int_{\mathbb{R}^N} \operatorname{card}(A \cap f^{-1}(y)) \, d\mathcal{H}^M y \qquad (3.130)$$

holds for each Lebesgue measurable subset A of $\mathbb{R}^M$.

Here $\mathcal{H}^M$ denotes the M-dimensional Hausdorff measure and $J_M f$ denotes the M-dimensional Jacobians of f which is defined in Definition 3.8.3. In case $M = N$, the M-dimensional Jacobian agrees with the usual Jacobian, $|\det(Df)|$.

The proof of the area formula separates into three fundamental parts. The first is to understand the situation for linear maps. The second is to extend our understanding to the behavior of maps that are well approximated by linear maps. This second part of the proof is essentially multi-variable calculus, and the area formula for C^1 maps follows readily. The third part of the proof brings in the measure theory that allows us to reduce the behavior of Lipschitz maps to that of maps that are well approximated by linear maps.

In the next section we will treat the co-area formula that applies to a Lipschitz map $f : \mathbb{R}^M \to \mathbb{R}^N$, but with $M \geq N$ instead of $M \leq N$. The proof of the co-area formula is similar to the proof of the area formula in that the same three steps of understanding linear maps, understanding maps well approximated by linear maps, and applying measure theory are fundamental. The discussion of linear maps in this section will be applicable to both the area formula and the co-area formula.

Linear Maps

A key ingredient in the area formula is the K-dimensional Jacobian which is a measure of how K-dimensional area transforms under the differential of a mapping. Since a linear map sends one parallelepiped into another, the fundamental question is "What is the K-dimensional area of the parallelepiped determined by a set of K vectors in $\mathbb{R}^N$?" Of course the answer is known, but recently G. J. Porter [1] has discovered a particularly lucid derivation that we give below.

Proposition 3.8.2 *If*

$$\vec{v}_i = \begin{pmatrix} v_{1\,i} \\ v_{2\,i} \\ \vdots \\ v_{N\,i} \end{pmatrix}, \ for\ i = 1, 2, \ldots, K, \qquad (3.131)$$

are vectors in $\mathbb{R}^N$, then the parallelepiped determined by those vectors has K-dimensional area

$$\sqrt{\det\left(V^{\mathbf{t}} \cdot V\right)}, \qquad (3.132)$$

where V is the $N \times K$ matrix with $\vec{v}_1, \vec{v}_2, \ldots, \vec{v}_K$ as its columns.

Proof: If the vectors $\vec{v}_1, \vec{v}_2, \ldots, \vec{v}_k$ are orthogonal, then the result is immediate. Thus we will reduce the general case to this special case.

Notice that Cavalieri's Principle shows us that adding a multiple of $\vec{v}_i$ to another vector $\vec{v}_j$, $j \neq i$, does not not change the K-dimensional area of the parallelepiped determined by the vectors. But also notice that such an operation on the vectors $\vec{v}_i$ is equivalent to multiplying V on the right by a $K \times K$ triangular matrix with 1's on the diagonal (upper triangular if $i < j$ and lower triangular if $i > j$). The Gram-Schmidt procedure is effected by a sequence of operations of precisely this type. Thus we see that there is an upper triangular matrix A with 1's on the diagonal such that VA has orthogonal columns and the columns of VA determine a parallelepiped with the same K-dimensional area as the parallelepiped determined by $\vec{v}_1, \vec{v}_2, \ldots, \vec{v}_N$. Since the columns of VA are orthogonal, we know that $\sqrt{\det\left((VA)^{\mathbf{t}} \cdot (VA)\right)}$ equals the K-dimensional area of the parallelepiped determined by its columns, and thus equals the K-dimensional area as the parallelepiped determined by $\vec{v}_1, \vec{v}_2, \ldots, \vec{v}_N$. Finally, we compute

$$\begin{aligned} \det\left((VA)^{\mathbf{t}} \cdot (VA)\right) &= \det\left(A^{\mathbf{t}} \cdot V^{\mathbf{t}} \cdot V \cdot A\right) \\ &= \det\left(A^{\mathbf{t}}\right) \det\left(V^{\mathbf{t}} \cdot V\right) \det(A) \\ &= \det\left(V^{\mathbf{t}} \cdot V\right). \end{aligned}$$

∎

3.8. THE AREA FORMULA

Definition 3.8.3

(i) Suppose $U \subset \mathbb{R}^M$ and $f : U \to \mathbb{R}^N$. We say f is **differentiable** at $a \in U$ if there exists a linear map $L : \mathbb{R}^M \to \mathbb{R}^N$ such that

$$\lim_{x \to a} \frac{|f(x) - f(a) - L(x - a)|}{|x - a|} = 0. \tag{3.133}$$

In case f is differentiable at a, we denote the linear map L such that (3.133) holds by $Df(a)$ and call it the **differential of f** at a.

(ii) Suppose that $f : \mathbb{R}^M \to \mathbb{R}^N$ is differentiable at a and $K \leq M$. We define the K-**dimensional Jacobian of f** at a, denoted $J_K f(a)$, by setting

$$J_K f(a) = \sup \left\{ \frac{\mathcal{H}^K[Df(a)(P)]}{\mathcal{H}^K[P]} : \right.$$

$$\left. P \text{ is a } K\text{-dimensional parallelepiped contained in } \mathbb{R}^M \right\}. \tag{3.134}$$

The conventional situation considered in elementary multi-variable calculus is that in which $K = M = N$. In that case, it is easily seen from Proposition 3.8.2 that one may choose P to be the unit M-dimensional cube and that $J_K f(a) = |\det(Df(a))|$.

Two other special cases are of interest: They are when $K = M < N$ and when $M > N = K$. When $K = M < N$, again one can choose P to be the unit M-dimensional cube in $\mathbb{R}^M$. The image of P under $Df(a)$ is the parallelepiped determined by the columns of the matrix representing $Df(a)$. It follows from Proposition 3.8.2 that $J_K f(a) = \sqrt{\det\left[(Df(a))^{\mathsf{t}} \cdot (Df(a))\right]}$.

When $M > N = K$, then P should be chosen to lie in the orthogonal complement of the kernel of $Df(a)$. This follows from Corollary 3.1.17 which tells us that orthogonal projection onto an affine subspace of $\mathbb{R}^N$ cannot increase Hausdorff measure. So, if we let L be the orthogonal complement of the kernel of $Df(a)$, then if P' is any parallelepiped, the image of the orthogonal projection of P' onto L is the same as the image of P'; but the orthogonal projection of P' has smaller K-dimensional area, giving a larger ratio of areas.

It is plain to see that the orthogonal complement of the kernel of $Df(a)$ is the span of the columns of $(Df(a))^{\mathsf{t}}$. If we begin with the parallelepiped determined by the columns of $(Df(a))^{\mathsf{t}}$, then that parallelepiped maps onto the parallelepiped determined by the columns of $(Df(a)) \cdot (Df(a))^{\mathsf{t}}$. By Proposition 3.8.2, the N-dimensional area of the first parallelepiped is

$$\sqrt{\det\left[(Df(a)) \cdot (Df(a))^{\mathsf{t}}\right]}$$

and the N-dimensional area of the second parallelepiped is

$$\sqrt{\det\left[((Df(a))\cdot(Df(a))^{\mathsf{t}})^{\mathsf{t}}\cdot((Df(a))\cdot(Df(a))^{\mathsf{t}})\right]}$$
$$= \det\left[(Df(a))\cdot(Df(a))^{\mathsf{t}}\right],$$

so the ratio is $J_K f(a) = \sqrt{\det\left[(Df(a))\cdot(Df(a))^{\mathsf{t}}\right]}$.

We summarize the above facts in the following lemma.

Lemma 3.8.4 *Suppose $f : \mathbb{R}^M \to \mathbb{R}^N$ is differentiable at a.*

(i) *If $K = M = N$, then*

$$J_K f(a) = |\det(Df(a))|. \tag{3.135}$$

(ii) *If $K = M \leq N$, then*

$$J_K f(a) = \sqrt{\det\left[(Df(a))^{\mathsf{t}}\cdot(Df(a))\right]}. \tag{3.136}$$

(iii) *If $M \geq N = K$, then*

$$J_K f(a) = \sqrt{\det\left[(Df(a))\cdot(Df(a))^{\mathsf{t}}\right]}. \tag{3.137}$$

Remark 3.8.5 The Generalized Pythagorean Theorem from Porter [1] allows one to see that the right-hand side of either (3.136) or (3.137) is equal to the square root of the sum of the squares of the $K \times K$ minors of $Df(a)$, where $K = \min\{M, N\}$. This is the form one is naturally led to if one develops the K-dimensional Jacobian via the alternating algebra over $\mathbb{R}^M$ and $\mathbb{R}^N$ as in Federer [4].

We will also need to make use of the polar decomposition of linear maps.

Theorem 3.8.6 (Polar Decomposition)

(i) *If $M \leq N$ and $T : \mathbb{R}^M \to \mathbb{R}^N$ is linear, then there exists a symmetric linear map $S : \mathbb{R}^M \to \mathbb{R}^M$ and an orthogonal linear map $U : \mathbb{R}^M \to \mathbb{R}^N$ such that $T = U \circ S$.*

(ii) *If $M \geq N$ and $T : \mathbb{R}^M \to \mathbb{R}^N$ is linear, then there exists a symmetric linear map $S : \mathbb{R}^N \to \mathbb{R}^N$ and an orthogonal linear map $U : \mathbb{R}^N \to \mathbb{R}^M$ such that $T = S \circ U^{\mathsf{t}}$.*

Proof:
(i) For convenience, let us first suppose that T is of full rank. The $M \times M$ matrix $T^{\mathsf{t}} \cdot T$ is easily seen to be symmetric and positive definite. So

3.8. THE AREA FORMULA

$T^t \cdot T$ has a complete set of M orthonormal eigenvectors $\vec{v}_1, \vec{v}_2, \ldots, \vec{v}_M$ associated with the positive eigenvalues $\lambda_1, \lambda_2, \ldots, \lambda_M$.

We define $S: \mathbb{R}^M \to \mathbb{R}^M$ by setting

$$S(\vec{v}_i) = \sqrt{\lambda_i}\, \vec{v}_i.$$

Using the orthonormal basis $\vec{v}_1, \vec{v}_2, \ldots, \vec{v}_M$, we see that S is represented by a diagonal matrix, thus S is symmetric.

We define $U: \mathbb{R}^M \to \mathbb{R}^N$ by setting

$$U(\vec{v}_i) = \frac{1}{\sqrt{\lambda_i}} T(\vec{v}_i).$$

We calculate

$$\begin{aligned} U(\vec{v}_i) \cdot U(\vec{v}_j) &= \frac{1}{\sqrt{\lambda_i}} \frac{1}{\sqrt{\lambda_j}} T(\vec{v}_i) \cdot T(\vec{v}_j) \\ &= \frac{1}{\sqrt{\lambda_i}} \frac{1}{\sqrt{\lambda_j}} \vec{v}_i \cdot (T^t \cdot T)(\vec{v}_j) \\ &= \frac{1}{\sqrt{\lambda_i}} \frac{1}{\sqrt{\lambda_j}} \lambda_j\, \vec{v}_i \cdot \vec{v}_j = \delta_{ij}. \end{aligned}$$

Thus U is an orthogonal map.

In case T is not of full rank, it follows that some of the λ_i's may be zero. For such i's we may choose $U(\vec{v}_i)$ arbitrarily, subject only to the requirement that $U(\vec{v}_1), U(\vec{v}_2), \ldots, U(\vec{v}_n)$ be an orthonormal set.

(ii) We apply (i) to the mapping T^t to obtain a symmetric S and orthogonal U so that $T^t = U \circ S$, but then $T = (U \circ S)^t = S \circ U^t$. ∎

Remark 3.8.7 The equations in Lemma 3.8.4 can be obtained using the Polar Decomposition of linear maps. For example, if we are considering $f: \mathbb{R}^M \to \mathbb{R}^N$ with $M \leq N$, then we apply the Polar Decomposition to write $Df(a) = U \circ S$. Since an orthogonal map clearly preserves Hausdorff measure, the entire effect of $Df(a)$ on the Hausdorff measure of a parallelepiped in $\mathbb{R}^M$ must be due to the non-negative definite symmetric map S, so we have

$$J_M f(a) = \det(S).$$

But then we can compute

$$\begin{aligned} \det(S) &= \sqrt{\det(S^t)\det(S)} \\ &= \sqrt{\det(S^t \cdot U^t \cdot U \cdot S)} = \sqrt{\det\left[(Df(a))^t \cdot (Df(a))\right]}. \end{aligned}$$

The first application of the Jacobian is in the following basic lemma concerning the behavior of Lebesgue measure under a linear map.

Lemma 3.8.8 *If $A \subset \mathbb{R}^M$ is Lebesgue measurable and $T : \mathbb{R}^M \to \mathbb{R}^M$ is linear, then*
$$\mathcal{L}^M(T(A)) = |\det(T)|\,\mathcal{L}^M(A).$$

Proof: Given $\epsilon > 0$, we can find an open U with $A \subset U$ and $\mathcal{L}^M(U \setminus A) < \epsilon$. We subdivide U into cubes and the image of each cube is a parallelepiped. So
$$\mathcal{L}^M(T(A)) \leq \mathcal{L}^M(T(U)) \leq |\det(T)|\,\mathcal{L}^M(U) \leq |\det(T)|\,[\epsilon + \mathcal{L}^M(A)].$$

Letting $\epsilon \downarrow 0$, we see that
$$\mathcal{L}^M(T(A)) \leq |\det(T)|\,\mathcal{L}^M(A).$$

Now we need to prove the reverse inequality. Note that if $\det(T) = 0$, then we are done. Assuming $\det(T) \neq 0$, we apply the case already proved to $T(A)$ and T^{-1} to see that
$$\mathcal{L}^M(A) = \mathcal{L}^M(T^{-1}(T(A))) \leq |\det(T^{-1})|\,\mathcal{L}^M(T(A)).$$

The result follows since $\det(T^{-1}) = (\det(T))^{-1}$. ∎

The Main Estimates for the Area Formula

Lemma 3.8.9 (Main Estimates for the Area Formula)
Suppose $T : \mathbb{R}^M \to \mathbb{R}^N$ is of full rank and $0 < \epsilon < \frac{1}{2}$. Let Π be orthogonal projection onto the image of T. Set
$$\lambda = \inf\{\langle T, v\rangle : |v| = 1\}. \tag{3.138}$$

If the Lebesgue measurable set $A \subset \mathbb{R}^M$ is such that

(i) *$Df(a)$ exists for $a \in A$,*

(ii) *$\|Df(a) - T\| < \epsilon$ holds for $a \in A$,*

(iii) *$|f(y) - f(a) - \langle Df(a), y - a\rangle| < \epsilon|y - a|$ holds for $y, a \in A$,*

(iv) *$\Pi|_{f(A)}$ is one-to-one,*

then
$$\begin{aligned}(1 - 3\epsilon\lambda^{-1})^M \cdot J_M T \cdot \mathcal{L}^M(A) &\leq \mathcal{H}^M(f(A)) \\ &\leq (1 + 2\epsilon\lambda^{-1})^M \cdot J_M T \cdot \mathcal{L}^M(A).\end{aligned} \tag{3.139}$$

3.8. THE AREA FORMULA

Proof: First we bound $\mathcal{H}^M(f(A))$ from above. We use the polar decomposition to write $T = U \circ S$, where $S : \mathbb{R}^M \to \mathbb{R}^M$ is symmetric and $U : \mathbb{R}^M \to \mathbb{R}^N$ is orthogonal, and we note that S is non-singular with $J_M S = J_M T$ and with $\lambda^{-1} = \|S^{-1}\|$.

Set $B = S(A)$ and $g = f \circ S^{-1}$. We know that

$$\mathcal{L}^M(B) = J_M S \cdot \mathcal{L}^M(A) = J_M L \cdot \mathcal{L}^M(A).$$

We claim that

$$\text{Lip}(g|_B) \leq 1 + 2\epsilon \lambda^{-1}.$$

To see this, suppose $z, b \in B$. Then with $a = S^{-1}(b)$, $y = S^{-1}(z)$, it follows that $|y - a| \leq \lambda^{-1}|z - b|$. Therefore we have

$$\begin{aligned}
|g(z) - g(b)| &\leq |g(z) - g(b) - \langle Dg(b), z - b\rangle| + |\langle Dg(b) - U, z - b\rangle| + |\langle U, z - b\rangle| \\
&= |f(y) - f(a) - \langle Df(a), y - a\rangle| \\
&\quad + |\langle (Df(a) - T) \circ S^{-1}, z - b\rangle| + |z - b| \\
&\leq \epsilon |y - a| + \|Df(a) - T\| \cdot \|S^{-1}\| \cdot |z - b| + |z - b| \\
&\leq \left(1 + 2\epsilon\lambda^{-1}\right) |z - b|.
\end{aligned} \quad (3.140)$$

Finally, we have

$$\begin{aligned}
\mathcal{H}^M(f(A)) &= \mathcal{H}^M(g(B)) \\
&\leq \left(1 + 2\epsilon\lambda^{-1}\right)^M \cdot \mathcal{L}^M(B) \\
&= \left(1 + 2\epsilon\lambda^{-1}\right)^M \cdot J_M T \cdot \mathcal{L}^M(A).
\end{aligned}$$

Next we bound $\mathcal{H}^M(f(A))$ from below. We continue to use the same notation for the polar decomposition. Set $C = \Pi(f(A)) = \Pi(g(B))$ and $h = (\Pi \circ g|_B)^{-1}$. We claim that

$$\text{Lip}(h|_C) \leq \left(1 - 3\epsilon\lambda^{-1}\right)^{-1}.$$

To see this, suppose $w, c \in C$. Let $b \in B$ be such that $\Pi \circ g(b) = c$ and $z \in B$ be such that $\Pi \circ g(z) = w$. Arguing as we did to obtain the upper bound (3.140), but with some obvious changes, we see that

$$|g(z) - g(b)| \geq \left(1 - 2\epsilon\lambda^{-1}\right)|z - b|.$$

Also, we have

$$\begin{aligned}
\epsilon \lambda^{-1}|z - b| &\geq |g(z) - g(b) - \langle Dg(b), z - b\rangle| \\
&= |\Pi(g(z) - g(b) - \langle Dg(b), z - b\rangle) \\
&\quad + \Pi^\perp(g(z) - g(b) - \langle Dg(b), z - b\rangle)| \\
&\geq |\Pi^\perp(g(z) - g(b) - \langle Dg(b), z - b\rangle)| \\
&= |\Pi^\perp(g(z) - g(b))|.
\end{aligned}$$

Thus we have

$$|\Pi(g(z)) - \Pi(g(b))| \geq |g(z) - g(b)| - |\Pi^\perp(g(z) - g(b))|$$
$$\geq (1 - 2\epsilon\lambda^{-1})|z - b| - \epsilon\lambda^{-1}|z - b|.$$

Finally, we have

$$J_M T \cdot \mathcal{H}^M(A) = \mathcal{L}^M(B)$$
$$\leq (1 - 3\epsilon\lambda^{-1})^M \cdot \mathcal{L}^M(C)$$
$$\leq (1 - 3\epsilon\lambda^{-1})^M \cdot \mathcal{H}^M(f(A)).\quad\blacksquare$$

Rademacher's Theorem

Theorem 3.8.10 (Rademacher's Theorem) *A Lipschitz function $f: \mathbb{R}^M \to \mathbb{R}^N$ is differentiable $\mathcal{L}^M$-almost everywhere and the differential is a measurable function.*

Proof: We may assume $N = 1$. We use induction on M. In case $M = 1$, the result follows from the classical theorem that an absolutely continuous function from $\mathbb{R}$ to $\mathbb{R}$ is differentiable $\mathcal{L}$-almost everywhere.

For the induction step $M > 1$, we see by Fubini's Theorem and the theorem about absolutely continuous functions that all M partial derivatives of f exist $\mathcal{L}^M$-almost everywhere and are measurable functions. The goal is to show that these partial derivatives actually represent the differential at almost every point.

Let us write $\mathbb{R}^M = \mathbb{R}^{M-1} \times \mathbb{R}$ and denote points $x \in \mathbb{R}^M$ by $x = (y, z)$, $y \in \mathbb{R}^{M-1}$, $z \in \mathbb{R}$. We consider a point $x_0 = (y_0, z_0)$ where the following two conditions are satisfied:

(i) As a function of the first $M - 1$ variables, f is differentiable.

(ii) All M partial derivatives of f exist and are approximately continuous (see Definition 3.5.7).

For convenience of notation, we assume $x_0 = (0, 0)$ and that all the partial derivatives at x_0 vanish. It suffices to show that, for each $\epsilon > 0$, there exists $r > 0$ such that

$$|f(y, z)| \leq \epsilon r$$

holds whenever $|(y, z)| < r$.

Let $\epsilon > 0$ be arbitrary. We can choose $r_0 > 0$ so that, for $|y| < r_0$, $|f(y, 0)| \leq \epsilon|y|$. Choosing r_0 smaller if necessary, we can also assume that, for $0 < r < r_0$, the set of points in the cube $[-r, r]^M$ for which the absolute value of the Mth partial derivative exceeds ϵ is less than $\epsilon^M r^M$.

3.8. THE AREA FORMULA

Inside any $(M-1)$-dimensional cube $\prod_{i=1}^{M-1}[y_i - \epsilon r, y_i]$ of side ϵr there must be a vertical line, say at y', in which the Mth partial derivative is bounded by ϵ except for a set of length ϵr. So

$$\begin{aligned} |f(y',z)| &\leq |f(y',0)| + \int_0^z \left|\frac{\partial f}{\partial z}(y',\zeta)\right| d\mathcal{L}\zeta \\ &\leq \epsilon|y'| + \epsilon|z| + M\epsilon r \end{aligned}$$

and thus

$$\begin{aligned} |f(y,z)| &\leq |f(y',z)| + M\sqrt{M-1}\,\epsilon r \\ &\leq (2 + (1+\sqrt{M-1})M)\,\epsilon r. \quad \blacksquare \end{aligned}$$

Measure Theory of Lipschitz Maps

We need to know that the images of Lipschitz maps are measurable. Recall from the discussion of Borel sets versus Suslin sets in Section 1.3 that the properties of a set can be significantly altered by applying a function to it.

Lemma 3.8.11 *Suppose $K < M$ and suppose $f : \mathbb{R}^M \to \mathbb{R}^N$ is a Lipschitz function. If $A \subset \mathbb{R}^M$ is $\mathcal{H}^K$-measurable and if $\mathcal{H}^K(A) < \infty$, then $f(A)$ is $\mathcal{H}^K$-measurable.*

Proof: Since A can be written as the union of countably many bounded measurable sets, it will suffice to prove the lemma under the additional assumption that A is bounded.

Note that sets of $\mathcal{H}^K$ measure zero are preserved under f. Since Hausdorff measure is Borel regular (by Lemma 3.1.2), it follows that A is contained in a Borel set B with $\mathcal{H}^K(A) = \mathcal{H}^K(B)$. Note that $\bigl(f(B) \setminus f(A)\bigr) \subset f(B \setminus A)$ and $\mathcal{H}^K\bigl(f(B \setminus A)\bigr) = 0$. Thus, it will suffice to show that $f(B)$ is $\mathcal{H}^K$-measurable whenever B is a bounded Borel set with $\mathcal{H}^K(B) < \infty$.

Also because $\mathcal{H}^K$ is Borel regular, a bounded Borel set B with $\mathcal{H}^K(B) < \infty$ is $\mathcal{H}^K$-almost equal to a countable union of compact subsets of B. So the f-image of B is $\mathcal{H}^K$-almost equal to a countable union of f-images of compact sets. Of course, the f-image of a compact set is compact, hence $\mathcal{H}^K$-measurable, and the result follows. $\blacksquare$

Lemma 3.8.12 *Suppose $K < M$ and suppose $f : \mathbb{R}^M \to \mathbb{R}^N$ is a Lipschitz function. If $A \subset \mathbb{R}^M$ is $\mathcal{H}^K$-measurable and if $\mathcal{H}^K(A) < \infty$, then the function on $\mathbb{R}^N$ defined by*

$$y \mapsto \operatorname{card}(A \cap f^{-1}(y)) \tag{3.141}$$

is $\mathcal{H}^K$ measurable, and

$$\int_{\mathbb{R}^N} \operatorname{card}(A \cap f^{-1}(y))\, d\mathcal{H}^K \leq [\operatorname{Lip} f]^K \mathcal{H}^K(A) \tag{3.142}$$

holds.

Proof: For each $i > 0$, write

$$A = \bigcup_j A_{i,j}$$

where the sets $A_{i,j}$, $j = 1, 2, \ldots$, are pairwise disjoint $\mathcal{H}^K$-measurable sets all having diameter less than $1/i$. Then

$$\sum_j \chi_{f(A_{i,j})}$$

is $\mathcal{H}^K$-measurable by Lemma 3.8.11 and

$$\sum_j \chi_{f(A_{i,j})}(y) \leq \operatorname{card}(A \cap f^{-1}(y)) \quad \text{for each } y \in \mathbb{R}^N.$$

Here

$$\sum_j \chi_{f(A_{i,j})}(y) \uparrow \operatorname{card}(A \cap f^{-1}(y)) \quad \text{for each } y \in \mathbb{R}^N, \quad \text{as } i \to \infty,$$

proving the $\mathcal{H}^K$-measurability of the function in (3.141).

Finally, (3.142) follows from the Monotone Convergence Theorem and the estimate

$$\begin{aligned}
\int_{\mathbb{R}^N} \sum_j \chi_{f(A_{i,j})} \, d\mathcal{H}^K &= \sum_j \mathcal{H}^K[f(A_{i,j})] \\
&\leq [\operatorname{Lip} f]^K \sum_j \mathcal{H}^K(A_{i,j}) \\
&= [\operatorname{Lip} f]^K \mathcal{H}^K(A).
\end{aligned}$$
∎

Corollary 3.8.13 *For f and A as in the preceding lemma,*

$$\mathcal{H}^K(\{y : f^{-1}(y) \cap A \text{ is infinite}\}) = 0. \tag{3.143}$$

Proof: Equation (3.143) follows immediately from (3.142). ∎

In the next few lemmas, we will show how to use the lexicographic ordering on $\mathbb{R}^M$ to divide a set into pieces on which the restriction of f is one-to-one.

Definition 3.8.14 *The lexicographic ordering, $\prec_\ell$, on $\mathbb{R}^M$ is given by defining*

$$(x_1, x_2, \ldots, x_M) \prec_\ell (y_1, y_2, \ldots, y_M)$$

if and only if, for some $j \in \{1, 2, \ldots, M\}$, $x_1 = y_1$, $x_2 = y_2$, $\ldots$, $x_{j-1} = y_{j-1}$, and $x_j < y_j$. For $x, y \in \mathbb{R}^M$, we will write $x \preceq_\ell y$ if $x \prec_\ell y$ or $x = y$.

3.8. THE AREA FORMULA

Lemma 3.8.15 *Suppose* $f : \mathbb{R}^M \to \mathbb{R}^N$ *is a Lipschitz function. If* $A \subset \mathbb{R}^M$ *is* $\mathcal{L}^M$*-measurable, then, for each* $j = 1, 2, \ldots$,

$$\begin{aligned} A_j &= \{x \in A : \operatorname{card}(f^{-1}[f(x)] \cap A) = j\,\} \\ &= f^{-1}\Big(\{y \in \mathbb{R}^N : \operatorname{card}(f^{-1}(y) \cap A) = j\,\}\Big) \end{aligned}$$

is $\mathcal{L}^M$*-measurable.*

Proof: It will suffice to show that, for each positive integer $j \geq 2$,

$$A'_j = \{x \in A : \operatorname{card}(f^{-1}[f(x)] \cap A) \geq j\,\}$$

is $\mathcal{L}^M$-measurable. It is also sufficient to prove this assertion under the additional assumption that A is bounded.

Fix $j \geq 2$. Define the Borel set Ξ by setting

$$\Xi = \left(\mathbb{R}^M\right)^{2j} \bigcap \{(u_1, u_2, \ldots, u_j, x_1, x_2, \ldots, x_j) :$$
$$u_{i_1} \neq u_{i_2},\ \text{for } 1 \leq i_1 < i_2 \leq j,$$
$$x_1 = x_2 = \cdots = x_j\}.$$

Define $\mathcal{A} \subset \left(\mathbb{R}^M\right)^{2j}$ by setting

$$\mathcal{A} = \{(u_1, u_2, \ldots, u_j, f(u_1), f(u_2), \ldots, f(u_j)) : u_1, u_2, \ldots, u_j \in A\}.$$

Since $\mathcal{A}$ is the image of A under a Lipschitz mapping, it follows that $\mathcal{A}$ has finite $\mathcal{H}^M$ measure and is $\mathcal{H}^M$-measurable by Lemma 3.8.11. Now notice that $A'_j = \Pi(\Xi \cap \mathcal{A})$ where $\Pi : \left(\mathbb{R}^M\right)^{2j} \to \mathbb{R}^M$ is projection onto the first factor. Since Π is also a Lipschitz map, we see that Lemma 3.8.11 implies that A'_j is $\mathcal{L}^M$-measurable. ∎

Remark 3.8.16 Note that the preceding lemma gives us an alternative proof that if $f : \mathbb{R}^M \to \mathbb{R}^N$ is a Lipschitz function and $A \subset \mathbb{R}^M$ is $\mathcal{L}^M$-measurable, then the function on $\mathbb{R}^N$ defined by $y \mapsto \operatorname{card}(A \cap f^{-1}(y))$ is $\mathcal{H}^M$-measurable.

Lemma 3.8.17 *Suppose* $f : \mathbb{R}^M \to \mathbb{R}^N$ *is a Lipschitz function. If* $A \subset \mathbb{R}^M$ *is* $\mathcal{L}^M$*-measurable, then there exist* $\mathcal{L}^M$*-measurable sets* $A_{j,k} \subset A$, $j = 1, 2, \ldots$, $k = 1, 2, \ldots, j$ *such that*

$$A_j = \bigcup_{k=1}^{j} A_{j,k}.$$

Here A_j *is as in Lemma 3.8.15; also, for each* $k \in \{1, 2, \ldots, j\}$, $f|_{A_{j,k}}$ *is one-to-one, and*

$$f(A_{j,k}) = \{y \in \mathbb{R}^N : \operatorname{card}(f^{-1}(y) \cap A) = j\,\}.$$

Proof: It is no loss of generality to assume that
$$A = f^{-1}\Big(\{y \in \mathbb{R}^N : \text{card}(f^{-1}(y) \cap A) = j \}\Big).$$
We can then argue by induction on j. The case $j = 1$ is, of course, trivial.
Suppose $j \geq 2$. Define the Borel sets Ξ_i, $i = 1, 2, \ldots, M$, by setting
$$\Xi_i = \left(\mathbb{R}^M\right)^4 \cap \{(u, v, x, y) : u_1 = v_1, \ldots, u_{i-1} = v_{i-1}, u_i < v_i, x = y\},$$
for $i = 1, 2, \ldots, M$. Let $\Pi : \left(\mathbb{R}^M\right)^4 \to \mathbb{R}^M$ be projection on the first factor. Define $\mathcal{A} \subset \left(\mathbb{R}^M\right)^4$ by setting
$$\mathcal{A} = \{(u, v, f(u), f(v)) : u, v \in A\}.$$
Since $\mathcal{A}$ is the image of A under a Lipschitz map, $\mathcal{A}$ is $\mathcal{H}^M$-measurable by Lemma 3.8.11. Set
$$B = \bigcup_{i=1}^{M} \Pi(\Xi_i \cap \mathcal{A}).$$
Each set $\Pi(\Xi_i \cap \mathcal{A})$ is $\mathcal{L}^M$-measurable because it is the orthogonal projection of a measurable set. Observe that B consists of all points $u \in A$ for which there exists a point $v \in A$ that is larger than u in the lexicographic ordering on $\mathbb{R}^M$. Setting
$$A_{j,1} = A \setminus B,$$
we obtain an $\mathcal{L}^M$-measurable set containing the lexicographically largest point of $A \cap f^{-1}(y)$ for each $y \in f(A)$. Notice also that, on B, f is a $(j-1)$-to-one function, so the induction hypothesis applies to B. ∎

Lemma 3.8.18 *Suppose $f : \mathbb{R}^M \to \mathbb{R}^N$ is a Lipschitz function and $A \subset \mathbb{R}^M$. If $y \in \mathbb{R}^N$ is such that $f^{-1}(y) \cap A$ is infinite, then every point of*
$$f^{-1}(y) \cap \{x : Df(x) \text{ exists and is non-singular }\} \cap A$$
is an isolated point of that set.

Proof: Obvious. ∎

Lemma 3.8.19 *Suppose $f : \mathbb{R}^M \to \mathbb{R}^N$ is a Lipschitz function. If $A \subset \mathbb{R}^M$ is $\mathcal{L}^M$-measurable and*
$$B = f^{-1}\Big(\{y : f^{-1}(y) \cap A \text{ is infinite }\}\Big)$$
$$\setminus \Big(\{x : J_M f(x) = 0\} \cup \{x : Df(x) \text{ does not exist }\}\Big),$$
has positive $\mathcal{L}^M$ measure, then there exists $B' \subset B$ with $\mathcal{L}^M(B') > 0$ such that $f|_{B'}$ is one-to-one.

3.8. THE AREA FORMULA

Proof: Choose a compact $C \subset B$ with positive measure. Then, by Lemma 3.8.18, $f^{-1}[f(x)]$ is finite for each $x \in C$. We can apply Lemma 3.8.17 to decompose C into countably many sets $C_{j,k}$, at least one of which must have positive measure. ∎

Proof of the Area Formula

We first prove the result under the additional assumptions that $Df(a)$ exists at every point $a \in A$, that $J_M f(a) > 0$ at every point of A, and that $f|_A$ is one-to-one. By Lusin's Theorem '1.3.12 we may assume that $Df(a)$ is the restriction to A of a continuous function. By Egoroff's Theorem 1.3.13 we may suppose that

$$\frac{|f(y) - f(a) - \langle Df(a), y - a \rangle|}{|y - a|}$$

converges uniformly to 0 as $y \in A$ approaches $a \in A$. It is plain that, for any $\epsilon > 0$, every subset of A with sufficiently small diameter satisfies conditions (i)–(iii) of Lemma 3.8.9 for some full rank linear $T : \mathbb{R}^M \to \mathbb{R}^N$—namely, we can choose T to be Df at any point in such a sufficiently small set. Since we are assuming that Df on A is the restriction of a continuous function, we can find a positive lower bound for λ in (3.138). To see that condition (iv) of Lemma 3.8.9 is also satisfied on a subset of A of small enough diameter, we suppose $\Pi \circ f(y) = \Pi \circ f(z)$; we show that, in this case, $\epsilon > 0$ can be chosen small enough compared to λ that conditions (i)–(iii) lead to a contradiction. Using (i)–(iii), we estimate

$$
\begin{aligned}
|\langle T, y - z \rangle| &= |\Pi \langle T, y - z \rangle)| \\
&\leq |\Pi \langle T - Df(a), y - z \rangle| + |\Pi \langle Df(a) - Df(z), y - z \rangle| \\
&\quad + |\Pi \langle Df(z), y - z \rangle| \\
&\leq \|T - Df(a)\| \, |y - z| + \|Df(a) - Df(z)\| \, |y - z| \\
&\quad + |\Pi \langle Df(z), y - z \rangle| \\
&= \|T - Df(a)\| \, |y - z| + \|Df(a) - Df(z)\| \, |y - z| \\
&\quad + |\Pi(f(y) - f(z) - \langle Df(z), y - z \rangle)| \\
&\leq \|T - Df(a)\| \, |y - z| + \|Df(a) - Df(z)\| \, |y - z| \\
&\quad + |f(y) - f(z) - \langle Df(a), y - z \rangle|.
\end{aligned}
$$

By choosing a, y, z in a small enough set we can bound the right-hand side of the preceding inequality above by $3\epsilon |y - z|$, while the left-hand side is bounded below by $\lambda |y - z|$. Choosing ϵ smaller than $\frac{1}{3}\lambda$ gives a contradiction. Thus (iv) also must hold on subsets of small enough diameter.

In case f is finite-to-one almost everywhere, we use Lemma 3.8.17 to decompose A into the sets $A_{j,k}$ on which the previous case applies.

For the part of A on which f is infinite-to-one we know the right-hand side of (3.130) is zero by Corollary 3.8.13. If the left-hand side of (3.130) were not zero, we would contradict Lemma 3.8.19.

Finally, to complete the proof, we need to show that the image of a set on which $J_M f = 0$ has measure zero. That follows by defining $f_\epsilon : \mathbb{R}^M \to \mathbb{R}^{M+N}$ by
$$x \mapsto \big(\epsilon x, f(x)\big).$$
This gives us the full rank hypothesis, but only increases the Jacobian by a bounded multiple of ϵ. The image of f is the orthogonal projection of the image of f_ϵ and thus its Hausdorff measure is no larger than the Hausdorff measure of the image of f_ϵ. We conclude as $\epsilon \downarrow 0$ that the Hausdorff measure of the image of f is 0. ∎

3.9 The Co-Area Formula

The main result of this section is the following theorem.

Theorem 3.9.1 (The Co-Area Formula) *If $f : \mathbb{R}^M \to \mathbb{R}^N$ is a Lipschitz function and $M \geq N$, then*

$$\int_A J_N f(x) \, d\mathcal{L}^M x = \int_{\mathbb{R}^N} \mathcal{H}^{M-N}(A \cap f^{-1}(y)) \, d\mathcal{L}^N y \qquad (3.144)$$

holds for each Lebesgue measurable subset A of $\mathbb{R}^M$.

Here $J_N f$ denotes the N-dimensional Jacobian of f which was defined in the previous section in Definition 3.8.3, and which was seen by (3.137) to be given by
$$J_N f(a) = \sqrt{\det\big[(Df(a)) \cdot (Df(a))^{\mathrm{t}}\big]}.$$
In case $M = N$, the N-dimensional Jacobian agrees with the usual Jacobian, $|\det(Df)|$, and the area and co-area formulas coincide. In case $M > N$, and $f : \mathbb{R}^M = \mathbb{R}^N \times \mathbb{R}^{M-N} \to \mathbb{R}^N$ is orthogonal projection onto the first factor, then the co-area formula simplifies to Fubini's theorem, thus one can think of the co-area formula as a generalization of Fubini's theorem to functions more complicated than orthogonal projection. The co-area formula was first proved in Federer [3].

As in the proof of the area formula, the proof of the co-area formula separates into three fundamental parts. The first is to understand the situation for linear maps. This was done in the previous section. The second part is to extend our understanding to the behavior of maps that are well approximated by linear maps. The third part of the proof brings in the measure theory that allows us to reduce the behavior of Lipschitz maps to that of maps that are well approximated by linear maps.

3.9. THE CO-AREA FORMULA

Main Estimates for the Co-Area Formula

Lemma 3.9.2 (Main Estimates for the Co-Area Formula)
Suppose $M > N$, $U : \mathbb{R}^N \to \mathbb{R}^M$ is orthogonal, and $0 < \epsilon < \frac{1}{2}$. If the Lebesgue measurable set $A \subset \mathbb{R}^M$ is such that

(i) $Df(a)$ exists for $a \in A$,

(ii) $\|Df(a) - U^{\mathrm{t}}\| < \epsilon$ holds for $a \in A$,

(iii) $|f(y) - f(a) - \langle Df(a), y - a \rangle| < \epsilon |y - a|$ holds for $y, a \in A$,

then

$$(1 - 2\epsilon)^M \int_{\mathbb{R}^N} \mathcal{H}^{M-N}(A \cap f^{-1}(y)) \, d\mathcal{L}^N y \le \int_A J_M f(a) \, d\mathcal{L}^M a$$

$$\le \int_{\mathbb{R}^N} \mathcal{H}^{M-N}(A \cap f^{-1}(y)) \, d\mathcal{L}^N y. \quad (3.145)$$

Proof: Let $V : \mathbb{R}^{M-N} \to \mathbb{R}^M$ be an orthogonal map such that $\ker(U^{\mathrm{t}})$ and $\ker(V^{\mathrm{t}})$ are orthogonal complements. Define $F : \mathbb{R}^M \to \mathbb{R}^N \times \mathbb{R}^{M-N}$ by setting

$$F(x) = (f(x), V^{\mathrm{t}}(x)),$$

and let $\Pi : \mathbb{R}^N \times \mathbb{R}^{M-N} \to \mathbb{R}^N$ be projection on the first factor. It is easy to see that

$$J_M F = J_N f.$$

Subsequently, we will show that $F|_A$ is one-to-one so that, by the area formula,

$$\mathcal{L}^M[F(A)] = \int_A J_M F \, d\mathcal{L}^M = \int_A J_N f \, d\mathcal{L}^M.$$

Thus, using Fubini's Theorem, we have

$$\int_A J_N f \, d\mathcal{L}^M = \mathcal{L}^M[F(A)]$$
$$= \int_{\mathbb{R}^N} \mathcal{H}^{M-N}[F(A) \cap \Pi^{-1}(z)] \, d\mathcal{L}^N z$$
$$= \int_{\mathbb{R}^N} \mathcal{H}^{M-N}[F(A \cap f^{-1}(z))] \, d\mathcal{L}^N z.$$

To complete the proof, we show $F|_A$ to be one-to-one and estimate the Lipschitz constant of F on $A \cap f^{-1}(z)$ and the Lipschitz constant of F^{-1} on $F(A \cap f^{-1}(z))$. Suppose $a, y \in A \cap f^{-1}(z)$. Then $F(a) = (f(a), V^{\mathrm{t}}(a)) =$

$(z, V^t(a))$ and $F(y) = (f(y), V^t(y)) = (z, V^t(y))$. We should like to compare $|a - y|$ and $|F(a) - F(y)|$. But the first components are the same, so
$$|F(a) - F(y)| = |V^t(a) - V^t(y)|.$$
On the one hand, V^t is distance decreasing, so
$$|F(a) - F(y)| \leq |a - y|.$$
On the other hand,
$$\begin{aligned}|\langle U^t, y - a\rangle| &\leq |\langle Df(a), y - a\rangle| + \|Df(a) - U^t\| \, |y - a| \\ &= |f(y) - f(a) - \langle Df(a), y - a\rangle| + \|Df(a) - U^t\| \, |y - a| \\ &< 2\epsilon |y - a|,\end{aligned}$$
and
$$|y - a|^2 = |V^t(a) - V^t(y)|^2 + |\langle U^t, y - a\rangle|^2,$$
so
$$|V^t(a) - V^t(y)|^2 \geq |y - a|^2 (1 - 4\epsilon^2).$$
Thus we have
$$\sqrt{1 - 4\epsilon^2} \, |y - a| \leq |F(y) - F(a)| \leq |y - a|. \qquad \blacksquare$$

Corollary 3.9.3 *Suppose $M > N$, $T : \mathbb{R}^M \to \mathbb{R}^N$ is of rank N, and $0 < \epsilon < \frac{1}{2}$. If the Lebesgue measurable set $A \subset \mathbb{R}^M$ is such that*

(i) *$Df(a)$ exists for $a \in A$,*

(ii) *$\|Df(a) - T\| < \epsilon$ holds for $a \in A$,*

(iii) *$|f(y) - f(a) - \langle Df(a), y - a\rangle| < \epsilon |y - a|$ holds for $y, a \in A$,*

then

$$(1 - 2\epsilon)^M \int_{\mathbb{R}^N} \mathcal{H}^{M-N}\left(A \cap f^{-1}(y)\right) d\mathcal{L}^N y \leq \int_A J_M f(a) \, d\mathcal{L}^M a$$
$$\leq \int_{\mathbb{R}^N} \mathcal{H}^{M-N}\left(A \cap f^{-1}(y)\right) d\mathcal{L}^N y. \qquad (3.146)$$

Proof: By the Polar Decomposition Theorem 3.8.6, there exists a symmetric linear map $S : \mathbb{R}^N \to \mathbb{R}^N$ and an orthogonal map $U : \mathbb{R}^N \to \mathbb{R}^M$ such

3.9. THE CO-AREA FORMULA

that $T = S \circ U^t$. Set $g = S^{-1} \circ f$. Then we apply the lemma to g and U to obtain

$$(1-2\epsilon)^M \int_{\mathbb{R}^N} \mathcal{H}^{M-N}(A \cap g^{-1}(z)) \, d\mathcal{L}^N z \leq \int_A J_M g(a) \, d\mathcal{L}^M a$$
$$\leq \int_{\mathbb{R}^N} \mathcal{H}^{M-N}(A \cap g^{-1}(z)) \, d\mathcal{L}^N z. \quad (3.147)$$

Notice that if $y = S(z)$, then

$$A \cap g^{-1}(z) = A \cap f^{-1}(y),$$

so by the change of variables formula in $\mathbb{R}^N$ applied to the mapping S, we have

$$\int_{\mathbb{R}^N} \mathcal{H}^{M-N}(A \cap g^{-1}(z)) J_N S \, d\mathcal{L}^N z = \int_{\mathbb{R}^N} \mathcal{H}^{M-N}(A \cap f^{-1}(y)) \, d\mathcal{L}^N y.$$

Also we have $J_N S\, J_M g = J_M f$, so

$$\int_A J_N g\, J_M g(a) \, d\mathcal{L}^M a = \int_A J_M f(a) \, d\mathcal{L}^M a$$

holds. Thus if we multiply all three terms in (3.147) by $J_N S$, we obtain (3.146). ∎

Measure Theory of Lipschitz Maps

We need to verify that the integrand on the right-hand side of (3.144) is measurable. (The measurability of the integrand on the left-hand side of (3.144) is given by Rademacher's Theorem 3.8.10.) First we obtain a useful preliminary estimate originally proved by Eilenberg and Harrold.

Lemma 3.9.4 *Suppose $0 \leq N \leq M < \infty$. There exists a constant $C(M,N)$ such that the following statement is true: If $f : \mathbb{R}^M \to \mathbb{R}^N$ is a Lipschitz function and $A \subset \mathbb{R}^M$ is $\mathcal{L}^M$-measurable, then*

$$\int_{\mathbb{R}^N}^* \mathcal{H}^{M-N}(A \cap f^{-1}(y)) \, d\mathcal{L}^N y \leq C(M,N) \, [\mathrm{Lip}(f)]^N \, \mathcal{L}^M(A) \quad (3.148)$$

holds.

Proof: We may assume the right-hand side of (3.148) is finite.

Fix $\sigma > 0$. By the definition of Hausdorff measure, there exists a cover of A by closed sets $S_1, S_2, \ldots$, all having diameter less than σ, such that

$$\sum_i \Upsilon_M \left(\frac{\mathrm{diam}(S_i)}{2} \right)^M \leq \mathcal{H}^M(A) + \sigma.$$

For $y \in \mathbb{R}^N$ we observe that

$$\mathcal{H}_\sigma^{M-N}(A \cap f^{-1}(y)) \leq \sum_{\{i: S_i \cap f^{-1}(y) \neq \emptyset\}} \Upsilon_{M-N} \left(\frac{\text{diam}(S_i)}{2}\right)^{M-N}$$

$$= 2^{N-M} \Upsilon_{M-N} \sum_i \left(\text{diam}(S_i)\right)^{M-N} \chi_{f(S_i)}(y).$$

Note also that if $p \in S_i$, then

$$f(S_i) \subset \overline{\mathbb{B}}\left(f(p), \text{Lip}(f)\, \text{diam}(S_i)\right),$$

so

$$\int_{\mathbb{R}^N} \chi_{f(S_i)} \, d\mathcal{L}^N \leq [\text{Lip}(f)]^N \, \Upsilon_N \left(\text{diam}(S_i)\right)^N.$$

Thus we have

$$\int_{\mathbb{R}^N}^* \mathcal{H}_\sigma^{M-N}\left(A \cap f^{-1}(y)\right) d\mathcal{L}^N y$$

$$\leq 2^{N-M} \Upsilon_{M-N} \sum_i \left(\text{diam}(S_i)\right)^{M-N} \int_{\mathbb{R}^N} \chi_{f(S_i)} \, d\mathcal{L}^N$$

$$\leq 2^{N-M} \Upsilon_{M-N} \Upsilon_N [\text{Lip}(f)]^N \sum_i \left(\text{diam}(S_i)\right)^N$$

$$\leq 2^N \frac{\Upsilon_{M-N}\, \Upsilon_N}{\Upsilon_M} \left(\mathcal{H}^M(A) + \sigma\right).$$

The result follows by letting $\sigma \downarrow 0$. ∎

Lemma 3.9.5 *Suppose $f : \mathbb{R}^M \to \mathbb{R}^N$ is a Lipschitz function. Then the mapping*

$$y \mapsto \mathcal{H}^N\left(A \cap f^{-1}(y)\right)$$

is $\mathcal{H}^M$-measurable.

Proof: By the previous lemma, we can ignore sets of arbitrarily small measure, hence we may assume that A is compact.

If A is compact, then

$$\{y : \mathcal{H}^{M-N}(A \cap f^{-1}(y)) \leq t\} = \bigcap_j V_j,$$

here V_j is the open subset of $\mathbb{R}^N$ consisting of all points y for which $A \cap f^{-1}(y)$ has a finite cover by open sets U_i of diameter less than $1/j$ such that

$$\sum_i \Upsilon_{M-N} \left(\frac{\text{diam}(U_i)}{2}\right)^{M-N} \leq t + \frac{1}{j}.$$
∎

3.9. THE CO-AREA FORMULA

Proof of the Co-area Formula

By Theorem 3.8.10 and (3.148), we may assume that $Df(a)$ exists at every point $a \in A$. First we prove the result under the additional assumption that $J_N f(a) > 0$ at every point of A. By Lusin's Theorem 1.3.12 we may assume that $Df(a)$ is the restriction to A of a continuous function. By Egoroff's Theorem 1.3.13 we may suppose that

$$\frac{|f(y) - f(a) - \langle Df(a), y - a \rangle|}{|y - a|}$$

converges uniformly to 0 as $y \in A$ approaches $a \in A$. It is plain that, for any $\epsilon > 0$, conditions (i)–(iii) of Lemma 3.9.3 are satisfied in any subset of A that has small enough diameter.

Finally, to complete the proof, we need to consider the case in which $J_N f = 0$ holds on all of A. In that case, the left-hand side of (3.144) is 0. We need to show that the right-hand side of (3.144) also equals 0. To this end, consider $f_\epsilon : \mathbb{R}^{M+N} \to \mathbb{R}^N$ defined by

$$(x, y) \mapsto f(x) + \epsilon x.$$

We can apply what has already been proved to the set

$$A \times [-1, 1]^N \subset \mathbb{R}^M \times \mathbb{R}^N.$$

We have $\mathcal{L}^{M+N}(A \times [-1, 1]^N) = 2^N \mathcal{L}^M(A)$, $J_N f_\epsilon \leq \epsilon [\epsilon + \mathrm{Lip}(f)]^{N-1}$, and

$$\int_{A \times [-1,1]^N} J_N f_\epsilon \, d\mathcal{L}^{M+N} = \int_{\mathbb{R}^N} \mathcal{H}^M \left[(A \times [-1,1]^N) \cap f_\epsilon^{-1}(z) \right] d\mathcal{L}^N z.$$

By (3.148) observe that

$$C(M, N) \mathcal{H}^M \left[(A \times [-1, 1]^N) \cap f_\epsilon^{-1}(z) \right]$$
$$\geq \int_{\mathbb{R}^N} \mathcal{H}^{M-N} \left[(A \times [-1,1]^N) \cap f_\epsilon^{-1}(z) \cap \Pi^{-1}(y) \right] d\mathcal{L}^N y$$
$$= \int_{[-1,1]^N} \mathcal{H}^{M-N} [A \cap f^{-1}(z - \epsilon y)] \, d\mathcal{L}^N y.$$

Thus

$$2^N \mathcal{L}^M(A) \epsilon [\epsilon + \mathrm{Lip}(f)]^{N-1}$$
$$\geq \int_{A \times [-1,1]^N} J_N f_\epsilon \, d\mathcal{L}^{M+N}$$
$$\geq \frac{1}{C(M,N)} \int_{\mathbb{R}^N} \int_{[-1,1]^N} \mathcal{H}^{M-N}[A \cap f^{-1}(z - \epsilon y)] \, d\mathcal{L}^N y \, d\mathcal{L}^N z$$
$$= \frac{1}{C(M,N)} \int_{[-1,1]^N} \int_{\mathbb{R}^N} \mathcal{H}^{M-N}[A \cap f^{-1}(z - \epsilon y)] \, d\mathcal{L}^N z \, d\mathcal{L}^N y$$
$$= \frac{2^N}{C(M,N)} \int_{\mathbb{R}^N} \mathcal{H}^{M-N}[A \cap f^{-1}(z)] \, d\mathcal{L}^N z$$

holds, where the last equation holds by translation invariance. Letting $\epsilon \downarrow 0$, we see that

$$\int_{\mathbb{R}^N} \mathcal{H}^{M-N}\left[A \cap f^{-1}(z)\right] d\mathcal{L}^N z = 0.$$ ∎

Chapter 4

Sobolev Spaces

4.1 Basic Definitions and Results

This book has given substantial attention to the C^k spaces. Such spaces are suitable classes from which to select the defining function for a domain, and the C^k spaces are natural in a number of other geometric contexts. However, in the study of partial differential equations and Fourier analysis, the Sobolev spaces are more convenient. The definition of the Sobolev spaces is less near the surface than that of the C^k spaces, but theorems about Sobolev spaces are *more accessible*. In the end, the Sobolev imbedding theorem allows one to pass back and forth between the C^k spaces and the Sobolev spaces (however one must pay with a certain lack of precision that is present in the imbedding theorem).

We will need the following basic definitions.

Definition 4.1.1

(i) *$\mathcal{D}$ is used to denote the space of infinitely differentiable functions on $\mathbb{R}^N$ having compact support and topologized by uniform convergence on compact sets.*

(ii) *The **Fourier transform** of $f \in L^1(\mathbb{R}^N)$ is denoted by $\hat{f}$ and is defined by*

$$\hat{f}(\xi) = \int_{\mathbb{R}^N} f(x) \, e^{-2\pi i x \cdot \xi} \, d\mathcal{L}^N x. \tag{4.1}$$

(iii) *The **Fourier Inversion Theorem** (e.g. see Stein and Weiss [1]) states that if $f, \hat{f} \in L^1(\mathbb{R}^N)$, then*

$$f(x) = \int_{\mathbb{R}^N} \hat{f}(\xi) \, e^{2\pi i x \cdot \xi} \, d\mathcal{L}^N \xi \tag{4.2}$$

holds for almost every $x \in \mathbb{R}^N$.

For simplicity, we begin our discussion of Sobolev spaces on the domain $\mathbb{R}^N$, or all of space. We follow the discussion in Krantz [4].

Definition 4.1.2 *If $\phi \in \mathcal{D}$ and $s \in \mathbb{R}$ then we define the norm*

$$\|\phi\|_{H^s} = \|\phi\|_s \equiv \left(\int |\hat{\phi}(\xi)|^2 (1 + |\xi|^2)^s \, d\xi \right)^{1/2}. \tag{4.3}$$

We let $H^s(\mathbb{R}^N)$ be the closure of $\mathcal{D}$ with respect to $\|\cdot\|_s$.

In the case that s is a non-negative integer, then

$$(1 + |\xi|^{2s}) \approx (1 + |\xi|^2)^s \approx \sum_{|\alpha| \leq 2s} |\xi|^\alpha \approx \left(\sum_{|\alpha| \leq s} |\xi|^\alpha \right)^2.$$

Therefore

$$\phi \in H^s \quad \text{if and only if} \quad \hat{\phi} \cdot [\textstyle\sum_{|\alpha| \leq s} |\xi|^\alpha] \in L^2.$$

This last condition means that $\hat{\phi} \xi^\alpha \in L^2$ for all multi-indices α with $|\alpha| \leq s$. By the Plancherel theorem (see Stein and Weiss [1]), we have

$$\left(\frac{\partial}{\partial x} \right)^\alpha \phi \in L^2 \quad \forall \alpha \text{ such that } |\alpha| \leq s.$$

Thus we have

Proposition 4.1.3 *If s is a non-negative integer, then*

$$H^s = \left\{ f \in L^2 : \frac{\partial^\alpha}{\partial x^\alpha} f \in L^2 \text{ for all } \alpha \text{ with } |\alpha| \leq s \right\}.$$

Here derivatives are interpreted in the sense of distributions.

Notice that if $s > r$ then $H^s \subset H^r$ because

$$|\hat{\phi}|^2 (1 + |\xi|^2)^r \leq C \cdot |\hat{\phi}|^2 (1 + |\xi|^2)^s.$$

The Sobolev spaces turn out to be easy to work with because they are modeled on L^2—indeed each H^s is canonically isomorphic as a Hilbert space to L^2 (exercise). But they are important because they can be related to the more classical spaces of smooth functions. That is the content of the Sobolev Imbedding Theorem:

Theorem 4.1.4 (Sobolev) *Let $s > N/2$. If $f \in H^s(\mathbb{R}^N)$, then f can be corrected, uniquely, on a set of measure zero to be continuous.*

More generally, if $k \in \{0, 1, 2, \ldots\}$ and if $f \in H^s$, $s > N/2 + k$, then f can be corrected on a set of measure zero to be C^k.

4.1. BASIC DEFINITIONS AND RESULTS

Proof: For the first part of the theorem, let $f \in H^s$. By definition, there exist $\phi_j \in \mathcal{D}$ such that $\|\phi_j - f\|_{H^s} \to 0$. Then

$$\|\phi_j - f\|_{L^2} = \|\phi_j - f\|_0 \le \|\phi_j - f\|_s^2 \to 0. \tag{4.4}$$

Our plan is to show that $\{\phi_j\}$ is an equibounded, equicontinuous family of functions. Then the Ascoli-Arzelá theorem (*e.g.* see Rudin [1]) will imply that there is a subsequence converging uniformly on compact sets to a (continuous) function g. But (4.4) guarantees that a subsequence of this subsequence converges pointwise to the function f. So $f = g$ almost everywhere and the required assertion follows.

To see that $\{\phi_j\}$ is equibounded, we calculate that

$$\begin{aligned} |\phi_j(x)| &= \left| \int e^{-2\pi i x \cdot \xi} \hat{\phi}_j(\xi) \, d\xi \right| \\ &\le \int |\hat{\phi}_j(\xi)| (1+|\xi|^2)^{s/2} (1+|\xi|^2)^{-s/2} \, d\xi \\ &\le \left(\int |\hat{\phi}_j(\xi)|^2 (1+|\xi|^2)^s \, d\xi \right)^{1/2} \cdot \left(\int (1+|\xi|^2)^{-s} \, d\xi \right)^{1/2}. \end{aligned}$$

Using polar coordinates, we may see easily that, for $s > N/2$,

$$\int (1+|\xi|^2)^{-s} \, d\xi < \infty.$$

Therefore

$$|\phi_j(x)| \le C\|\phi_j\|_{H^s} \le C'$$

and $\{\phi_j\}$ is equibounded.

To see that $\{\phi_j\}$ is equicontinuous, we write

$$|\phi_j(x) - \phi_j(y)| = \left| \int \hat{\phi}_j(\xi) \left(e^{-2\pi i x \cdot \xi} - e^{-2\pi i y \cdot \xi} \right) d\xi \right|.$$

Observe that $|e^{-2\pi i x \cdot \xi} - e^{-2\pi i y \cdot \xi}| \le 2$ and, by the mean value theorem,

$$|e^{-2\pi i x \cdot \xi} - e^{-2\pi i y \cdot \xi}| \le 2\pi |x-y| |\xi|.$$

Then, for any $0 < \epsilon < 1$,

$$\begin{aligned} |e^{-2\pi i x \cdot \xi} - e^{-2\pi i y \cdot \xi}| &= |e^{-2\pi i x \cdot \xi} - e^{-2\pi i y \cdot \xi}|^{1-\epsilon} |e^{-2\pi i x \cdot \xi} - e^{-2\pi i y \cdot \xi}|^\epsilon \\ &\le 2\pi^\epsilon |x-y|^\epsilon |\xi|^\epsilon. \end{aligned}$$

Therefore

$$\begin{aligned} |\phi_j(x) - \phi_j(y)| &\le C \int |\hat{\phi}_j(\xi)| |x-y|^\epsilon |\xi|^\epsilon \, d\xi \\ &\le C|x-y|^\epsilon \int |\hat{\phi}_j(\xi)| (1+|\xi|^2)^{\epsilon/2} \, d\xi \\ &\le C|x-y|^\epsilon \|\phi_j\|_{H^s} \left(\int (1+|\xi|^2)^{-s+\epsilon} \, d\xi \right)^{1/2}, \end{aligned}$$

where the last inequality is a consequence of the Schwarz inequality. If we select $0 < \epsilon < 1$ such that $-s + \epsilon < -N/2$ then we find that

$$\int (1 + |\xi|^2)^{-s+\epsilon} d\xi$$

is finite. It follows that the sequence $\{\phi_j\}$ is equicontinuous and we are done.

The second assertion of the theorem may be derived from the first by a simple inductive argument. We leave the details as an exercise. ∎

Remark 4.1.5

(i) If $s = N/2$, then the first part of the theorem is false (exercise). In fact the sharp statement that one can make is that in this, case, f lies in the space of functions of bounded mean oscillation or that f is exponentially integrable (see Stein [1], Folland and Stein [1], Adams [1]).

(ii) Theorem 4.1.4 may be interpreted as saying that $H^s \subset C^k$ for $s > k + N/2$. In other words, the identity provides a continuous imbedding of H^s into C^k. A converse is also true. Namely, if $H^s \subset C^k$ for some non-negative integer k, then $s > k + N/2$.

To see this, notice that the hypotheses $u_j \to u$ in H^s and $u_j \to v$ in C^k imply that $u = v$. Therefore the inclusion of H^s into C^k is a closed map. It is therefore continuous by the closed graph theorem. Thus there is a constant C such that

$$\|f\|_{C^k} \leq C\|f\|_{H^s}. \tag{4.5}$$

Now, for $x \in \mathbb{R}^N$ fixed and α a multi-index with $|\alpha| \leq k$, the tempered distribution e_x^α defined by

$$e_x^\alpha(\phi) = \left(\frac{\partial^\alpha}{\partial x^\alpha}\right)\phi(x)$$

is bounded in $(C^k)^*$ with bound independent of x and α (but depending on k). (See Stein and Weiss [1] or Krantz [4] for a discussion of tempered distributions.) Hence, by (4.5), $\{e_x^\alpha\}$ forms a bounded set in $(H^s)^* \equiv H^{-s}$. As a result, for $|\alpha| \leq k$ we have that

$$\begin{aligned}
\|e_x^\alpha\|_{H^{-s}} &= \left(\int |\widehat{(e_x^\alpha)}(\xi)|^2 (1+|\xi|^2)^{-s} d\xi\right)^{1/2} \\
&= \left(\int |(-2\pi i\xi)^\alpha e^{2\pi i x \cdot \xi}|^2 (1+|\xi|^2)^{-s} d\xi\right)^{1/2} \\
&\leq C\left(\int (1+|\xi|^2)^{-s+|\alpha|} d\xi\right)^{1/2}
\end{aligned}$$

4.1. BASIC DEFINITIONS AND RESULTS

is finite, independent of x and α. But this can only happen if $2(k-s) < -N$, that is if $s > k + N/2$.

We cannot proceed any further without giving a precise definition of Sobolev spaces on a domain. In fact we shall give three such definitions, and in the next section we will compare and contrast them.

Definition 4.1.6

(i) Let $\Omega \subset \mathbb{R}^N$ be any domain. Let $s \geq 0$ be an integer. Define the norm

$$\|f\|_{H^s(\Omega)} = \sum_{|\alpha| \leq s} \left\| \frac{\partial^\alpha f}{\partial x^\alpha} \right\|_{L^2(\Omega)}.$$

Let $\mathcal{P}^s$ consist of those $f \in C^s(\Omega)$ for which $\|f\|_{H^s(\Omega)}$ is finite. This is clearly a linear space. Define $H^s(\Omega)$ to be the closure of $\mathcal{P}^s$ in the $\|\cdot\|_{H^s(\Omega)}$ norm.

(ii) Let $\Omega, \|\cdot\|_{H^s(\Omega)}$ be as in the last definition. Let $W^s(\Omega)$ consist of those $f \in L^2(\Omega)$ such that $(\partial^\alpha/\partial x^\alpha)f \in L^2$ for all $|\alpha| \leq s$. Here derivatives are, perforce, interpreted in the weak sense.

(iii) Let $\Omega, \|\cdot\|_{H^s(\Omega)}$ be as in the preceding two definitions. We define $W_0^s(\Omega)$ to be the closure in the $\|\cdot\|_{H^s}$ topology of the space $C_c^\infty(\Omega)$.

Remark 4.1.7

(i) It is clear from the definitions that $W_0^s \subset W^s$ and $W_0^s(\Omega) \subset H^s(\Omega)$. That these inclusions are proper when s is large enough is clear from the Sobolev imbedding theorem. For if $s > N/2$ then an element of $W_0^s(\Omega)$ would necessarily vanish on $\partial \Omega$. And density in the $\|\cdot\|_{H^s(\Omega)}$ norm would imply uniform density. Thus $W_0^s(\Omega)$ could not be dense in either of the first two spaces. However more is true. In fact if $s > 1/2$, then $W_0^s(\Omega)$ is not dense in $W^s(\Omega)$, nor is it dense in $H^s(\Omega)$. The reader may try his hand at this as an exercise, or consult Taylor [1].

(ii) It is a striking fact, discovered as recently as 1964 (see Meyers and Serrin [1]) that, *for any domain* $\Omega \subset \mathbb{R}^N$, $H^s(\Omega) = W^s(\Omega)$. The proof of this statement is a standard, but tricky, adaptation of the usual approximation by cutting off and convolving with a family of Friedrichs mollifiers. We omit the details, but refer the reader to Adams [1].

Exercise: Imitate the proof of the Sobolev Imbedding Theorem 4.1.4 to prove Rellich's lemma: If $s > r$ and if $\Omega_2 \subset\subset \Omega_1$ then the inclusion map

$$i : H^s(\Omega_1) \to H^r(\Omega_2)$$

given by
$$f \mapsto f|_{\Omega_2}$$
is a compact operator (see Krantz [4] for the details).

4.2 Restriction and Trace Theorems for Sobolev Spaces

In the preceding section, we discussed the most basic facts about the Sobolev spaces. In this section, we will adapt some of our earlier results to bounded domains in space. First, we discuss the significance of the so-called "restriction theorem:"

Let $S = \{(x_1, \ldots, x_{N-1}, 0)\} \subset \mathbb{R}^N$. Then S is a hypersurface, and is the boundary of $\{(x_1, \ldots, x_N) \in \mathbb{R}^N : x_N > 0\}$. It is the simplest example of the type of geometric object that arises as the boundary of a domain. It is natural to want to be able to restrict a function f defined on a neighborhood of S, or on one side of S, to S. If f is continuous on a neighborhood of S, then the restriction of f to S is trivially and unambiguously defined, simply because a continuous function is well-defined at every point.

If, instead of being continuous, f is an element of the Sobolev space $H^{N/2+\epsilon}$, then we may apply the Sobolev imbedding theorem to correct f on a set of measure zero (in a unique manner) to obtain a continuous function. The corrected f may then be restricted to S. By contrast, restriction *prior* to the correction on the set of measure zero is *prima facie* ambiguous. This is because a Sobolev space "function" is really an equivalence class of functions any pair of which need only agree up to a set of measure zero. The set S itself has measure zero, and thus two different elements of the equivalence class of f may have different restrictions to S.

We wish to develop a notion of calculating the restriction or "trace" of a Sobolev space function on a hypersurface that applies to H^r for $r \ll N/2$ and is such that the restriction operation works naturally in the context of Sobolev classes without relying on the Sobolev imbedding theorem. Our first result is a bound on norms. This bound, given in estimate (4.6), shows us when the trace of a Sobolev function is defined, thus the theorem is known as the Sobolev trace theorem.

Theorem 4.2.1 *Identify S with $\mathbb{R}^{N-1}$ in the natural way. Let $s > 1/2$. Then the mapping from $C_c^\infty(\mathbb{R}^N)$ to $C_c^\infty(\mathbb{R}^{N-1})$ defined by*
$$\phi \mapsto \phi|_S$$
extends to a bounded linear operator from $H^s(\mathbb{R}^N)$ to $H^{s-1/2}(\mathbb{R}^{N-1})$. That is, there exists a constant $C = C(s) > 0$ such that

$$\left\| \phi|_S \right\|_{H^{s-1/2}(\mathbb{R}^{N-1})} \le C(s) \|\phi\|_{H^s(\mathbb{R}^N)}. \tag{4.6}$$

4.2. SOBOLEV SPACE RESTRICTION AND TRACE THEOREMS

Remark 4.2.2

(i) Since $H^s(\mathbb{R}^N)$ and $H^{s-1/2}(\mathbb{R}^{N-1})$ are defined to be the closures of $C_c^\infty(\mathbb{R}^N)$ and $C_c^\infty(\mathbb{R}^{N-1})$ respectively, we may use the theorem to conclude the following: If T is any $(N-1)$- dimensional affine subspace of $\mathbb{R}^N$, then a function $f \in H^s(\mathbb{R}^N)$ has a well-defined trace in $H^{s-1/2}$ on T, provided that $s > 1/2$. Conversely, we shall see that if $g \in H^{s-1/2}(T), s > 1/2$, then there is a function $\tilde{g} \in H^s(\mathbb{R}^N)$ such that $\tilde{g}$ has trace g on T.

(ii) It is a bit awkward to state the theorem as we have (that is, as an *a priori* estimate on C_c^∞ functions). As an exercise, the reader should attempt to reformulate the theorem directly in terms of the H^s spaces to see that in fact the statement of Theorem 4.2.1 is as simple as it can be made.

Proof of Theorem 4.2.1: We introduce the notation

$$(x', x_N) = (x_1, \ldots, x_N)$$

for an element of $\mathbb{R}^N$. If $u \in C_c^\infty(\mathbb{R}^N)$, then we will use the notation $u^r(x')$ to denote $u(x', 0)$. Now we have

$$\|u^r\|^2_{H^{s-1/2}(\mathbb{R}^{N-1})} = \int_{\mathbb{R}^{N-1}} |\widehat{u^r}(\xi')|^2 (1+|\xi'|^2)^{s-1/2} \, d\xi'$$

$$= \int_{\mathbb{R}^{N-1}} \left[-\int_0^\infty \frac{\partial}{\partial x_N} |\widehat{u_1}(\xi', x_N)|^2 (1+|\xi'|^2)^{s-1/2} \, dx_N \right] d\xi'. \quad (4.7)$$

Here $\widehat{u}_1$ denotes the partial Fourier transform in the variable x'. The product rule yields that, if D is a first derivative, then $D(|h|^2) \leq 2|h| \cdot |Dh|$. Therefore (4.7) does not exceed

$$2 \int_{\mathbb{R}^{N-1}} \int_0^\infty |\widehat{u_1}(\xi', x_N)| \cdot \left| \frac{\partial}{\partial x_N} \widehat{u_1}(\xi', x_N) \right| \cdot (1+|\xi'|^2)^{s-1/2} \, dx_N \, d\xi'.$$

Since $2\alpha\beta \leq \alpha^2 + \beta^2$, the last line is

$$\leq \int_{\mathbb{R}^{N-1}} \int_0^\infty \left| \frac{\partial}{\partial x_N} \widehat{u_1}(\xi', x_N) \right|^2 \, dx_N (1+|\xi'|^2)^{s-1} \, d\xi'$$

$$+ \int_{\mathbb{R}^{N-1}} \int_0^\infty |\widehat{u_1}(\xi', x_N)|^2 \, dx_N (1+|\xi'|^2)^s \, d\xi'$$

$$\equiv I + II.$$

Now apply Plancherel's theorem to term II in the x_N variable. The result is

$$II \leq C \int_{\mathbb{R}^{N-1}} \int_\mathbb{R} |\hat{u}(\xi', \xi_N)|^2 \, d\xi_N (1+|\xi'|^2)^s \, d\xi'$$

$$\leq C\|u\|^2_{H^s(\mathbb{R}^N)}.$$

Plancherel's theorem, applied in the x_N variable to the term I, yields

$$\begin{aligned} I &\leq C \int_{\mathbb{R}^{N-1}} \int_{\mathbb{R}} \left| \frac{\partial}{\partial x_N} \widehat{u_1}(\xi', x_N) \right|^2 dx_N (1+|\xi'|^2)^{s-1} d\xi' \\ &\leq C \int_{\mathbb{R}^{N-1}} \int_{\mathbb{R}} |\hat{u}(\xi', \xi_N)|^2 \cdot |\xi_N|^2 d\xi_N (1+|\xi'|^2)^{s-1} d\xi' \\ &\leq C \int_{\mathbb{R}^N} |\hat{u}(\xi)|^2 \cdot (1+|\xi|^2)^s d\xi \\ &= C \|u\|_{H^s(\mathbb{R}^N)}. \quad \blacksquare \end{aligned}$$

As an immediate corollary we have:

Corollary 4.2.3 *Let $u \in H^s(\mathbb{R}^N)$ and let α be a multi-index such that $s > |\alpha| + 1/2$. Then $D^\alpha u$ has trace in $H^{s-|\alpha|-1/2}(\mathbb{R}^{N-1})$.*

The reader will note that, in the trace theorem, it is only necessary to have $u \in H^{s+1/2}(\mathbb{R}^N_+)$ in order to obtain a well-defined restriction to $\mathbb{R}^{N-1}$. In other words, the function to be restricted need only be defined on *one* side of the boundary.

Now we present a converse to the Sobolev trace theorem. This converse is the extension theorem for Sobolev spaces. We continue to use the notation $S = \{(x', 0) \in \mathbb{R}^N\}$. In this theorem it is convenient to let D_N denote $i(\partial/\partial x_N)$.

Theorem 4.2.4 *Let k be a non-negative integer, and suppose $s > 1/2 + k$. If $\phi_0, \ldots, \phi_k$ are defined on S and satisfy $\phi_j \in H^{s-j-1/2}(\mathbb{R}^{N-1})$, then there exists a function $f \in H^s(\mathbb{R}^N)$ such that $D_N^j f$ has trace ϕ_j on S, for $j = 0, \ldots, k$. Moreover*

$$\|f\|_{H^s(\mathbb{R}^N)} \leq C_{s,k} \sum_{j=0}^k \|\phi_j\|^2_{H^{s-j-1/2}(\mathbb{R}^{N-1})}.$$

Proof: Let $h \in C_c^\infty(\mathbb{R}), h \equiv 1$ in a neighborhood of the origin, $0 \leq h \leq 1$. We define

$$\widehat{u_1}(\xi', x_N) = \sum_{j=0}^k \frac{1}{j!} (-ix_N)^j h\left(x_N (1+|\xi'|^2)^{1/2}\right) \widehat{\phi_j}(\xi')$$

and

$$f(x) = \int e^{-2\pi i x' \cdot \xi'} \widehat{u_1}(\xi', x_N) d\xi'.$$

This is the function f that we seek. For if m is any integer, $0 \leq m \leq k$, then

$$D_N^m f(x', 0) = \int e^{-2\pi i x' \cdot \xi'} D_N^m \widehat{u_1}(x', x_N)\big|_{x_N = 0} d\xi'$$

4.2. SOBOLEV SPACE RESTRICTION AND TRACE THEOREMS

$$
\begin{aligned}
&= \int e^{-2\pi i x' \cdot \xi'} i^m \sum_{j=0}^{k} \frac{1}{j!} \sum_{\ell=0}^{m} \binom{m}{\ell} \frac{\partial^\ell}{\partial x_N^\ell} (-ix_N)^j \bigg|_{x_N=0} \\
&\quad \times \frac{\partial^{m-\ell}}{\partial x_N^{m-\ell}} [h(x_N(1+|\xi'|^2)^{1/2})]\bigg|_{x_N=0} \cdot \widehat{\phi_j}(\xi')\, d\xi' \\
&= \int e^{-2\pi i x' \cdot \xi'} i^m \sum_{j=0}^{k} \frac{1}{j!} \binom{m}{j} j!(-i)^j \\
&\quad \frac{\partial^{m-j}}{\partial x_N^{m-j}} [h(x_N(1+|\xi'|^2)^{1/2})]\bigg|_{x_N=0} \cdot \widehat{\phi_j}(\xi')\, d\xi' \\
&= \int e^{-2\pi i x' \cdot \xi'} \widehat{\phi_m}(\xi')\, d\xi' \\
&= \hat{\phi}_m(x'),
\end{aligned}
$$

verifying our first assertion.

Now we need to check that f has the right Sobolev norm. We have

$$
\begin{aligned}
\|f\|_{H^2}^2 &= \int_{\mathbb{R}^N} |\hat{f}(\xi)|^2 (1+|\xi|^2)^s\, d\xi \\
&= \int_{\mathbb{R}^N} \left|\widehat{(u_1(\xi',\cdot))}_2 (\xi_N)\right|^2 (1+|\xi|^2)^s\, d\xi \\
&= \frac{1}{4\pi^2} \int_{\mathbb{R}^N} \left|\sum_{j=0}^{k} \frac{1}{j!} \frac{\partial^j}{\partial \xi_N^j} \hat{h}\left(\frac{\xi_N}{(1+|\xi'|^2)^{1/2}}\right) \frac{1}{(1+|\xi'|^2)^{1/2}}\right|^2 \\
&\quad \times |\widehat{\phi_j}(\xi')|^2 (1+|\xi|^2)^s\, d\xi. \qquad (4.8)
\end{aligned}
$$

Using the fact that $|\sum_{j=0}^{k} \zeta_j|^2 \leq C \cdot \left(\sum_{j=0}^{k} |\zeta_j|^2\right)$, where the constant C depends only on k, we see that (4.8) does not exceed

$$
\begin{aligned}
&C \int_{\mathbb{R}^N} \sum_{j=0}^{k} \frac{1}{(j!)^2} \left|\frac{\partial^j}{\partial \xi_N^j} \hat{h}\left(\frac{\xi_N}{(1+|\xi'|^2)^{1/2}}\right)\right|^2 \\
&\quad \times \frac{1}{1+|\xi'|^2} \cdot |\widehat{\phi_j}(\xi)|^2 (1+|\xi|^2)^s\, d\xi \\
&\leq C \int_{\mathbb{R}^N} \sum_{j=0}^{k} \frac{1}{(j!)^2} \left|\hat{h}^{(j)}\left(\frac{\xi_N}{(1+|\xi'|^2)^{1/2}}\right)\right|^2 \\
&\quad \times \frac{1}{(1+|\xi'|^2)^{j+1}} \cdot |\widehat{\phi_j}(\xi)|^2 (1+|\xi|^2)^s\, d\xi. \qquad (4.9)
\end{aligned}
$$

Now write $d\xi = d\xi' d\xi_N$. We make the change of variables

$$
\xi_N \mapsto \xi_N (1+|\xi'|^2)^{1/2}.
$$

Noting that

$$1 + |\xi|^2 = 1 + |\xi'|^2 + |\xi_N|^2 \mapsto 1 + |\xi'|^2 + |\xi_N|^2(1+|\xi'|^2)$$
$$= (1+|\xi'|^2)(1+|\xi_N|^2),$$

we see that (4.9) does not exceed

$$C \sum_{j=0}^{k} \frac{1}{(j!)^2} \int_{\mathbb{R}^{N-1}} \int_{\mathbb{R}} |\hat{h}^{(j)}(\xi_N)|^2 \frac{1}{(1+|\xi'|^2)^{j+1}} (1+|\xi'|^2)^{1/2}$$
$$\times |\widehat{\phi_j}(\xi')|^2 \left[(1+|\xi'|^2)(1+|\xi_N|^2)\right]^s d\xi' d\xi_N$$
$$= C \sum_{j=0}^{k} \frac{1}{(j!)^2} \int_{\mathbb{R}^{N-1}} |\hat{\phi}_j(\xi')|^2 (1+|\xi'|^2)^{s-j-1/2} d\xi'$$
$$\times \int_{\mathbb{R}} \left|\hat{h}^{(j)}(\xi_N)\right|^2 (1+|\xi_N|^2)^s d\xi_N$$
$$\leq C \sum_{j=0}^{k} \frac{1}{(j!)^2} \|\phi_j\|_{H^{s-j-1/2}(\mathbb{R}^{N-1})} \cdot \|h\|^2_{H^{s+k}(\mathbb{R})}$$
$$\leq C_{s,k} \sum_{j=0}^{k} \|\phi_j\|^2_{H^{s-j-1/2}(\mathbb{R}^{N-1})},$$

completing the proof of our extension theorem. ∎

Remark 4.2.5

(i) The extension $\widehat{u_1}$ was constructed by a scheme based on ideas that go back at least to A. P. Calderón—see Stein [1] and references therein.

(ii) Consider a smoothly bounded domain $\Omega \subset \mathbb{R}^N$. Let $\{U_j\}_{j=1}^k$ be a covering of $\partial\Omega$ with open subsets of $\mathbb{R}^N$. We may choose these open sets in such a way that there are diffeomorphisms $\Phi_j : U_j \to W_j \subset \mathbb{R}^N$ such that $\Phi_j(\partial\Omega) = W_j \cap \{x \in \mathbb{R}^N : x_N = 0\}$. This claim follows from the implicit function theorem. We may assume that there is a universal finite constant C such that $\|\nabla \Phi_j\|_{\sup} \leq C$ and $\|\nabla[\Phi_j]^{-1}\|_{\sup} \leq C$ for each j. Let ψ_j be a partition of unity subordinate to the covering $\{U_j\}$ of $\partial\Omega$ such that $\sum_j \psi_j \equiv 1$ in a neighborhood of $\partial\Omega$ in $\mathbb{R}^N$.

Suppose that $f \in H^s(\Omega)$. Fix an index j, $1 \leq j \leq k$. We write $\Phi_j(\Omega \cap U_j) \equiv V_j$. Then $g_j \equiv \psi_j \cdot f \circ \left(\Phi_j^{-1}\big|_{V_j}\right)$ lies in $H^s(V_j)$. According to our trace theorem for Sobolev functions on the upper half space $\mathbb{R}^N_+$, the function g_j has a trace on $\Phi_j(\partial\Omega)$. Unraveling all the notation, we find that $\psi_j \cdot f$ has a trace on $\partial\Omega$. Summing over j, we conclude that f itself has a trace on $\partial\Omega$. We shall not examine minimal hypotheses on the smoothness of $\partial \Omega$ that will guarantee the validity of the trace theorem for elements of H^s, $s > 1/2$. Clearly it suffices to assume that the boundary is C^k with $k > s$.

Remarks similar to those in the preceding two paragraphs apply to show that the extension theorem has a valid formulation for Sobolev functions on the boundary of a domain with sufficiently smooth boundary. We shall say no more about the matter here.

4.3 Domain Extension Theorems for Sobolev Spaces

Next we turn our attention to domain extension theorems for Sobolev functions. Such theorems are related to the converse of the trace theorem, but they have a slightly different formulation. We will be concerned here with extending a Sobolev function from a given domain in $\mathbb{R}^N$ to all of space.

It is easy to see that a successful extension theorem will necessitate *some* regularity condition on the boundary of the domain in question. For let us consider the situation in $\mathbb{R}^2$. Fix an integer $k > 0$. Let

$$\Omega_k = \{(x_1, x_2) \in \mathbb{R}^2 : 0 \leq x_1 \leq 1, 0 \leq x_2 \leq x_1{}^k\}.$$

Define $f(x_1, x_2) = 1/x_1$. If $k \geq 2$ then $f \in L^2(\Omega_k)$. If $k \geq 4$ then $f \in W^1(\Omega_k)$. More generally, if $k \geq 2s + 2$ then $f \in W^s(\Omega_k)$. But for $s > 1$ we cannot expect f to extend to a function in $W^s(\mathbb{R}^2)$. If it did, then the extended function would be (after correction on a set of measure zero) continuous—by the Sobolev theorem. Since f is unbounded at $(0, 0)$, that conclusion is clearly impossible.

Very sharp conditions on the boundary of a domain in Euclidean space, in order to guarantee a valid extension theorem, are known (see Jones [1]). We shall not attempt such a precise discussion, but shall instead treat some of the standard results.

Definition 4.3.1

(i) *Let $\Omega \subset \mathbb{R}^N$ be a domain. A linear operator*

$$E : W^m(\Omega) \to W^m(\mathbb{R}^N)$$

is said to be a **simple m-extension operator** *for Ω if*

1. *$Eu(x) = u(x)$ for almost every point $x \in \Omega$;*
2. *$\|Eu\|_{W^m(\mathbb{R}^N)} \leq K \cdot \|u\|_{W^m(\Omega)}$.*

(ii) With Ω as above, we say that

$$E : \{\text{functions on } \Omega\} \to \{\text{functions on } \mathbb{R}^N\}$$

is a **total extension operator** *if the restriction of E to $W^m(\Omega)$ is a simple m-extension operator for each non-negative integer m.*

We shall present three extension theorems for Sobolev spaces (and this three part unity is quite typical of extension questions for other function spaces too). In our discussion, we shall restrict attention to bounded domains in Euclidean space. This choice makes the statements of the theorems a bit cleaner. A treatment of unbounded domains may be found in Adams [1] and in Stein [1]. It is also possible to develop Sobolev space theory for domains in a manifold—see Hörmander [1].

Theorem 4.3.2 *If $\Omega \subset \mathbb{R}^N$ is a bounded domain with C^m boundary, then there exists a simple m-extension operator for each non-negative integer m.*

Proof: First let us assume that Ω is the upper half space

$$\mathbb{R}^N_+ = \{(x_1, x_2, \ldots, x_N) \in \mathbb{R}^N : x_N > 0\}.$$

(Of course this domain is not bounded, but that will be seen not to matter in what follows.) It is convenient to denote a point of $\mathbb{R}^N_+$ by (x', x_N), where x' is a shorthand for $(x_1, \ldots, x_{N-1})$.

Set

$$Eu(x) = \begin{cases} u(x) & \text{if } x_N > 0 \\ \sum_{j=1}^{m+1} \lambda_j u(x', -jx_N) & \text{if } x_N \leq 0, \end{cases}$$

where $\lambda_1, \ldots, \lambda_{m+1}$ are the unique solutions to the $m+1$ equations in $m+1$ unknowns given by

$$\sum_{j=1}^{m+1} (-j)^k \lambda_j = 1 \quad , \quad k = 0, 1, \ldots, m.$$

Also define the auxiliary operators

$$E_\alpha u(x) = \begin{cases} u(x) & \text{if } x_N > 0 \\ \sum_{j=1}^{m+1} (-j)^{\alpha_N} \lambda_j u(x', -jx_N) & \text{if } x_N \leq 0, \end{cases}$$

for $\alpha = (\alpha_1, \ldots, \alpha_N)$ any multi-index with $|\alpha| \leq m$.

If $u \in C^m(\overline{\mathbb{R}^N_+})$ then it is straightforward to calculate that

$$Eu \in C^m(\mathbb{R}^N),$$
$$D^\alpha E(u) = E_\alpha D^\alpha u(x) \quad , \quad |\alpha| \leq m,$$

4.3. EXTENSION THEOREMS FOR SOBOLEV SPACES

$$\int_{\mathbb{R}^N} |D^\alpha Eu(x)|^2 \, dx = \int_{\mathbb{R}^N_+} |D^\alpha u(x)|^2 \, dx$$

$$+ \int_{\mathbb{R}^N_-} \left| \sum_{j=1}^{m+1} (-j)^{\alpha_N} \lambda_j D^\alpha u(x', -jx_N) \right|^2 dx$$

$$\leq K \cdot \int_{\mathbb{R}^N_+} |D^\alpha u|^2 \, dx.$$

By the previously noted density result Remaark 4.1.5(ii)), i.e., $H^m = W^m$, these inequalities extend to all of W^m. This completes the proof of the theorem when the domain is the half space.

If now Ω is a bounded domain with C^m boundary, then we follow a by now familiar procedure of covering $\partial\Omega$ with finitely many open sets U_j, each diffeomorphic to a ball, and each equipped with a mapping $\Phi_j : U_j \to W_j \subset \mathbb{R}^N$ such that $\Phi(U_j \cap \partial\Omega) = W_j \cap \{x \in \mathbb{R}^N : x_N = 0\}$ and $\Phi_j(U_j \cap \Omega) = W_j \cap \mathbb{R}^N_+$. This, in effect, reduces the problem on Ω to that already treated on the upper half space. To wit, if $f \in W^m(\Omega)$ then one extends each of the functions $f \circ \Phi_j^{-1}$ from $W_j \cap \mathbb{R}^N_+$ to W^j, then pulls the extension back via Φ_j and patches all these together. Details are left to the reader. ∎

A refinement of the proof of the last theorem yields the following stronger result. We omit the proof, and refer the reader to Adams [1] for details.

Theorem 4.3.3 *If $\Omega \subset \mathbb{R}^N$ is a bounded domain with C^m boundary, then there is a total extension operator for Ω.*

Using deeper techniques, related to the Calderón-Zygmund singular integral theory (for which see Stein [1]), it is actually possible to prove the following still stronger result. Again we omit the proof. Proofs may be found in Adams [1] and Stein [1]. As previously noted, an even sharper result appears in Jones [1]. The result of Jones is precisely sharp in dimension two.

Theorem 4.3.4 *If $\Omega \subset \mathbb{R}^N$ is a bounded domain with Lipschitz boundary, then there is a total extension operator for Ω.*

Chapter 5

Smooth Mappings

5.1 Sard's Theorem

Sard's Theorem is an important tool in differential topology and in Morse theory. We shall begin our discussion of Sard's theorem by treating the most elementary form of the result, and then shall later develop generalizations and variants of the theorem. It is worth noting that the basic idea here was discovered by A. B. Brown in 1935 (see Brown [1]). Later, in 1939 and 1942, the result was rediscovered by A. P. Morse (see Morse [1]) and A. Sard (see Sard [1]). The method of proof of Sard's theorem is quite robust and can be modified to yield a number of interesting results. We shall close the section with a discussion of one of these variants that is known as "hard Sard."

Definition 5.1.1 *Let $U \subset \mathbb{R}^N, V \subset \mathbb{R}^M$ be open sets and let $f : U \to V$ be a C^1 function. We say that $y \in V$ is a **critical value** of f if there is an $x \in U$ with $f(x) = y$ and $\nabla f(x)$ has rank less than M. The point x is of course called a **critical point**.*

We begin with the following special case of the theorem of Brown, Morse, and Sard:

Theorem 5.1.2 *Let $U \subset \mathbb{R}^N, V \subset \mathbb{R}^M$ be open sets and let $f : U \to V$ be a C^2 function. Assume that $0 < N < 2M$. Let $\mathcal{C}$ denote those points in U where $\nabla f = 0$. Then the image of $\mathcal{C}$ under f has Lebesgue volume measure equal to zero.*

Proof: For each N-tuple of integers $k = (k_1, \ldots, k_N)$ let

$$Q_k = \{x = (x_1, \ldots, x_N) \in \mathbb{R}^N : k_j \leq x_j \leq k_j + 1, j = 1, \ldots, N\}.$$

Then $\mathbb{R}^N = \cup_k Q_k$ and the Q_k have disjoint interiors. Set

$$S_k = \{y \in V : y \in f(\mathcal{C}) \text{ and } \exists x \in Q_k \text{ such that } f(x) = y\}.$$

Then $\cup_k S_k$ equals $f(\mathcal{C})$, and it is enough for us to show that each S_k has measure zero. Given this reduction, we may as well suppose that f has as its domain the cube $Q = \{x \in \mathbb{R}^N : 0 \leq x_j \leq 1, j = 1, \ldots, N\}$.

Now the function ∇f is uniformly continuous on Q. Better yet, $\nabla^2 f$ is continuous on Q, and hence is bounded. It follows from the mean value theorem that ∇f itself is Lipschitz. Let K be the Lipschitz constant:

$$|\nabla f(x) - \nabla f(x+h)| \leq K|h|. \tag{5.1}$$

Let $\epsilon > 0$. Choose a $\delta > 0$ such that if $x, t \in Q$ and $|x - t| < N\delta$, then $|\nabla f(x) - \nabla f(t)| < \epsilon$. In fact $\delta = \epsilon/(KN)$ will do. We may certainly assume that $\delta < \epsilon$.

Select an integer p, $(1/\delta)+1 \geq p > 1/\delta$. Note that $1/\delta \sim (KN)/\epsilon$. Break Q up into p^N closed subcubes, with side length equal to $1/p$, sides parallel to the axes, and disjoint interiors.

Consider any one of the p^N subcubes, call it P, that contains a critical point. The diameter of this cube is less than $\sqrt{N}\delta$. And there is a point in the cube where ∇f vanishes. It follows that $|\nabla f| < \epsilon$ on the entire cube.

We conclude, from the mean value theorem applied to f restricted to one dimensional segments in P (for instance), that the *image* of the cube P under f has diameter at most $\epsilon \cdot \sqrt{N}\delta < \epsilon \cdot \sqrt{N}/p$. Thus it has volume at most

$$\left[2\sqrt{N}\epsilon/p\right]^M. \tag{5.2}$$

Summing over all subcubes P that contain critical points (there are at most p^N of these) we find that the set $f(\mathcal{C})$ has volume at most

$$p^N \cdot \left[2\sqrt{N}\epsilon/p\right]^M = \left(2\sqrt{N}\epsilon\right)^M \cdot p^{N-M}. \tag{5.3}$$

But now recall that $p \sim KN/\epsilon$. Thus (5.3) is essentially

$$\left(2\sqrt{N}\epsilon\right)^M (KN)^{N-M} \epsilon^{M-N} \leq \left(2\sqrt{N}\right)^M (KN)^{N-M} \epsilon^{2M-N}.$$

Since $\epsilon > 0$ was arbitrary, and since $2M - N > 0$, we conclude that the set $f(\mathcal{C})$ has measure zero. ∎

In fact Sard's theorem is true with no restrictions on the dimension of the domain and of the range, and it is true for the set of *all* critical values—not just the image of the points where the gradient vanishes. However a certain smoothness of the function f is necessary; examples that demonstrate the necessity of the smoothness requirements may be found in Federer [4], p. 317 *ff*. The complete statement of the classical Sard theorem is this:

Theorem 5.1.3 *Let $U \subset \mathbb{R}^N, V \subset \mathbb{R}^M$ be open sets and let $f : U \to V$ be a C^k function, where $k \geq N/M$. Then the set of critical values of f has Lebesgue M-dimensional volume measure equal to zero.*

5.1. SARD'S THEOREM

Proof: In this proof we shall assume that the map is C^∞, and we shall not give the more delicate proof that is valid for C^k functions, $k \geq N/M$. We refer the reader to Federer [4], p. 316 for a full account of these matters. The proof that we present comes from Milnor [1] and Sternberg [1].

We proceed by induction on N. Notice that the statement of the theorem makes sense, and is obviously true, for $N = 0$. That is the starting point of our induction. Assume now that the result has been proved for dimension $N - 1$.

Let C be the set of points where the rank of ∇f is less than M. Let $C_1 \subset C$ be the set of points where $\nabla f = 0$ (this set was denoted by $\mathcal{C}$ in Theorem 5.1.2). More generally, let C_i denote the set of points where $\nabla f = 0, \nabla^2 f = 0, \ldots \nabla^i f = 0$. Of course each C_i is closed and we have

$$C \supset C_1 \supset C_2 \supset C_3 \supset \cdots.$$

We shall prove the following three assertions:

1. The image of $C \setminus C_1$ under f has measure zero.

2. The image of $C_i \setminus C_{i+1}$ under f has measure zero, $i = 1, 2, \ldots$.

3. The image of C_j under f has measure zero if j is sufficiently large.

Step 1: If $M = 1$ then $C = C_1$ and there is nothing to prove. So assume that $M \geq 2$. Choose $t \in C \setminus C_1$. By definition, there is some partial derivative $\partial f_\ell / \partial x_j(t)$ that does not vanish. After juggling indices, we may assume that $\partial f_1 / \partial x_1(t) \neq 0$. Define

$$h(x) = (f_1(x), x_2, \ldots, x_N).$$

Then the inverse function theorem implies that there is a neighborhood V of t on which h is a diffeomorphism onto an open set $V' \subset \mathbb{R}^N$.

Now the composition $g = f \circ h^{-1}$ will map V' into $\mathbb{R}^M$, and the set C' of critical points of g is nothing other than $h(V \cap C)$. In other words, the set $g(C')$ of critical values of g is equal to $f(V \cap C)$. So it is enough for us to determine the measure of $g(C')$.

For a point of the form $(s, x_2, \ldots, x_N) \in V'$, observe that the image $g(s, x_2, \ldots, x_M)$ belongs to the hyperplane $\{s\} \times \mathbb{R}^{M-1}$. Therefore g maps hyperplanes into hyperplanes. Now define

$$g^s : \left[\{s\} \times \mathbb{R}^{N-1}\right] \cap V' \to \{s\} \times \mathbb{R}^{M-1}$$

to be the restriction of g to the hyperplane $\{s\} \times \mathbb{R}^{N-1}$. Notice that the Jacobian matrix of g has the form

$$\begin{pmatrix} 1 & 0 \\ * & (\partial g_i^s / \partial x_\ell) \end{pmatrix}.$$

Thus a point of $\{s\} \times \mathbb{R}^{N-1}$ is critical for g^s if and only if it is critical for the map g itself.

By the inductive hypothesis, the set of critical values of g^s has $M-1$ dimensional measure zero. By Fubini's theorem (taking the union over s), it follows that the set of critical values of g itself has M dimensional measure zero.

Step 2: Fix j, and consider a point $t \in C_j \setminus C_{j+1}$. By definition, some $(j+1)^{\text{st}}$ derivative of some component of f is not zero at t. Thus there is some k^{th} order partial derivative D, and index ℓ, and another index m, such that $w(t) \equiv Df_\ell(t) = 0$ but $\partial/\partial x_m Df_\ell(t) \neq 0$. We may take $m=1$.

Then the map $h : U \to \mathbb{R}^N$ defined by

$$h(x) = (w(x), x_2, \ldots, x_N)$$

is a diffeomorphism of some neighborhood V of t onto an open set $V' \subset \mathbb{R}^N$. The map h maps $C_j \cap V$ into the hyperplane $\{0\} \times \mathbb{R}^{N-1}$. Proceeding as in Step 1, we consider

$$g \equiv f \circ h^{-1} : V' \to \mathbb{R}^M.$$

We let

$$g^* : \left[\{0\} \times \mathbb{R}^{N-1}\right] \cap V' \to \mathbb{R}^M$$

denote the restriction of g to the hyperplane $\{0\} \times \mathbb{R}^{N-1}$.

By induction, the set of critical values of g^* has $(M-1)$ dimensional measure zero. But each point of $h(C_j \cap V)$ is a critical point of g^* (because all derivatives of order not exceeding j vanish by definition). Thus

$$[g^* \circ h](C_j \cap V) = f(C_j \cap V)$$

has measure zero. Since we may cover $C_j \setminus C_{j+1}$ by countably many such sets V, we may conclude that $f(C_j \setminus C_{j+1})$ has measure zero.

Step 3: Refer to the proof of Theorem 5.1.2 above. For a point $x \in C_j$ we may replace line (5.1) in that proof by

$$|f(x+h) - f(x)| \leq K'|h|^{j-1}.$$

Thus the estimate for the volume of the image of a cube in line (5.2) becomes

$$\left[2\sqrt{N}\delta^{j-1}/p\right]^M.$$

Since $\delta < \epsilon$, in the final calculation the term ϵ^{2M-N} is replaced by ϵ^{jM-N}. If $j > N/M$ then the rest of the proof goes through. ∎

5.1. SARD'S THEOREM

Implicit in the proof that we have just presented of Sard's theorem is the following remarkable lemma of A. P. Morse. We record it here for interest's sake, and invite the reader to give a proof (details may also be found in Sternberg [1], p. 49).

Lemma 5.1.4 (Morse) *Let $A \subset \mathbb{R}^N$ and let $q \geq 0$ be an integer. Then we may decompose A as $A = \cup_{i=0}^{\infty} A_i$ such that: Let U be a neighborhood of A and $f : U \to \mathbb{R}^M$ a C^q mapping. Suppose that every point of A is a critical point of f. Then there are functions $e_i(\epsilon)$ (these functions depend on f) such that each e_i satisfies $e_i(\epsilon) \to 0$ monotonically as ϵ decreases to 0 and*
$$|f(x) - f(y)| \leq e_i(\|x - y\|) \cdot \|x - y\|^k$$
for any $x, y \in A_i$.

Remark 5.1.5 We note explicitly that the sets A_i in the decomposition of A in Lemma 5.1.4 are independent of the choice of the function f.

Here is a typical application of Sard's theorem.

Example 5.1.6 Let $f : \mathbb{R}^N \to \mathbb{R}$ be a C^k function, $k \geq N$, that is **proper**, where "proper" means that the inverse image of any compact set is compact. Then for almost every $c \in \mathbb{R}$ the set
$$S_c \equiv \{x \in \mathbb{R}^N : f(x) = c\}$$
is either empty or is a C^k compact manifold.

To see this, notice that Sard's theorem guarantees that the set $\mathcal{C}$ of critical values for f has measure zero. If c is in the complement of $\mathcal{C}$, and is in the image of f, then every point of $f^{-1}(c)$ is *not* a critical point. Thus $\nabla f \neq 0$ in a neighborhood of each such point, and therefore $f^{-1}(c)$ is a C^k hypersurface in a neighborhood of each such point.

We have established that if $c \notin \mathcal{C}$ and if $f^{-1}(c)$ is nonempty, then it is a C^k manifold. The properness of f guarantees that it is compact.

Now we give a formulation of a version of "hard Sard." We shall only sketch the proof, and refer the reader to Hirsch [1] or Federer [4] for a more detailed treatment.

Theorem 5.1.7 *Let $U \subset \mathbb{R}^N, V \subset \mathbb{R}^M$ be open sets and let $f : U \to V$ be a C^k function. Let us say that a critical value w is of* **rank** *ℓ if $w = f(x)$ for at least one point x at which the Jacobian matrix has rank not exceeding ℓ. Denote by $\mathcal{C}_\ell$ the set of points in V that are critical values of rank ℓ. Then*
$$\mathcal{H}^{\ell + (N-\ell)/k}(\mathcal{C}_\ell) = 0.$$

In other words, the set of critical values of rank ℓ has Hausdorff dimension not exceeding $\ell + (N - \ell)/k$.

The spirit of the proof is that, near a point at which the Jacobian of f has rank ℓ, the mapping will collapse a small N-dimensional ball in U to an essentially ℓ- dimensional ball in the image. Since the latter balls cover the set $\mathcal{C}_\ell$, it is clear that this puts a bound on the Hausdorff dimension of $\mathcal{C}_\ell$.

5.2 Extension Theorems

We have already (in Section 1.1) defined what it means for a function $f : U \to \mathbb{R}^M$ to be C^k when U is an open subset of $\mathbb{R}^N$. A difficult question arises when f is not *a priori* defined on an open set. What should it mean in that case for f to be C^k? An easy definition, which is very often used, is to require the existence of an extension of f to a larger open domain that contains the original domain and to demand that the extension be C^k on that larger domain. While this definition is expedient, it really begs the question. One should be able to determine from information about f on its given domain whether or not f is in a particular smoothness class. It should not be necessary to appeal to an extension "oracle" to learn the answer. The replacement for an extension "oracle" is an extension theorem.

The following classical extension theorem is well-known and can be found in many references (*e.g.* see Dugundji [1]).

Theorem 5.2.1 (Tietze's Extension Theorem) *Let S be a closed subset of $\mathbb{R}^N$. A function $f : S \to \mathbb{R}^M$ is continuous if and only if f has a continuous extension to all of $\mathbb{R}^N$.*

Tietze's Extension Theorem is a model for what we would like to accomplish: A characterization of the functions that are restrictions of those in a certain category.

The preceding extension theorem is more topological than analytical, requiring little from the geometry of $\mathbb{R}^N$. To move into the arena of analysis we deal first with Lipschitz functions. Recall from Definition 3.1.15 that $f : X \to Y$, with X and Y subsets of Euclidean spaces, is a Lipschitz function if there exists $C < \infty$ such that

$$|f(x_1) - f(x_2)| \leq C|x_1 - x_2| \quad \text{holds for all} \quad x_1, x_2 \in X, \tag{5.4}$$

and the smallest such C is called the Lipschitz constant of f. The following theorem is known as Kirzbraun's Extension Theorem.

Theorem 5.2.2 (Kirzbraun [1])
Suppose $S \subset \mathbb{R}^N$. A function $f : S \to \mathbb{R}^M$ is Lipschitz if and only if f has a Lipschitz extension to all of $\mathbb{R}^N$. The extended function may be taken to have the same Lipschitz constant.

Proof: It is clear that the restriction of a Lipschitz function is Lipschitz. The heart of the matter is proving the existence of an extension. Let us

5.2. EXTENSION THEOREMS

use C to denote the Lipschitz constant for $f : S \to \mathbb{R}^M$. In case $M = 1$, we can simply use the following formula, due to H. Whitney, to set

$$F(x) = \sup\{f(z) - C\operatorname{dist}(x,z) : z \in S\}, \tag{5.5}$$

and easily check that this defines an extension of f with the same Lipschitz constant.

In case $M > 1$, the formula (5.5) can be applied to each component to obtain a Lipschitz function, but the Lipschitz constant is generally larger than C. To obtain a Lipschitz extension having the same Lipschitz constant C requires a more elaborate construction.

Consider the family $\mathcal{F}$ of Lipschitz extensions of f (to some set T with $S \subset T$) which also have the Lipschitz constant C. This family is non-empty because it at least contains the original function f.

We define a partial ordering on $\mathcal{F}$ as follows: Suppose $g_1 : T_1 \to \mathbb{R}^M$ and $g_2 : T_2 \to \mathbb{R}^M$ are both elements of $\mathcal{F}$, then we write $g_1 \preceq g_2$ if and only if g_2 is an extension of g_1, i.e. $S \subset T_1 \subset T_2$ and $g_1(x) = g_2(x)$, for all $x \in T_1$. (The same partial ordering is defined if we recall that a function from a subset of $\mathbb{R}^N$ to $\mathbb{R}^M$ is a set of elements of $\mathbb{R}^N \times \mathbb{R}^M$, and we partially order $\mathcal{F}$ by inclusion.) By Hausdorff's Maximal Principle, $\mathcal{F}$ has a maximal totally ordered sub-family $\widetilde{\mathcal{F}}$. Let $\widetilde{S}$ be the union of the domains of the functions in $\widetilde{\mathcal{F}}$. We define a function $F : \widetilde{S} \to \mathbb{R}^M$ by setting $F(x) = g(x)$ where $g \in \widetilde{\mathcal{F}}$ and $x \in \operatorname{dom}(g)$. One readily sees that F is Lipschitz and that $\operatorname{Lip}(F) = C$.

We claim that $\widetilde{S} = \mathbb{R}^N$. If not, then we can fix $x_0 \in \mathbb{R}^N \setminus \widetilde{S}$. A contradiction will be reached if we can show that there is $y_0 \in \mathbb{R}^M$ such that

$$|y - y_0| \leq C|x - x_0| \quad \text{whenever} \quad y = F(x) \quad \text{and} \quad x \in \widetilde{S},$$

that is, if we can show

$$\emptyset \neq \bigcap_{x \in \widetilde{S}} \mathbf{B}\bigl(F(x), C|x - x_0|\bigr). \tag{5.6}$$

Since (5.6) involves an intersection of compact sets, it suffices to show that any such finite intersection is non-empty. Accordingly, let $x_1, x_2, \ldots, x_n \in \widetilde{S}$ be fixed. Set $y_i = F(x_i)$, $r_i = |x_i - x_0|$, for $i = 1, 2, \ldots, n$, and $r^* = \sup\{r_1, r_2, \ldots, r_n\}$. We know that for any sufficiently large value of γ

$$K_\gamma \equiv \bigcap_{i=1}^n \mathbf{B}(y_i, \gamma r_i) \neq \emptyset.$$

Set

$$\gamma_0 = \inf\{\gamma : K_\gamma \neq \emptyset\}.$$

Since

$$K_{\gamma_0} = \bigcap_{\gamma > \gamma_0} K_\gamma \tag{5.7}$$

and the intersection of any finitely many of the sets on the right-hand side of (5.7) is non-empty, we see that

$$K_{\gamma_0} \neq \emptyset.$$

It will suffice to show $\gamma_0 \leq C$. We may, of course, assume $\gamma_0 > 0$. Note that K_{γ_0} must contain exactly one point, for if $y', y'' \in K_{\gamma_0}$, then we can use the following convexity argument:

$$\begin{aligned}
|(y'+y'')/2 - y_i|^2 &= |y'+y''|^2/4 + |y_i|^2 - (y'+y'') \cdot y_i \\
&= |y'|^2/2 + |y''|^2/2 - |y'-y''|^2/4 + |y_i|^2 \\
&\quad - y' \cdot y_i - y'' \cdot y_i \\
&= \left(|y'-y_i|^2 + |y''-y_i|^2\right)/2 - |y'-y''|^2/4 \\
&\leq \gamma_0^2 r_i^2 - \frac{|y'-y''|^2}{4(r^*)^2} r_i^2,
\end{aligned}$$

holds for $i = 1, 2, \ldots, n$, and $(y'+y'')/2 \in K_\gamma$ with

$$\gamma = \sqrt{\gamma_0^2 - \frac{|y'-y''|^2}{4(r^*)^2}} < \gamma_0,$$

contradicting the definition of γ_0.

By translating the coordinate system if necessary, we may assume $\{0\} = K_{\gamma_0}$. Consequently, we have $|y_i| \leq \gamma_0 r_i$, for $i = 1, 2, \ldots, n$. We now claim that 0 is in the convex hull of $\{y_i : |y_i| = \gamma_0 r_i\}$. Were that not the case, there would exist an $M - 1$ dimensional plane separating the origin from $\{y_i : |y_i| = \gamma_0 r_i\}$, but then for all sufficiently small $\epsilon > 0$ we would have $\mathbb{B}(0, \epsilon)$ on the opposite side of the plane from $\{y_i : |y_i| = \gamma_0 r_i\}$, again contradicting the definition of γ_0. Thus we can choose non-negative $\lambda_1, \lambda_2, \ldots, \lambda_n$ with

$$|y_i| < \gamma_0 r_i \quad \Rightarrow \quad \lambda_i = 0,$$

and

$$1 = \sum_{i=1}^n \lambda_i,$$

$$0 = \sum_{i=1}^n \lambda_i y_i.$$

It follows that

$$0 = 2\left|\sum_{i=1}^n \lambda_i y_i\right|^2$$

5.2. EXTENSION THEOREMS

$$= 2\sum_{i,j=1}^{n} \lambda_i \lambda_j y_i \cdot y_j$$

$$= \sum_{i,j=1}^{n} \lambda_i \lambda_j \left[|y_i|^2 + |y_j|^2 - |y_i - y_j|^2\right]$$

$$\geq \sum_{i,j=1}^{n} \lambda_i \lambda_j \left[\gamma_0^2 r_i^2 + \gamma_0^2 r_j^2 - C^2 |x_i - x_j|^2\right]$$

$$= \sum_{i,j=1}^{n} \lambda_i \lambda_j \left[2(x_i - x_0) \cdot \gamma_0 (x_j - x_0) + (\gamma_0^2 - C^2)|x_i - x_j|^2\right]$$

$$= 2\left|\gamma_0 \sum_{i=1}^{n} \lambda_i (x_i - x_0)\right|^2 + (\gamma_0^2 - C^2) \sum_{i,j=1}^{n} \lambda_i \lambda_j |x_i - x_j|^2. \quad (5.8)$$

If there were but one non-zero λ_i, then the second term in (5.8) would vanish and the first would be positive, a contradiction. Thus there are at least two non-zero λ_i's and the second term in (5.8) must be non-positive, forcing $\gamma_0 \leq C$, as desired. ∎

To discuss the extension of smoother functions, we first need to assess what the characterizing condition might be. To this end we consider Taylor's Theorem. It is elementary to see that, for a real-valued function ϕ of one real variable which is k-times continuously differentiable,

$$\frac{1}{k!} \int_0^1 \left[\phi^{(k)}(0) - \phi^{(k)}(t)\right] d(1-t)^k$$

$$= -\frac{1}{k!}\phi^{(k)}(0) + \frac{1}{(k-1)!} \int_0^1 \left[\phi^{(k-1)}(0) - \phi^{(k-1)}(t)\right] d(1-t)^{k-1}$$

must hold. Induction leads to

$$\frac{1}{k!} \int_0^1 \left[\phi^{(k)}(0) - \phi^{(k)}(t)\right] d(1-t)^k$$

$$= -\sum_{j=2}^{k} \frac{1}{j!}\phi^{(j)}(0) + \int_0^1 [\phi'(0) - \phi'(t)] d(1-t)$$

$$= -\sum_{j=2}^{k} \frac{1}{j!}\phi^{(j)}(0) - \phi'(0) - \phi(0) + \phi(1),$$

which (after rearranging the terms) is the standard Taylor formula with remainder. Now suppose U is a domain in $\mathbb{R}^N$ that contains the points a and x and the line segment connecting them, and suppose $f : U \to \mathbb{R}$ is C^k on an open set containing that line segment. Define the real-valued function of one variable, ϕ, by setting

$$\begin{aligned}\phi(t) &= f((1-t)a + tx) \\ &= f(a_1 + t(x_1 - a_1), \ldots, a_N + t(x_N - a_N)).\end{aligned}$$

It is easy to see that

$$\begin{aligned}\phi^{(j)}(t) &= \sum_{i_1=1}^N \cdots \sum_{i_j=1}^N (x_{i_1} - a_{i_1}) \ldots (x_{i_j} - a_{i_j}) \frac{\partial^j f}{\partial x_{i_1} \ldots \partial x_{i_j}}((1-t)a + tx) \\ &= \sum_{|\alpha|=j} \frac{j!}{\alpha!}(x-a)^\alpha \frac{\partial^{|\alpha|} f}{\partial x^\alpha}((1-t)a + tx),\end{aligned}$$

for $j = 0, 1, \ldots, k$. Thus we obtain

Theorem 5.2.3 (Taylor's Theorem) *If $f : U \to \mathbb{R}$ is C^k on an open set containing the closed line segment joining a and x, then*

$$\begin{aligned}f(x) &= f(a) + \sum_{j=1}^k \sum_{|\alpha|=j} \frac{(x-a)^\alpha}{\alpha!} \frac{\partial^{|\alpha|} f}{\partial x^\alpha}(a) \\ &+ \int_0^1 \sum_{|\alpha|=k} \frac{(x-a)^\alpha}{\alpha!} \left[\frac{\partial^{|\alpha|} f}{\partial x^\alpha}(a) - \frac{\partial^{|\alpha|} f}{\partial x^\alpha}((1-t)a + tx) \right] d(1-t)^k.\end{aligned}$$

Definition 5.2.4 *When*

$$f(a) + \sum_{j=1}^k \sum_{|\alpha|=j} \frac{(x-a)^\alpha}{\alpha!} \frac{\partial^{|\alpha|} f}{\partial x^\alpha}(a) = P_a(x)$$

*is considered as a polynomial in $x_1, \ldots, x_N$ it is called the **Taylor polynomial** for f at a or, synonymously, the **k-jet** of f at a.*

Remark 5.2.5 *It is easy to see that if $j \leq k$ and β is a multi-index with $|\beta| = j$, then*

$$\frac{\partial^\beta P_a}{\partial x^\beta}(x)$$

is the $(k-j)$-jet of

$$\frac{\partial^\beta f}{\partial x^\beta}$$

at a.

From Taylor's theorem follows:

5.3. PROOF OF THE WHITNEY EXTENSION THEOREM

Corollary 5.2.6 *If $f : U \to \mathbb{R}$ is C^k on the open set U, A is a compact subset of U, $P_a(x)$ denotes the k-jet of f at a, and β is a multi-index with $|\beta| \leq k$, then*

$$\lim_{x \to a} \left| \frac{\partial^\beta f}{\partial x^\beta}(x) - \frac{\partial^\beta P_a}{\partial x^\beta}(x) \right| |x - a|^{|\beta|-k} = 0$$

holds uniformly on A.

In the early 1930's H. Whitney proved the converse of this corollary to Taylor's theorem:

Theorem 5.2.7 (Whitney Extension Theorem) *Suppose k is a nonnegative integer, A is a closed subset of $\mathbb{R}^N$, and to each $a \in A$ there corresponds a polynomial in $x_1, \ldots, x_N$ of degree not exceeding k such that for each multi-index β with $|\beta| \leq k$*

$$\lim_{b \to a} \left| \frac{\partial^\beta P_b}{\partial x^\beta}(b) - \frac{\partial^\beta P_a}{\partial x^\beta}(b) \right| |b - a|^{|\beta|-k} = 0 \tag{5.9}$$

holds uniformly on compact subsets of A. Then there exists a C^k map $g : \mathbb{R}^N \to \mathbb{R}$ with

$$\frac{\partial^\alpha g}{\partial x^\alpha}(a) = \frac{\partial^\alpha P_a}{\partial x^\alpha}(a) \tag{5.10}$$

for each multi-index α with $|\alpha| \leq k$ and each $a \in A$.

We will give the proof of the Whitney Extension Theorem in the next section.

5.3 Proof of the Whitney Extension Theorem

The Whitney Extension Theorem 5.2.7 allows one to extend a given function on a closed set A to a C^k function on all of $\mathbb{R}^N$, provided there exist prospective Taylor polynomials for the function that satisfy the *a priori* bounds that Taylor's Theorem requires of a C^k function. The fundamental geometric tool needed for the proof of the Whitney Extension Theorem is the **Whitney decomposition** of $\mathbb{R}^N \setminus A$. A Whitney decomposition of an open set U is a family of closed cubes, with disjoint interiors, the union of which equals U. In addition, each cube in a Whitney decomposition of U must have its diameter bounded above and below by constant multiples of the distance from the cube to the complement of U. Whitney's original construction of what is now called a Whitney decomposition was a paragon of simplicity (see item 8 on page 67 of Whitney [1]), but not *all* the cubes had their diameters bounded from below by a multiple of the distance from the cube to the complement of the open set. Our version of the proof retains the simplicity of Whitney's construction while obtaining the lower bound for all cubes.

Theorem 5.3.1 (Whitney Decomposition) *Let $1 \le \Gamma_1$ and $2\Gamma_1 + 1 \le \Gamma_2$ be fixed. For any non-empty closed subset A of $\mathbb{R}^N$, there exists a family $\mathcal{W}$ of closed cubes with sides parallel to the coordinate axes such that*

(i) $\mathbb{R}^N \setminus A = \cup_{C \in \mathcal{W}} C$,

(ii) $\mathring{C} \cap \mathring{C}' = \emptyset$, *if $C, C' \in \mathcal{W}$ with $C \ne C'$*,

(iii) $\Gamma_1 \cdot \mathrm{diam}(C) \le \mathrm{dist}(C, A) \le \Gamma_2 \cdot \mathrm{diam}(C)$, *for $C \in \mathcal{W}$*.

Proof: Assume without loss of generality that $0 \in A$. Let $\mathcal{R}_0$ denote the family of $3^N - 1$ cubes, all of side length $\frac{2}{3}$, obtained by taking all possible cartesian products of N intervals chosen from the set

$$\left\{ [-1, -\tfrac{1}{3}],\ [-\tfrac{1}{3}, \tfrac{1}{3}],\ [\tfrac{1}{3}, 1] \right\},$$

but omitting the N-fold cartesian product of $[-\tfrac{1}{3}, \tfrac{1}{3}]$ with itself. For k any integer, let $\mathcal{R}_k$ denote the family

$$\{ 3^k C : C \in \mathcal{R}_0 \}$$

of cubes. For $k > 0$, the cubes in $\mathcal{R}_k$ are dilations of the cubes in $\mathcal{R}_0$; for $k < 0$, the cubes in $\mathcal{R}_k$ are contractions of the cubes in $\mathcal{R}_0$. Let $\mathcal{R}$ denote the union of all the families $\mathcal{R}_k$ (see Figure 5.1), so

$$\mathcal{R} = \bigcup_{k \in \mathbb{Z}} \mathcal{R}_k.$$

Note that if C is a cube in $\mathcal{R}_k$, then

$$\begin{aligned}
\mathrm{diam}(C) &= 2\sqrt{N} 3^{k-1}, \\
\mathrm{dist}(C, A) &\le \mathrm{dist}(C, 0) \le \sqrt{N} 3^{k-1},
\end{aligned}$$

thus any cube $C \in \mathcal{R}$ satisfies

$$\mathrm{dist}(C, A) \le \tfrac{1}{2} \mathrm{diam}(C) \le \Gamma_2 \cdot \mathrm{diam}(C). \tag{5.11}$$

Of course, the cubes in $\mathcal{R}$ have pairwise disjoint interiors and sides parallel to the coordinate axes.

Let $\mathcal{W}_0$ denote the family consisting of all cubes C in $\mathcal{R}$ that satisfy

$$\Gamma_1 \cdot \mathrm{diam}(C) \le \mathrm{dist}(C, A), \tag{5.12}$$

in addition to (5.11). Set

$$\mathcal{B}_0 = \mathcal{R} \setminus \mathcal{W}_0.$$

Let m be a non-negative integer. Suppose the family of cubes $\mathcal{B}_m$ has already been formed, and suppose that for each cube $C \in \mathcal{B}_m$ the inequalities (5.11) and

$$\mathrm{dist}(C, A) < \Gamma_1 \cdot \mathrm{diam}(C) \tag{5.13}$$

5.3. PROOF OF THE WHITNEY EXTENSION THEOREM

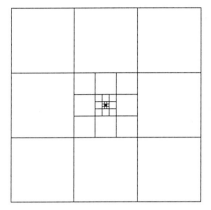

Figure 5.1: **Initial Decomposition**

both hold. This is the case for $\mathcal{B}_0$. Let $\mathcal{B}'_m$ be the set of cubes formed by subdividing each cube of $\mathcal{B}_m$ into 2^N congruent subcubes with pairwise disjoint interiors and sides parallel to the axes. If C' is one of the cubes of $\mathcal{B}'_m$ formed by subdividing the cube C of $\mathcal{B}_m$, we see that

$$\begin{aligned}
\operatorname{diam}(C') &= \tfrac{1}{2}\operatorname{diam}(C), \\
\operatorname{dist}(C', A) &\leq \operatorname{dist}(C, A) + \operatorname{diam}(C') \\
&< \Gamma_1 \cdot \operatorname{diam}(C) + \operatorname{diam}(C') \\
&= 2\Gamma_1 \cdot \operatorname{diam}(C') + \operatorname{diam}(C') \\
&\leq \Gamma_2 \cdot \operatorname{diam}(C').
\end{aligned}$$

So every cube in $\mathcal{B}'_m$ satisfies (5.11). Let $\mathcal{W}_{m+1}$ be the family consisting of all the cubes in $\mathcal{B}'_m$ that satisfy (5.12), and let $\mathcal{B}_{m+1}$ be the family consisting of all the cubes in $\mathcal{B}'_m$ that satisfy (5.13). Thus $\mathcal{B}_{m+1}$ satisfies the same condition assumed for $\mathcal{B}_m$, and the construction can continue inductively.

Set

$$\mathcal{W} = \bigcup_{m=0}^{\infty} \mathcal{W}_m.$$

It is clear from the construction that every cube in $\mathcal{W}$ satisfies the required inequalities, but also $\cup \mathcal{W}$ exhausts $\mathbb{R}^N \setminus A$ because

$$\mathbb{R}^N \setminus A \subset \bigcup \mathcal{R}$$

and $\mathbb{R}^N \setminus A$ is open. ∎

Definition 5.3.2 *For a non-empty closed set $A \subset \mathbb{R}^N$, a collection, $\mathcal{W}$, of closed cubes with sides parallel to the coordinate axes and satisfying (i)–(iii)* of Theorem 5.3.1 will be called a **Whitney decomposition of** $\mathbb{R}^N \setminus A$ or a **Whitney family for** A.

Whitney's definition of the extension of a function from A to $\mathbb{R}^N \setminus A$ is given by a summation over all the cubes of the Whitney decomposition of $\mathbb{R}^N \setminus A$: For each cube the summand is a cut-off function times one of the hypothesized polynomials $P_\xi(x)$, where $\xi \in A$ is a point as near as possible to the cube. The cut-off function associated with a given cube is a translation of one fixed function with support scaled to match the diameter of the cube. To see that the extension formula comes out correctly, we will need to establish some facts about that cut-off function and its derivatives. Such is our next task.

Definition 5.3.3 *For $C \subset \mathbb{R}^N$ a cube, let $\widetilde{C}$ denote the cube with the same center as C, with sides parallel to the sides of C, but with side-length double that of C.*

Lemma 5.3.4 *Let $1 \leq \Gamma_1$ and $2\Gamma_1 + 1 \leq \Gamma_2$ be fixed. There exists a constant $K = K_{\Gamma_1, \Gamma_2}$ such that whenever A is a non-empty closed subset of $\mathbb{R}^N$, $\mathcal{W}$ is a Whitney decomposition of $\mathbb{R}^N \setminus A$ and $p \in \mathbb{R}^N \setminus A$, then*

(i) $C \in \mathcal{W} \Longrightarrow \widetilde{C} \cap A = \emptyset$,

(ii) $\operatorname{card}\{C \in \mathcal{W} : p \in \widetilde{C}\} \leq K$.

Proof: (i) Arguing by contradiction, if we suppose $a \in \widetilde{C} \cap A$, then we can consider the line through a and the center, c, of C and let x be the point of ∂C nearest to a and x' the point of ∂C farthest from a. Then we have

$$\begin{aligned} |a - x| + |x - c| &= |a - c| \\ &\leq 2|x - c| \\ &= |x - x'| \\ &\leq \operatorname{diam}(C), \end{aligned}$$

hence
$$\operatorname{dist}(A, C) \leq |a - x| < \operatorname{diam}(C) \leq \Gamma_1 \operatorname{diam}(C),$$
contradicting condition (iii) of Theorem 5.3.1.

(ii) Fix $p \in \mathbb{R}^N \setminus A$ and set $\delta = \operatorname{dist}(p, A) > 0$. We will establish the existence of constants c_1 and c_2 such that if $p \in \widetilde{C}$, then

$$C \subset \overline{\mathbb{B}}(p, c_1 \delta) \quad \text{and} \quad \operatorname{diam}(C) \geq c_2 \delta.$$

Since the cubes in $\mathcal{W}$ are pairwise disjoint, we see by considering the N-dimensional volumes that it suffices to take

$$K = \Omega(N) \left(\frac{c_1}{c_2}\right)^N N^{N/2},$$

5.3. PROOF OF THE WHITNEY EXTENSION THEOREM

where Υ_N denotes the N-dimensional volume of the unit ball in $\mathbb{R}^N$.

Assuming $p \in \tilde{C}$, we easily estimate

$$\operatorname{dist}(p, C) \leq \tfrac{1}{2} \operatorname{diam}(C). \tag{5.14}$$

Also, it is clear that

$$\operatorname{dist}(p, A) - \operatorname{dist}(p, C) \leq \operatorname{dist}(C, A) \leq \operatorname{dist}(p, A) + \operatorname{dist}(p, C)$$

holds. By condition (iii) of Theorem 5.3.1, we also have

$$\operatorname{diam}(C) \leq \Gamma_1 \operatorname{diam}(C) \leq \operatorname{dist}(C, A) \leq \Gamma_2 \operatorname{diam}(C).$$

First we obtain

$$\begin{aligned} \operatorname{diam}(C) &\leq \operatorname{dist}(C, A) \\ &\leq \operatorname{dist}(p, A) + \operatorname{dist}(p, C) \\ &\leq \delta + \tfrac{1}{2} \operatorname{diam}(C), \end{aligned}$$

so that

$$\operatorname{diam}(C) \leq 2\delta$$

and, by (5.14),

$$\operatorname{dist}(p, C) \leq \delta.$$

Clearly if $r = \operatorname{diam}(C) + \operatorname{dist}(p, C)$, then $C \subset \overline{\mathbb{B}}(p, r)$, so we can take $c_1 = 3$ and obtain $C \subset \overline{\mathbb{B}}(p, c_1 \delta)$.

Next we estimate

$$\begin{aligned} \delta &= \operatorname{dist}(p, A) \\ &\leq \operatorname{dist}(p, C) + \operatorname{dist}(C, A) \\ &\leq \tfrac{1}{2} \operatorname{diam}(C) + \Gamma_2 \cdot \operatorname{diam}(C) \\ &= (\tfrac{7}{2} + \Gamma_2) \operatorname{diam}(C), \end{aligned}$$

so we can take $c_2 = \tfrac{1}{2} + \Gamma_2$. ∎

Definition 5.3.5

(i) *For the remainder of this section, fix some C^∞ function $\phi : \mathbb{R} \to \mathbb{R}$ with the following properties*

$$0 \leq \phi(t) \leq 1 \text{ for all } t \in \mathbb{R}, \tag{5.15}$$

$$[-\tfrac{1}{2} - \epsilon, \tfrac{1}{2} + \epsilon] \subset \{t : \phi(t) = 1\} \quad \text{for some } \epsilon > 0, \tag{5.16}$$

$$\operatorname{supp} \phi \subset (-1, 1). \tag{5.17}$$

(ii) For a cube C with sides parallel to the coordinate axes centered at $p = (p_1, p_2, \ldots, p_N)$ and with side length $s > 0$, that is

$$C = \prod_{i=1}^{N} [-\tfrac{s}{2} + p_i, \tfrac{s}{2} + p_i],$$

we set

$$\widetilde{\phi}_C(x) = \prod_{i=1}^{N} \phi\left(\frac{x_i - p_i}{s}\right).$$

Based on Definition 5.3.5, the following result is plainly true.

Lemma 5.3.6

(i) For any closed cube C with sides parallel to the coordinate axes and having non-empty interior, the function $\widetilde{\phi}_C$ is supported in the interior of $\widetilde{C}$ and is identically 1 on C.

(ii) If $\mathcal{W}$ is a Whitney family for the non-empty closed subset A, then

$$\sum_{C \in \mathcal{W}} \widetilde{\phi}_C(x) \geq 1 > 0 \text{ for all } x \in \mathbb{R}^N \setminus A, \qquad (5.18)$$

$$\sum_{C \in \mathcal{W}} \widetilde{\phi}_C(x) = 0 \text{ for all } x \in A, \qquad (5.19)$$

and there exists a family of constants $\{\widetilde{K}_\alpha\}$ (determined by the particular choice of ϕ) such that

$$\left| D^\alpha \widetilde{\phi}_C \right| \leq \widetilde{K}_\alpha \left[\text{diam}(C)\right]^{-|\alpha|} \qquad (5.20)$$

holds for each multi-index α.

Lemma 5.3.7 If $\mathcal{W}$ is a Whitney family for the non-empty closed subset A and

$$h(x) = 1 \bigg/ \left(\sum_{C \in \mathcal{W}} \widetilde{\phi}_C(x)\right),$$

then there exists a family of constants $\{\widehat{K}_\alpha\}$ such that

$$\left|\frac{\partial^{|\alpha|} h}{\partial x^\alpha}(x)\right| \leq \widehat{K}_\alpha \qquad (5.21)$$

holds on $\mathbb{R}^N \setminus A$.

5.3. PROOF OF THE WHITNEY EXTENSION THEOREM

For the proof of Lemma 5.3.7, we will use the formula for the derivatives of the composition of two functions. In the one variable case, this is known as the Formula of Faà de Bruno (see Krantz and Parks [2], Lemma 1.3.1). For the situation at hand, we will not need the actual values of the coefficients. We leave it to the reader to use induction to verify the formula.

Proposition 5.3.8 (Generalized Formula of Faà de Bruno)
There are positive integer coefficients $C_{[\mathbf{A},\alpha]}$ such that, for each multi-index $\lambda \in \Lambda_N$, if f and g are functions of class $C^{|\lambda|}$ and $h = g \circ f$, then

$$D^\lambda h(x) = \sum_{\mathbf{A}} \prod_{\alpha \in \Lambda_N} C_{[\mathbf{A},\alpha]} \, g^{(k)}[f(x)] \, (D^\alpha f(x))^{\mathbf{A}(\alpha)},$$

holds, where the sum extends over functions $\mathbf{A} : \Lambda_N \to \mathbb{Z}^+$ such that

$$\sum_{\alpha \in \Lambda_N} \mathbf{A}(\alpha)\, \alpha = \lambda,$$

and where $k = k(\mathbf{A})$ is defined by

$$k = \sum_{\alpha \in \Lambda_N} \mathbf{A}(\alpha).$$

Proof of Lemma 5.3.7: Set $g(t) = 1/t$ and

$$\Phi(x) = \sum_{C \in \mathcal{W}} \widetilde{\phi}_C(x),$$

then $h = g \circ \Phi$. Now, since Φ is bounded below on $\mathbb{R} \setminus A$ by 1, we see quite simply that

$$\left| g^{(k)}[\Phi(x)] \right| \leq k!$$

holds on $\mathbb{R} \setminus A$. By the upper bound (5.20) and the bound on the number of non-vanishing $\widetilde{\phi}_C$ at any point of $\mathbb{R} \setminus A$, we have a bound on $D^\alpha \Phi(x)$, and the lemma follows from the Generalized Formula of Faà de Bruno. ∎

In the next definition, we construct a partition of unity for $\mathbb{R}^N \setminus A$. It is crucial that we can estimate the size of the partial derivatives of the functions making up this partition of unity and that all those functions vanish on A.

Definition 5.3.9 (Whitney Partition of Unity) *Suppose $\mathcal{W}$ is a Whitney decomposition for the non-empty closed set A.*

(i) *For $C \in \mathcal{W}$, we set*

$$\phi_C(x) = \widetilde{\phi}_C(x) \bigg/ \left(\sum_{C' \in \mathcal{W}} \widetilde{\phi}_{C'}(x) \right) \text{ for } x \in \mathbb{R}^N \setminus A,$$

so

$$\sum_{C \in \mathcal{W}} \phi_C(x) = 1 \text{ for all } x \in \mathbb{R}^N \setminus A, \qquad (5.22)$$

$$\sum_{C \in \mathcal{W}} \phi_C(x) = 0 \text{ for all } x \in A. \qquad (5.23)$$

There exists a family of constants $\{K_\alpha\}$ such that

$$|D^\alpha \phi_C(x)| \le K_\alpha \bigl(\mathrm{diam}(C)\bigr)^{-|\alpha|} \qquad (5.24)$$

holds for each multi-index α.

(ii) Let $\xi : \mathcal{W} \to A$ be such that

$$\mathrm{dist}(C, A) = \mathrm{dist}\,(\xi(C), A) \qquad (5.25)$$

holds for each $C \in \mathcal{W}$. Of course, the choice of $\xi(C)$ need not be unique; any point meeting the condition may be chosen.

Remark 5.3.10 The constants K_α in (i) depend not only on the choice of ϕ, but also on the value of K_{Γ_1, Γ_2} from Lemma 5.3.4.

Now that we have dealt with the preliminaries regarding the Whitney decomposition and the associated partition of unity, we can proceed to the actual proof.

Proof of the Whitney Extension Theorem 5.2.7: Let k be a nonnegative integer and let A be a non-empty closed set. Suppose that to each $a \in A$ there corresponds a polynomial

$$P_a(x_1, x_2, \ldots, x_N),$$

of degree not exceeding k, such that the limits in the hypotheses of the Whitney Extension Theorem are attained uniformly on compact subsets of A. Recall those limits are Eq. (5.9)

$$\lim_{b \to a} \left| \frac{\partial^\beta P_b}{\partial x^\beta}(b) - \frac{\partial^\beta P_a}{\partial x^\beta}(b) \right| |b - a|^{|\beta| - k} = 0.$$

Choosing $\Gamma_1 > 2$ in Theorem 5.3.1, we let $\mathcal{W}$ be a Whitney decomposition for $\mathbb{R}^N \setminus A$.

The desired function g can simply be written down by setting

$$g(x) = \begin{cases} P_x(x) & \text{for } x \in A, \\ \sum_{C \in \mathcal{W}} \phi_C(x) P_{\xi(C)}(x) & \text{for } x \in \mathbb{R}^N \setminus A. \end{cases} \qquad (5.26)$$

5.3. PROOF OF THE WHITNEY EXTENSION THEOREM

The difficulty lies in showing that g is C^k and has the polynomial P_a as its Taylor polynomial of degree k at $a \in A$. Obviously, g is infinitely differentiable on $\mathbb{R}^N \setminus A$ and

$$D^\alpha g(x) = \sum_{C \in W} \sum_{\beta \leq \alpha} \binom{\alpha}{\beta} D^\beta \phi_C(x) D^{\alpha-\beta} P_{\xi(C)}(x) \quad (5.27)$$

for $x \in \mathbb{R}^N \setminus A$ and any multi-index α.

For any multi-index α, with $|\alpha| \leq k$, we define a function $T_\alpha : \mathbb{R}^N \to \mathbb{R}$ by setting

$$T_\alpha(x) = \begin{cases} (D^\alpha P_x)(x) & \text{for } x \in A, \\ (D^\alpha g)(x) & \text{for } x \in \mathbb{R}^N \setminus A. \end{cases}$$

Also, for any any multi-index α, with $|\alpha| < k$, and any $x \in \mathbb{R}^N$, we define a linear function $L_\alpha(x) : \mathbb{R}^N \to \mathbb{R}$ by setting

$$\langle L_\alpha(x), v \rangle = \sum_{i=1}^{N} (v \cdot \mathbf{e}_i)\, T_{\alpha^{i+}}(x),$$

where

$$\left(\alpha^{i+}\right)_j = \begin{cases} \alpha_i + 1 & \text{for } j = i, \\ \alpha_j & \text{for } j \neq i. \end{cases}$$

We will show that, if α is a multi-index with $|\alpha| \leq k$, then T_α is continuous at every point of A, and if $|\alpha| < k$ and $a \in A$, then

$$\lim_{x \to a} \frac{|T_\alpha(x) - T_\alpha(a) - \langle L_\alpha(a), x - a \rangle|}{|x - a|} = 0.$$

This will verify that g is C^k on all of $\mathbb{R}^N$ and that the Taylor polynomial of g at any point $a \in A$ coincides with P_a.

For notational convenience, in that which follows let us assume that A is compact. Then we may set

$$\epsilon(\delta) = \sup\left\{ \frac{|D^\alpha P_{\hat{a}}(a) - D^\alpha P_a(a)|}{|\hat{a} - a|^{|\alpha|-k}} : \hat{a}, a \in A,\ 0 < |\hat{a} - a| \leq \delta,\ 0 \leq |\alpha| \leq k \right\}.$$

By hypothesis, $\epsilon(\delta) \downarrow 0$ as $\delta \downarrow 0$.

We have broken the details of the proof up into six steps. The first step will show that T_α is continuous on A.

Step 1: Suppose $|\alpha| \leq k$, and consider $\hat{a} \to a$ with $\hat{a}, a \in A$.

We have

$$|T_\alpha(\hat{a}) - T_\alpha(a)| = |D^\alpha P_{\hat{a}}(\hat{a}) - D^\alpha P_a(a)|$$
$$\leq |D^\alpha P_{\hat{a}}(\hat{a}) - D^\alpha P_a(\hat{a})| + |D^\alpha P_a(\hat{a}) - D^\alpha P_a(a)|. \quad (5.28)$$

As $\hat{a} \to a$, the first term in line (5.28) approaches 0 because of (5.9) while the second term approaches 0 because of the continuity of $D^\alpha P_a$.

The second step will show that the difference quotient goes to zero when attention is restricted to A. The estimate is almost as simple as in Step 1.

Step 2: Suppose $|\alpha| < k$, and consider $\hat{a} \to a$ with $\hat{a}, a \in A$.

Similarly to Step 1, we have

$$\frac{|T_\alpha(\hat{a}) - T_\alpha(a) - \langle L_\alpha(a), \hat{a} - a \rangle|}{|\hat{a} - a|}$$

$$\leq \frac{|D^\alpha P_{\hat{a}}(\hat{a}) - D^\alpha P_a(\hat{a})|}{|\hat{a} - a|} \qquad (5.29)$$

$$+ \frac{|D^\alpha P_a(\hat{a}) - D^\alpha P_a(a) - \langle L_\alpha(a), \hat{a} - a \rangle|}{|\hat{a} - a|}. \qquad (5.30)$$

As $\hat{a} \to a$, the term in line (5.29) approaches 0 again because of (5.9) while the term in line (5.30) approaches 0 by Taylor's Theorem applied to $D^\alpha P_a$.

When we consider approaching the point $a \in A$ through points lying outside of A, the situation becomes more complicated. In Steps 3 and 4 we establish a pair of preliminary estimates that will facilitate the rest of the argument.

Step 3: Suppose $|\alpha| \leq k$, and consider $\hat{a}, a \in A$ and $x \in \mathbb{R}^N$.

We apply Taylor's Theorem to the polynomials $D^\alpha P_{\hat{a}}$ and $D^\alpha P_a$ at the point a to estimate

$$|D^\alpha P_{\hat{a}}(x) - D^\alpha P_a(x)|$$

$$\leq \sum_{0 \leq |\gamma| \leq k - |\alpha|} |D^{\gamma+\alpha} P_{\hat{a}}(a) - D^{\gamma+\alpha} P_a(a)| \frac{|x-a|^{|\gamma|}}{\gamma!}$$

$$\leq \epsilon(|\hat{a} - a|) \sum_{0 \leq |\gamma| \leq k - |\alpha|} |\hat{a} - a|^{k - |\gamma| - |\alpha|} \frac{|x-a|^{|\gamma|}}{\gamma!}. \qquad (5.31)$$

Comparing the summation in line (5.31) with the binomial formula, we conclude that there exists a constant Γ_3 such that

$$|D^\alpha P_{\hat{a}}(x) - D^\alpha P_a(x)| \leq \Gamma_3 \epsilon(|\hat{a} - a|) \left(|\hat{a} - a| + |x - a|\right)^{k - |\alpha|} \qquad (5.32)$$

holds.

Step 4: Suppose $|\alpha| \leq k$, and consider $x \in \mathbb{R}^N \setminus A$ and $\hat{a} \in A$ such that

$$|x - \hat{a}| = \text{dist}(x, A).$$

5.3. PROOF OF THE WHITNEY EXTENSION THEOREM

Noting that

$$D^\alpha P_{\hat{a}}(x) = D^\alpha \left[\sum_{C \in W} \phi_C(x) P_{\hat{a}}(x) \right]$$

$$= \sum_{C \in W} \sum_{\beta \leq \alpha} \binom{\alpha}{\beta} D^\beta \phi_C(x) D^{\alpha-\beta} P_{\hat{a}}(x),$$

we see that

$$|T_\alpha(x) - D^\alpha P_{\hat{a}}(x)|$$

$$= \left| \sum_{C \in W} \sum_{\beta \leq \alpha} \binom{\alpha}{\beta} D^\beta \phi_C(x) D^{\alpha-\beta} P_{\xi(C)}(x) - D^\alpha P_{\hat{a}}(x) \right|$$

$$\leq \sum_{C \in W} \sum_{\beta \leq \alpha} \binom{\alpha}{\beta} \left| D^\beta \phi_C(x) \right| \left| D^{\alpha-\beta} P_{\xi(C)}(x) - D^{\alpha-\beta} P_{\hat{a}}(x) \right|. \quad (5.33)$$

We will estimate the term in line (5.33). Consider

$$\left| D^\beta \phi_C(x) \right| \left| D^{\alpha-\beta} P_{\xi(C)}(x) - D^{\alpha-\beta} P_{\hat{a}}(x) \right|,$$

where we may assume $x \in \widetilde{C}$, since otherwise $D^\beta \phi_C(x) = 0$. By (5.24), we have

$$\left| D^\beta \phi_C(x) \right| \leq K_\beta \left(\mathrm{diam}(C) \right)^{-|\beta|}. \quad (5.34)$$

We now need to show that $|x - \hat{a}|$ is comparable to $\mathrm{diam}(C)$ and that $|\xi(C) - x|$ and $|\xi(C) - \hat{a}|$ can be bounded by a constant multiple of $\mathrm{diam}(C)$. To see this, let $p \in C$ be such that $|\xi(C) - p| = \mathrm{dist}(C, A)$. We have

$$\mathrm{dist}(C, A) \leq |p - \hat{a}| \leq |p - x| + |x - \hat{a}|,$$

so

$$\begin{aligned} |x - \hat{a}| &\geq \mathrm{dist}(C, A) - |p - x| \\ &\geq \mathrm{dist}(C, A) - \mathrm{diam}(\widetilde{C}) \\ &\geq \mathrm{dist}(C, A) - 2\,\mathrm{diam}(C) \\ &\geq (\Gamma_1 - 2)\,\mathrm{diam}(C). \qquad \text{(Recall we chose } \Gamma_1 > 2.\text{)} \end{aligned}$$

Also, we estimate

$$\begin{aligned} \Gamma_2 \cdot \mathrm{diam}(C) &\geq \mathrm{dist}(C, A) \\ &= |p - \xi(C)| \\ &\geq |x - \xi(C)| - |p - x| \\ &\geq |x - \xi(C)| - 2\,\mathrm{diam}(C) \\ &\geq |x - \hat{a}| - 2\,\mathrm{diam}(C). \end{aligned}$$

Thus we have

$$(\Gamma_2 + 2)\mathrm{diam}(C) \geq |x - \hat{a}| \geq (\Gamma_1 - 2)\mathrm{diam}(C). \tag{5.35}$$

Next, we can estimate

$$\begin{aligned} |\xi(C) - \hat{a}| &\leq |\xi(C) - p| + |p - x| + |x - \hat{a}| \\ &\leq \mathrm{dist}(C, A) + \mathrm{diam}(\widetilde{C}) + \mathrm{dist}(x, A) \\ &\leq \Gamma_2 \cdot \mathrm{diam}(C) + 2\,\mathrm{diam}(C) \\ &\quad + \mathrm{dist}(x, C) + \mathrm{dist}(C, A) \\ &\leq (3 + 2\Gamma_2)\,\mathrm{diam}(C). \end{aligned}$$

Thus we also have

$$|\xi(C) - x| \leq |\xi(C) - \hat{a}| + |\hat{a} - x| \leq (5 + 3\Gamma_2)\mathrm{diam}(C). \tag{5.36}$$

Returning to the estimation of the term in line (5.33), we can apply (5.32) and (5.36) to conclude that

$$\left|D^{\alpha-\beta} P_{\xi(C)}(x) - D^{\alpha-\beta} P_{\hat{a}}(x)\right|$$
$$\leq \Gamma_3\, \epsilon(\xi(C), \hat{a}) \left[\mathrm{diam}(C)\right]^{k-|\alpha|+|\beta|}.$$

Combined with (5.34) this shows us that there exist constants Γ_4 and Γ_5 such that

$$|T_\alpha(x) - D^\alpha P_{\hat{a}}(x)| \leq \Gamma_4\, \epsilon(\Gamma_5\, |x - \hat{a}|)\, |x - \hat{a}|^{k-\alpha}. \tag{5.37}$$

In Step 5, we will complete the demonstration of continuity of T_α.

Step 5: Suppose $|\alpha| \leq k$, and consider $x \to a$ with $x \in \mathbb{R}^N \setminus A$.
Let $\hat{a} \in A$ be such that

$$|x - \hat{a}| = \mathrm{dist}(x, A).$$

Clearly, we have $|x - \hat{a}| \leq |x - a|$, so $|x - \hat{a}| \to 0$ and $\hat{a} \to a$ as $x \to a$.
We estimate

$$\begin{aligned} &|T_\alpha(x) - T_\alpha(a)| \\ &= |T_\alpha(x) - D^\alpha P_{\hat{a}}(x) + D^\alpha P_{\hat{a}}(x) \\ &\quad - D^\alpha P_a(x) + D^\alpha P_a(x) - T_\alpha(a)| \\ &\leq |T_\alpha(x) - D^\alpha P_{\hat{a}}(x)| + |D^\alpha P_{\hat{a}}(x) - D^\alpha P_a(x)| \quad (5.38)\\ &\quad + |D^\alpha P_a(x) - D^\alpha P_a(a)|. \quad (5.39) \end{aligned}$$

5.4. APPLICATION OF THE WHITNEY THEOREM

As $x \to a$ the term in line (5.38) approaches 0 by the estimate (5.37) in Step 4, the second term in line (5.38) approaches 0 by the estimate (5.32) in Step 3, and the term in line (5.39) approaches 0 by the continuity of $D^\alpha P_a$.

Step 6 will complete the demonstration of differentiablity of T_α.

Step 6: Suppose $|\alpha| < k$.

Again let $\hat{a} \in A$ be such that

$$|x - \hat{a}| = \text{dist}(x, A).$$

We estimate

$$\frac{|T_\alpha(x) - T_\alpha(a) - \langle L_\alpha(a), x - a\rangle|}{|x - a|}$$

$$\leq \frac{|T_\alpha(x) - D^\alpha P_{\hat{a}}(x)|}{|x - a|} \qquad (5.40)$$

$$+ \frac{|D^\alpha P_{\hat{a}}(x) - D^\alpha P_a(x)|}{|x - a|} \qquad (5.41)$$

$$+ \frac{|D^\alpha P_a(x) - D^\alpha P_a(a) - \langle L_\alpha(a), x - a\rangle|}{|x - a|}. \qquad (5.42)$$

The term in line (5.40) approaches 0 by the estimate (5.37) in Step 4, the term in line (5.41) approaches 0 by the estimate (5.32) in Step 3, and the term in line (5.42) approaches 0 by Taylor's Theorem applied to the polynomial $D^\alpha P_a$. ∎

5.4 Application of the Whitney Extension Theorem

The so-called Cantor function provides a well-known example of a function on the unit interval that is non-decreasing and non-constant, but for which the derivative vanishes almost everywhere. By a similar construction H. Whitney was able to construct a function on an arc in the plane that can be extended to a C^1 function in the whole plane with the amazing property that the entire arc consists of critical points of the function, but the function is not constant on the arc. The extension of the function from the arc to the whole plane is accomplished by the Whitney Extension Theorem 5.2.7. We base our construction on Whitney [2].

The Cantor Set

We form a Cantor set C in the closed unit square $[0, 1] \times [0, 1]$ as follows: Fix $0 < \lambda < 1/2$. Let C_1 consist of four closed squares of side length λ

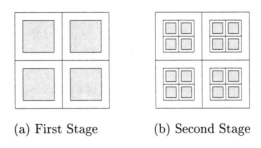

(a) First Stage   (b) Second Stage

Figure 5.2: **Forming the Cantor Set**

centered in the four quarters of the unit square, as in Figure 5.2(a). Let us denote these squares by S_0, S_1, S_2, S_3 in positive cyclical order and, to be precise, let S_0 be the square in the bottom left quarter of the unit square. Notice that the distance from any square S_i to the exterior of its quarter of the unit square is $\frac{1}{2}(\frac{1}{2} - \lambda)$ so, if $i \neq j$, then $\text{dist}(S_i, S_j) \geq (\frac{1}{2} - \lambda)$. Let C_2 consist of sixteen closed squares of side length λ^2 formed by applying the preceding construction with each square S_i taking the role of the unit square. These sixteen squares are illustrated in Figure 5.2(b). The subsquares of S_i will be denoted by $S_{i0}, S_{i1}, S_{i2}, S_{i3}$ again labeled in cyclic order, but we will specify later whether this cycle in the positive or negative direction and we will specify later which subsquare is designated S_{i0}. As before, we can estimate the distance between subsquares: if $i_2 \neq i_2'$, then $\text{dist}(S_{i_1 i_2}, S_{i_1 i_2'}) \geq \lambda(\frac{1}{2} - \lambda)$. Continue this construction to form $C_3, C_4, \ldots$ where C_k consists of 4^j squares of side length λ^j, with those squares denoted by $S_{i_1 i_2 \ldots i_k}$. The distance estimate extends to

$$\text{dist}(S_{i_1 i_2 \ldots i_{k-1} i_k}, S_{i_1 i_2 \ldots i_{k-1} i_k'}) \geq \lambda^{k-1}(\tfrac{1}{2} - \lambda) \quad \text{if} \quad i_k \neq i_k'. \tag{5.43}$$

The Cantor set is then
$$C = \bigcap_{k=1}^{\infty} C_k.$$

Any point $p \in C$ is associated with a unique sequence $(i_1, i_2, \ldots)$ such that
$$\{p\} = \bigcap_{k=1}^{\infty} S_{i_1 i_2 \ldots i_k}. \tag{5.44}$$

Thus we can easily define a function $f : C \to \mathbb{R}$ by setting
$$f(p) = \sum_{k=1}^{\infty} i_k \mu^k. \tag{5.45}$$

Lemma 5.4.1 *Suppose $p, q \in C$ with corresponding sequences $(i_1, i_2, \ldots)$ and $(j_1, j_2, \ldots)$. If $i_1 = j_1$, $i_2 = j_2$, $\ldots$, $i_k = j_k$ and $i_{k+1} \neq j_{k+1}$, then*
$$|p - q| \geq (\tfrac{1}{2} - \lambda)\lambda^k \tag{5.46}$$

5.4. APPLICATION OF THE WHITNEY THEOREM

and
$$|f(p) - f(q)| \leq \frac{3\mu^{k+1}}{1-\mu}. \tag{5.47}$$

Proof: This follows easily from (5.43) and (5.45). ∎

Corollary 5.4.2 *There exist constants $0 < \sigma$ and $\Gamma < \infty$ such that if $p, q \in C$ with $p \neq q$, then*
$$|f(p) - f(q)| \leq \Gamma |p-q|^\sigma. \tag{5.48}$$

Proof: The corollary follows by solving for k in both (5.46) and (5.47) and comparing the results. We find
$$\sigma = \frac{\log \mu}{\log \lambda}, \qquad \Gamma = \frac{1-\mu}{3\mu} \left(\tfrac{1}{2} - \lambda\right)^\sigma. \qquad \blacksquare$$

Remark 5.4.3 The estimate in Corollary 5.4.2 shows that, if we let K be the greatest integer strictly less than σ, then f can be extended from C to $\mathbb{R}^2$ as a C^K function with every point of C a critical point. This is a consequence of the Whitney Extension Theorem 5.2.7 applied with the polynomial P_a at $a \in C$ chosen to be the constant polynomial $f(a)$. While K can be made as large as we wish by choosing μ small, the resulting set of critical points is totally disconnected. Also, if we consider the set of numbers $f(C)$, all of which are critical values, this is also generally a disconnected set. A crucial observation is that if $\mu = \tfrac{1}{4}$, then $f(C) = [0, 1]$. This choice of μ is governed by fact that we have put four subsquares in each square at every stage of the construction.

The Arc

From now on we set
$$\mu = \tfrac{1}{4} \qquad \text{and choose} \qquad \tfrac{1}{4} < \lambda < \tfrac{1}{2}. \tag{5.49}$$

We describe how to run an arc A given by $\varphi : [0, 1] \to \mathbb{R}^2$ through the Cantor set C. The construction proceeds in stages in concert with the formation of the Cantor set. Simultaneously, we will extend f from the Cantor set C to the arc.

First Stage: The following construction is illustrated in Figure 5.3(a).

(i) Starting at the point $(0, 0)$, run a line segment to the bottom left corner of the square S_0. Define φ on the interval $[0, \tfrac{1}{9}]$ to map linearly onto this line segment, with 0 mapping to $(0, 0)$. The value of f on this part of the arc will equal the minimum value of f on the set $S_0 \cap C$, i.e. 0.

182   CHAPTER 5. SMOOTH MAPPINGS

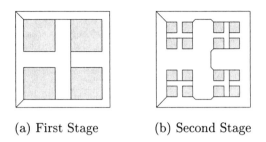

(a) First Stage   (b) Second Stage

Figure 5.3: **Running an Arc through the Cantor Set**

(ii) From the bottom right corner of S_0 run a line segment to the bottom left corner of S_1. Define φ on the interval $[\frac{2}{9}, \frac{3}{9}]$ to map linearly onto this line segment, with $\frac{2}{9}$ mapping to the bottom right corner of S_0. The value of f on this segment will be the maximum of f on the set $S_0 \cap C$, which also equals the minimum value of f on the set $S_1 \cap C$. That value is $\frac{1}{4}$. (This coincidence of the maximum of f on $S_0 \cap C$ and minimum of f on $S_1 \cap C$ is a result of choosing $\mu = \frac{1}{4}$.)

(iii) From the top left left corner of S_1 run a line segment to the bottom left corner of S_2. Define φ on the interval $[\frac{4}{9}, \frac{5}{9}]$ to map linearly onto this line segment, with $\frac{4}{9}$ mapping to the top left corner of S_1. The value of f on this segment is $\frac{1}{2}$ which is the maximum of f on $S_1 \cap C$ and the minimum of f on $S_2 \cap C$.

(iv) From the top left corner of S_2 run a line segment to the right corner of S_3. Define φ on the interval $[\frac{6}{9}, \frac{7}{9}]$ to map linearly onto this line segment, with $\frac{6}{9}$ mapping to the top left corner of S_2. On this segment, set f equal to $\frac{3}{4}$, the maximum of f on $S_2 \cap C$ and the minimum of f on $S_3 \cap C$.

(v) From the top left corner of S_3 run a line segment to the point $(0, 1)$. Define φ on the interval $[\frac{8}{9}, 1]$ to map linearly onto this line segment, with $\frac{8}{9}$ mapping to the top left corner of S_3. On this segment, set f equal to 1, the maximum of f on $S_3 \cap C$.

The arc φ has been defined on

$$[0, \tfrac{1}{9}] \cup [\tfrac{2}{9}, \tfrac{3}{9}] \cup [\tfrac{4}{9}, \tfrac{5}{9}] \cup [\tfrac{6}{9}, \tfrac{7}{9}] \cup [\tfrac{8}{9}, 1].$$

Define $\varphi_1 : [0, 1] \to \mathbb{R}^2$ by extending φ linearly to the missing intervals

$$[\tfrac{1}{9}, \tfrac{2}{9}], \ [\tfrac{3}{9}, \tfrac{4}{9}], \ [\tfrac{5}{9}, \tfrac{6}{9}], \ [\tfrac{7}{9}, \tfrac{8}{9}].$$

Second Stage: Scale the previous construction by a factor of λ and rotate and translate to fit it in the subsquares S_i and to connect with the part of the arc already constructed. At this time, we can determine the ordering of the labeling of the squares S_{ij}, so the second index increases as we proceed along the arc. In the process, the definition of φ is extended to the intervals

$$\left[\tfrac{9\ell}{81}, \tfrac{1+9\ell}{81}\right], \left[\tfrac{2+9\ell}{81}, \tfrac{3+9\ell}{81}\right], \left[\tfrac{4+9\ell}{81}, \tfrac{5+9\ell}{81}\right], \left[\tfrac{6+9\ell}{81}, \tfrac{7+9\ell}{81}\right], \left[\tfrac{8+9\ell}{81}, \tfrac{9+9\ell}{81}\right]$$

which are subsets of $[\tfrac{\ell}{9}, \tfrac{1+\ell}{9}]$, for $\ell = 1, 3, 5, 7$. This is illustrated in Figure 5.3(b). The values of f are again set by the maximum and minimum process used in the first stage.

Define $\varphi_2 : [0,1] \to \mathbb{R}^2$ by extending linearly to the intervals on which φ is not yet defined.

k^{th} **Stage:** Scale the first stage construction and rotate and translate to fit it in the subsquares $S_{i_1 i_2 \ldots i_{k-1}}$. Again this determines the indexing of the subsquares $S_{i_1 i_2 \ldots i_k}$. Again φ is extended into the subintervals where it has not previously been defined and φ_k is defined by extending linearly to the intervals on which φ is still not defined. The function values are assigned by the maximum/minimum process as before.

The sequence of functions $\varphi_k : [0,1] \to \mathbb{R}^2$ converges uniformly to complete the definition of φ and to thus define the arc.

Theorem 5.4.4 *Let A be the arc constructed above and let f be the function defined on A as above. Then the function f can be extended to a C^1 function on all of $\mathbb{R}^2$ with every point of A a critical point and with $f(A) = [0,1]$.*

Proof: Since $\mu < \lambda$ we have $1 < \sigma$ in Corollary 5.4.2. The distance between points of $A \setminus C$ that have differing values is at least as large as for points of C, so the estimate (5.48) extends to all of A. ∎

5.5 Multidimensional Versions of the Fundamental Theorem of Calculus

Introductory Remarks

A cornerstone of multivariable real analysis is the generalization, in its various guises, of the Fundamental Theorem of Calculus from a bounded interval in $\mathbb{R}^1$ to a (smoothly) bounded domain in $\mathbb{R}^N$. Already in this informal statement we can begin to sense a profound difference between the one dimensional setting and the multi-dimensional setting. For in one dimension any open set is a disjoint union of intervals; therefore it is reasonable to take a bounded open interval as a "typical" open set. In several real variables there is no typical open set. In many applications, the Whitney decomposition (Theorem 5.3.1) serves in several variables as a substitute for the easy structure theorem for open sets in $\mathbb{R}^1$. In the present section, however, we will deal with a smoothly bounded domain in $\Omega \subset \mathbb{R}^N$ just as it is given.

Let f be a C^1 function defined on $\overline{\Omega}$. The goal of any multidimensional Fundamental Theorem of Calculus is to relate the integration of f on the boundary of Ω to the integral of some derivative of f on the interior of Ω. Recall that, in dimension one, the first half of this syllogism is almost trivial: It consists of evaluating f (with appropriate signs, or orientation) at the endpoints of the interval. In higher dimensions, the boundary will be a smooth manifold of dimension at least one, and the boundary integration becomes a substantive consideration.

At the end of the section we shall briefly describe de Rham's theorem for a smooth manifold. In a sense, de Rham's theorem puts into the natural context of cohomology theory what the multidimensional Fundamental Theorem only begins to suggest: Namely that the operation ∂ of calculating the boundary of a geometric object and the operation $\partial/\partial x_j$ of taking the partial derivative of a smooth function are in fact obverse facets of the very same phenomenon. One is the topological notion of differentiation and the other is the analytic notion. The multidimensional Fundamental Theorem of Calculus—in its many different manifestations—makes the connection between these two concepts both explicit and concrete.

In some sense, Stokes's theorem is the paradigm that sits above all the versions of the Fundamental Theorem in any dimension. Thus, speaking strictly logically, one should first derive the theorem of Stokes and then derive the Gauss-Green theorem, the divergence theorem, and other versions of the Fundamental Theorem from it.[1] However the strictly logical approach is not always the most didactically useful. In fact, it will be more convenient for us to first describe, state, and prove the divergence theorem in any dimension. This is so because this particular version of the Fundamental Theorem may be formulated and proved with a minimum of notation and fuss. And all the essential ideas are already present in this result.

One of the key issues in formulating the Fundamental Theorem is that of orienting the domain and the boundary. Recall that, in one real dimension, the domain is an interval that is oriented according to the natural ordering of the real numbers. The boundary is oriented in (what we will see to be) a compatible fashion by assigning to the right-hand endpoint of $[a,b]$ a plus sign and to the left-hand endpoint a minus sign. Thus the one dimensional Fundamental Theorem takes the form

$$\int_a^b f(x)\,dx = f(b) - f(a).$$

---

[1] The Gauss-Green Theorem can also be generalized in another direction by considering sets that are not smoothly bounded; that topic was treated in the earlier discussion of domains with finite perimeter in Section 3.7.

5.5. A MULTIDIMENSIONAL FUNDAMENTAL THEOREM

Of course orienting a boundary of dimension one or greater, in a fashion that is compatible with the orientation of the domain that it bounds, is a more complex matter. Fortunately, we will be assisted by the fact that we are dealing with a domain sitting in space.

The Concept of Orientation

The standard way to impose an orientation on an N-dimensional manifold M is to do so pointwise: If $P \in M$ and U is a small topologically trivial neighborhood of P in M, then choosing an orientation at P amounts to selecting a generator for the homology group $H_N(U \setminus \{P\})$ (the rigorous definition involves *relative homology*, but what we have said captures the spirit of the idea). In a two dimensional manifold M, for example $H_2(U \setminus \{P\})$ is just $\mathbb{Z}$ and there are two choices for the generator (corresponding to ± 1)—geometrically, choosing an orientation for M at $P \in M$ then amounts to choosing a direction of rotation.

Since we will be orienting a domain, and its boundary, in Euclidean space, we can dispense with technical machinery such as homology groups. However it will be useful to keep the ideas in the last paragraph in mind as motivation.

Now, orienting the domain Ω itself amounts to choosing a "standard" order for the differentials $dx_1, dx_2, \ldots, dx_N$. Of course we will follow custom and select $dx_1 \, dx_2 \cdots dx_N$, or more precisely $dx_1 \wedge dx_2 \wedge \cdots \wedge dx_N$, as our standard order.[2] The compatible orientation at any boundary point P is determined by selecting the outward unit normal vector ν to the boundary at P. If ρ is a C^1 defining function for $\partial\Omega$ near P then this normal is given by

$$\nu(P) = \frac{\mathbf{grad}\,\rho}{|\mathbf{grad}\,\rho|}.$$

For example, if the dimension $N = 3$, then specifying the outward unit normal ν determines, by way of the "right hand rule," a direction of rotation in the two dimensional boundary of Ω. According to our earlier discussion, this is the way that we specify an orientation.

If $\Omega \subset\subset \mathbb{R}^N$ is a domain with C^1 boundary, then we let $d\sigma$ denote the $N-1$ dimensional Hausdorff measure on the boundary. We let dV be the standard Lebesgue N-dimensional volume measure in $\mathbb{R}^N$.

Now if

$$\mathbf{X} = \langle X_1, X_2, \ldots, X_N \rangle = \sum_{j=1}^{N} X_j \frac{\partial}{\partial x_j}$$

---

[2]The differentials $dx_1, dx_2, \ldots, dx_N$ are the dual basis to the standard basis $\mathbf{e}_1, \mathbf{e}_2, \ldots, \mathbf{e}_N$ for $\mathbb{R}^N$. The symbol "$\wedge$" used here denotes exterior multiplication which is a part of the subject of Grassmann algebra. A thorough treatment of Grassmann algebra can be found in Federer [4], Chapter 1. Differentials will be essential for our discussion of Stoke's Theorem later in this section.

is a vector field, defined on $\overline{\Omega}$ and having C^1 coefficients X_j, then the **divergence** of $\mathbf{X}$ is defined to be

$$\mathrm{div} \mathbf{X} = \sum_{j=1}^{N} \frac{\partial X_j}{\partial x_j}.$$

It is a straightforward exercise with the chain rule to see that the divergence is invariant under coordinate changes.

Now we can state our theorem.

Theorem 5.5.1 (The Divergence Theorem) *Let $\Omega \subset \mathbb{R}^N$ be a bounded domain with a C^1 defining function. Define the unit outward normal vector field to the boundary ν as above. Let*

$$\mathbf{X} = \langle X_1, X_2, \ldots, X_N \rangle = \sum_{j=1}^{N} X_j \frac{\partial}{\partial x_j} \tag{5.50}$$

be a vector field, defined on $\overline{\Omega}$ and having C^1 coefficients. Then

$$\int_{\partial \Omega} \mathbf{X} \cdot \nu \, d\sigma = \int_{\Omega} \mathrm{div} X \, dV. \tag{5.51}$$

Remark 5.5.2 We have already noted that the divergence $\mathrm{div} \mathbf{X}$ of a vector field $\mathbf{X}$ is a coordinate-free quantity. Now let us look at the left hand side of (5.51). It is not true that ν is coordinate-free, nor is it true that $d\sigma$ is coordinate-free. However the expression $\nu d\sigma$ (which is essentially a volume frame) *is* coordinate-free. These observations will be useful in the proof of the divergence theorem, for we will perform the calculations on local coordinate patches. These patches will be chosen in such a fashion as to simplify the proof. We follow the ideas in Loomis and Sternberg [1].

Proof of the Divergence Theorem: To simplify things a bit we shall assume that the boundary is C^2 (however only C^1 is really necessary). Let U be a tubular neighborhood of the boundary of Ω. Choose $\eta > 0$ such that $V_\eta = \{x \in \mathbb{R}^N : \mathrm{dist}(x, \partial \Omega) < \eta\}$ lies in U. Cover $\overline{\Omega}$ with open Euclidean balls of radius η. Pass to a finite subcover, and let ϕ_ℓ be a smooth partition of unity subordinate to this cover. In summary, we have a collection of pairs $\{(U_\ell, \phi_\ell)\}_{\ell=1}^m$, where each U_ℓ is an open ball of radius η, each $\phi_\ell \in C_c^\infty(U_\ell)$, and $\sum_{\ell=1}^m \phi_\ell \equiv 1$ on $\overline{\Omega}$.

In particular, $\mathbf{X} = \sum_\ell \phi_\ell \mathbf{X}$. Since the statement of the divergence theorem is linear, it is sufficient to prove the theorem for each $\mathbf{Y}^\ell \equiv \phi_\ell \mathbf{X}$. Of course we need only consider $\mathbf{Y}^\ell$ such that $\mathrm{supp}\, \phi_\ell \cap \overline{\Omega} \neq \emptyset$. We do so in what follows without further comment. We will write $\mathbf{Y}^\ell = \langle (\mathbf{Y}^\ell)_1, (\mathbf{Y}^\ell)_2, \cdots, (\mathbf{Y}^\ell)_N \rangle$.

The Case $U_\ell \subset \Omega$: This case is relatively trivial, and is not at all the point of the theorem. Clearly the left side of (5.51) is zero in this case. As for the right hand side, recall that if $\psi(x)$ any any $C_c^\infty(\mathbb{R})$ function then

$$\int_{\mathbb{R}} \frac{d}{dx} \psi(x)\, dx = 0 \tag{5.52}$$

(just use the Fundamental Theorem of Calculus). Now we analyze the term

$$\int_\Omega \frac{\partial}{\partial x_j}(\mathbf{Y}^\ell)_j\, dV$$

by integrating in the x_j variable first. But then (5.52) tells us that the integral is zero. This completes the treatment of the first case.

The Case $U_\ell \cap \partial\Omega \neq \emptyset$: We use the ideas in Section 1.2 to choose a local coordinate system on U_ℓ so that

$$U_\ell \cap \Omega = \{x \in U_\ell : x_N > 0\}.$$

In particular,
$$U_\ell \cap \partial\Omega = \{x \in U_\ell : x_N = 0\}.$$

Let us begin our study with the right side of (5.51). The integrals of $\frac{\partial}{\partial x_j}(\mathbf{Y}^\ell)_j$ for $1 \leq j < N$ vanish just as in the first case. Using the Fundamental Theorem of Calculus in the variable x_N, we see that the integral of $\frac{\partial}{\partial x_N}(\mathbf{Y}^\ell)_N$ equals $-\int_{\mathbb{R}^{N-1}}(\mathbf{Y}^\ell)_N\, dx_1\, dx_2 \cdots dx_{N-1}$. But we have normalized coordinates so that $(\mathbf{Y}^\ell)_N = \nu \cdot \mathbf{Y}^\ell$. Thus we see that, starting from the right side of (5.51), we have obtained the left side. ∎

Differential Forms and Stokes's Theorem

Differential forms were invented for the purpose of performing oriented integrations on manifolds. Masterful treatments appear in Federer [4] and Whitney [3]. A more summary treatment is in Krantz [2]. For the present discussion, suffice it to say that a differential form of degree k (or a k-form) in $\mathbb{R}^N$ is a linear combination, with coefficients that are smooth functions, of expressions of the form $dx^\alpha = dx_{\alpha_1} \wedge dx_{\alpha_2} \wedge \cdots dx_{\alpha_k}$; hence $\alpha_1, \alpha_2, \ldots, \alpha_k$ are elements of $\{1, 2, \ldots, N\}$.

Differential forms are alternating in their indices, so that repetition of an index causes the form to be zero. We integrate a form β on a coordinate patch U in a k-dimensional manifold M (or, more particularly, on an open subset U of $\mathbb{R}^N$) by using a smooth parametrization

$$\phi : W \to U$$

from an open set $W \subset \mathbb{R}^N$ to U. Write $\beta = \sum b_\alpha(x) dx^\alpha$. We assume that each multi-index α has length k, corresponding to the dimension of $U \subset M$. Then

$$\int_U \beta \equiv \int_W \phi^*\beta,$$

where $\phi^*\beta$ is the canonical pullback of β under ϕ (whose nature is forced by the functoriality of exterior algebra). To put it more prosaically, if we let $w_1, w_2, \ldots, w_N$ be the coordinates on W then $x = \phi(w)$ and

$$dx_j = \sum_{p=1}^{N} \frac{\partial x_j}{\partial w_p} dw_p;$$

the form (being made up of objects such as dx_j) transforms accordingly. Of course $\phi^*\beta$ ends up being a linear combination of expressions dw^α of length k with C^∞ coefficients. One performs the resulting integration by explicitly calculating the orientation:

$$\int b(w)\, dw^\alpha \equiv (-1)^\varepsilon \int b(w)\, dw^{\tilde\alpha},$$

where ε is the sign of the permutation of $\alpha = (\alpha_1, \alpha_2, \ldots, \alpha_k)$ to a degree k multi-index $\tilde\alpha$ with entries in increasing order.

The final notion that we need is that of exterior differentiation. If $\beta = \sum_\alpha b_\alpha dx^\alpha$ is a differential form, then we define its **exterior derivative**, $d\beta$, by setting[3]

$$d\beta = \sum_\alpha \sum_{j=1}^{N} \frac{\partial b_\alpha}{\partial x_j} dx_j \wedge dx^\alpha. \tag{5.53}$$

A simple calculation using the alternating property of exterior algebra establishes the fact that, for any form β, $d(d\beta) = 0$.

With these preliminaries out of the way, we may now formulate a version of Stokes's theorem.

Theorem 5.5.3 (Stokes's Theorem) *Let $\Omega \subset \mathbb{R}^N$ be a bounded domain with C^1 defining function. Let β be a differential form of degree $N-1$ with C^1 coefficients defined on $\overline\Omega$. Then*

$$\int_{\partial\Omega} \beta = \int_\Omega d\beta. \tag{5.54}$$

We shall not provide a proof of Stokes's theorem. The steps are just the same as in the proof of Green's theorem, and it is arguable that these two results are the same theorem stated with two different notations. The ambitious reader may wish, for instance, to derive Green's theorem from Stokes's. A reasonable facility with exterior algebra will prove to be requisite for this exercise. Details may be found in Krantz [2].

De Rham's Theorem

We conclude this section with a quick description of de Rham's theorem.

---
[3]See Federer [4] or Whitney [3]) for a coordinate-free definition of the exterior derivative.

5.5. A MULTIDIMENSIONAL FUNDAMENTAL THEOREM

This theorem has proved to be a major influence in modern mathematical analysis and geometry. Although the theorem is correctly attributed to G. de Rham, it is the culmination of ideas instigated by E. Cartan and others. And it has also proved to be a step along the way to other great insights, such as the Atiyah-Singer Index Theorem (see Gilkey [1]).

Let M be a compact manifold of dimension k. Let us denote the differential forms of degree p on M with the notation $\bigwedge^p(M)$. Then, with the aid of the exterior differentiation operator d, we have a complex (in the sense of algebraic topology):

$$\bigwedge\nolimits^0(M) \xrightarrow{d_0} \bigwedge\nolimits^1(M) \xrightarrow{d_1} \cdots \xrightarrow{d_{p-1}} \bigwedge\nolimits^p(M) \xrightarrow{d_p} \cdots \qquad (5.55)$$

By this we mean that the image of any given map is contained in the kernel of the map following. Thus it makes sense to form quotient groups as follows. Let d_p denote the exterior differentiation operator acting on p-forms. Then for $p = 1, 2, \ldots$, we define the **de Rham cohomology group**, $\mathcal{H}^p(M)$ by setting

$$\mathcal{H}^p(M) = \ker d_p \, / \, \operatorname{im} d_{p-1}. \qquad (5.56)$$

Here the quotient is taken in the sense of modules.

It seems plain that differential forms taken as a whole, and these quotient modules in particular, contain a lot of information about the differential geometric structure of M. As an example, suppose that M were the annulus in $\mathbb{R}^2$ (this is not a compact manifold, to be sure, but suits our temporary purpose). Consider the differential form

$$\beta = \frac{-y}{x^2 + y^2} \, dx + \frac{x}{x^2 + y^2} \, dy.$$

Then β is well-defined on M, $d\beta = 0$, yet β is not in the image of d_0 (exercise). As we know from advanced calculus, the only reason that such a β can be constructed is because M has a hole. And this hole is one that is detected by the first homotopy group, and also by the first singular homology group.

De Rham's theorem turns this example into a general phenomenon. We use $H^p(X)$ to denote the standard singular cohomology group of a topological space X (see Greenberg and Harper [1]).

Theorem 5.5.4 (de Rham) *Let M be a compact manifold. Then for each $p = 1, 2, \ldots$ we have the group identity*

$$H^p(M) = \mathcal{H}^p(M). \qquad (5.57)$$

In short, the de Rham theorem says that the structure of M, calculated from topological considerations, equals the structure of M when calculated from analytical considerations.

We shall provide no further details about de Rham's theorem, but instead refer the reader to de Rham [1] and Whitney [3].

Chapter 6

Convexity

6.1 The Classical Notion of Convexity

Let $S \subset \mathbb{R}^N$ be any set. In classical geometry, S is said to be **convex** if, whenever P and Q are points of S then the entire segment $\overline{PQ}$ lies in S. We will refer to this concept as **geometric convexity**.

Convex sets originally arose in Archimedes's axiomatic treatment of arc length. (Refer to Fenchel [1] for an authoritative discussion of this and other historical matters.) While geometric convexity was treated sporadically in the mathematics of the eighteenth and nineteenth centuries, it did not receive systematic treatment until the twentieth. Some of the modern treatises are Valentine [1], Bonneson and Fenchel [1], and Lay [1].

The hallmark of all the existing treatises on convexity is a strictly geometric approach. For tone, we record here the statements of two results that are part of the métier of classical treatments:

Theorem 6.1.1 (Helly) *Let* $\mathcal{S} = \{B_1, \ldots, B_k\}$ *be a family of geometrically convex sets in* $\mathbb{R}^N$ *with* $k \geq N+1$. *If every subfamily of* $N+1$ *geometrically convex sets in* $\mathcal{S}$ *has non-empty intersection, then the intersection of all the sets in* $\mathcal{S}$ *is non-empty.*

An interesting application of Helly's theorem is the following:

Theorem 6.1.2 (Kirchberger) *Let* K *and* L *be non-empty compact subsets of* $\mathbb{R}^N$. *Then* K *and* L *can be strictly separated by a hyperplane if and only if for each subset* $T \subset K \cup L$ *containing at most* $N+2$ *points there exists a hyperplane separating* $T \cap K$ *from* $T \cap L$.

Our viewpoint in this book is that the geometric approach to convex sets is not expedient for the purposes of analysis. The reason is that the geometric approach is not quantitative and not formulated in the language of functions. Our purpose here is to provide such a quantitative, function-theoretic development.

We begin by considering a domain Ω, i.e. a connected open set. We assume that Ω has C^2 boundary, and thus is given by a C^2 defining function ρ:

$$\Omega = \{x \in \mathbb{R}^N : \rho(x) < 0\}.$$

As usual, we require that $\nabla \rho \neq 0$ on $\partial \Omega$. What does it mean for such a set to be convex?

Let us approach this question in the language of freshman calculus. Let **p** be a two dimensional plane that intersects the interior of Ω. Since the present discussion is for motivational purposes only, let us suppose that at each point Q of $\mathbf{p} \cap \partial \Omega$ the intersection is transversal. This means that the tangent space to $\partial \Omega$ at Q *does not* contain the tangent space to **p**. In other words, the two tangent spaces span all of $\mathbb{R}^N$. If Ω is geometrically convex, then it follows *a fortiori* that $\Omega \cap \mathbf{p}$ is also geometrically convex. Now let us think of this intersection as a planar domain.

Imagine that the planar coordinate system in $\Omega \cap \mathbf{p}$ has been rotated and translated so that the tangent line to $\partial \Omega \cap \mathbf{p}$ at Q is in fact the x-axis and that the point Q corresponds to the origin. With these normalizations, we may think of the boundary, near the origin, as the graph of a function of the single variable x, that is, the boundary is given as the graph

$$y = f(x),$$

with $0 = f(0) = f'(0)$. If $\Omega \cap \mathbf{p}$ lies in the upper half-plane, then the restriction of ρ to this planar coordinate system has the form $r(x,y) = f(x) - y$, because ρ is negative on $\partial \Omega \cap \mathbf{p}$. Also the graph of f is "concave up," so calculus teaches us that $f'' \geq 0$, hence that

$$\frac{\partial^2 r}{\partial x^2}(x,y) \geq 0. \tag{6.1}$$

Similarly, if $\Omega \cap \mathbf{p}$ lies in the lower half-plane, then the restriction of ρ is $r(x,y) = y - f(x)$, and since the graph of f is "concave down," we have $f'' \leq 0$. Again (6.1) holds.

What we have just discovered, using elementary geometry, local coordinates, and calculus, is the following: if Ω has C^2 boundary and is geometrically convex, if $Q \in \partial \Omega$, and if ξ is a tangent direction at Q, then

$$\left. \frac{\partial}{\partial t^2} \rho(Q + t\xi) \right|_{t=0} \geq 0.$$

Using the chain rule, this condition may be rewritten as

$$\sum_{j,k=1}^{N} \frac{\partial^2 \rho}{\partial x_j \partial x_k}(Q) \xi_j \xi_k \geq 0.$$

This calculation will be the basis for our analytic definition of convexity.

6.1. THE CLASSICAL NOTION OF CONVEXITY

Definition 6.1.3 Let $\Omega = \{x \in \mathbb{R}^N : \rho(x) < 0\}$, where ρ is a C^2 defining function for the domain Ω.

(i) Let $Q \in \partial\Omega$. We say that $\partial\Omega$ is **analytically convex** at Q if for every choice of $\xi = (\xi_1, \ldots, \xi_N)$ in the tangent space to $\partial\Omega$ at Q it holds that

$$\sum_{j,k=1}^{N} \frac{\partial^2 \rho}{\partial x_j \partial x_k}(Q) \xi_j \xi_k \geq 0. \tag{6.2}$$

(ii) If every point of $\partial\Omega$ is analytically convex, then we say that the domain Ω is **analytically convex**.

(iii) If it further holds that whenever $0 \neq \xi = (\xi_1, \ldots, \xi_N)$ and ξ lies in the tangent space to $\partial\Omega$ at Q it follow that

$$\sum_{j,k=1}^{N} \frac{\partial^2 \rho}{\partial x_j \partial x_k}(Q) \xi_j \xi_k > 0, \tag{6.3}$$

then we say that Q is a point of **strong analytic convexity**.

(iv) If every point of $\partial\Omega$ is strongly analytically convex, then we say that the domain Ω is **strongly analytically convex**.

If a smooth boundary point is a point of analytic convexity, but not of strong analytic convexity, then we sometimes (for emphasis) refer to it as a point of **weak analytic convexity**.

Lemma 6.1.4 Let $\Omega \subset \mathbb{R}^N$ be strongly analytically convex. Then there is a constant $C > 0$ and a defining function $\tilde{\rho}$ for Ω such that

$$\sum_{j,k=1}^{N} \frac{\partial^2 \tilde{\rho}}{\partial x_j \partial x_k}(P) w_j w_k \geq C|w|^2, \quad \forall P \in \partial\Omega, w \in \mathbb{R}^N. \tag{6.4}$$

Proof: Let ρ be some fixed C^2 defining function for Ω. For $\lambda > 0$ define

$$\rho_\lambda(x) = \frac{\exp(\lambda \rho(x)) - 1}{\lambda}.$$

We shall select λ in a moment. Let $P \in \partial\Omega$ and set

$$X = X_P = \left\{ w \in \mathbb{R}^N : |w| = 1 \text{ and } \sum_{j,k} \frac{\partial^2 \rho}{\partial x_j \partial x_k}(P) w_j w_k \leq 0 \right\}.$$

Then no element of X could be a tangent vector at P, hence

$$X \subset \{w : |w| = 1 \text{ and } \sum_j \partial \rho/\partial x_j(P) w_j \neq 0\}.$$

Since X is defined by a non-strict inequality, it is closed; it is of course also bounded. Hence X is compact and

$$\mu \equiv \min\left\{\left|\sum_j \partial\rho/\partial x_j(P)\, w_j\right| : w \in X\right\}$$

is attained and is non-zero. Define

$$\lambda = \frac{-\min_{w \in X} \sum_{j,k} \frac{\partial^2 \rho}{\partial x_j \partial x_k}(P) w_j w_k}{\mu^2} + 1.$$

Set $\tilde{\rho} = \rho_\lambda$. Then for any $w \in \mathbb{R}^N$ with $|w| = 1$ we have (since $\exp(\rho(P)) = 1$) that

$$\sum_{j,k} \frac{\partial^2 \tilde{\rho}}{\partial x_j \partial x_k}(P)\, w_j w_k$$

$$= \sum_{j,k} \left\{ \frac{\partial^2 \rho}{\partial x_j \partial x_k}(P) + \lambda \frac{\partial \rho}{\partial x_j}(P) \frac{\partial \rho}{\partial x_k}(P) \right\} w_j w_k$$

$$= \sum_{j,k} \left\{ \frac{\partial^2 \rho}{\partial x_j \partial x_k}(P) \right\} w_j w_k + \lambda \left|\sum_j \frac{\partial \rho}{\partial x_j}(P)\, w_j\right|^2. \qquad (6.5)$$

If $w \notin X$, then the expression in (6.5) is positive by definition, while if $w \in X$ then the expression is positive by the choice of λ. Since $\{w \in \mathbb{R}^N : |w| = 1\}$ is compact, there is thus a $C > 0$ such that

$$\sum_{j,k} \frac{\partial^2 \tilde{\rho}}{\partial x_j \partial x_k}(P)\, w_j w_k \geq C, \quad \forall w \in \mathbb{R}^N \text{ such that } |w| = 1.$$

This establishes our inequality (6.4) for $P \in \partial\Omega$ fixed and w in the unit sphere of $\mathbb{R}^N$. For arbitrary w, we set $w = |w|\hat{w}$, with $\hat{w}$ in the unit sphere. Then (6.4) holds for $\hat{w}$. Multiplying both sides of the inequality for $\hat{w}$ by $|w|^2$ and performing some algebraic manipulations, we obtain the result for fixed P and all $w \in \mathbb{R}^N$.

Finally, notice that our estimates—in particular the existence of C, hold uniformly over points in $\partial\Omega$ near P. Since $\partial\Omega$ is compact, we see that the constant C may be chosen uniformly over all boundary points of Ω. ∎

Notice that the statement of Lemma 6.1.4 has two important features: (i) that the constant C may be selected uniformly over the boundary and (ii) that the inequality (6.4) holds for all $w \in \mathbb{R}^N$ (not just tangent vectors). In fact it is impossible to arrange for anything like (6.4) to be true at a weakly convex point. Also note that our proof shows that (6.4) is true not

6.1. THE CLASSICAL NOTION OF CONVEXITY

just for $P \in \partial \Omega$ but for P in a neighborhood of $\partial \Omega$. It is this sort of stability of the notion of strong convexity that makes it a more useful device than weak convexity.

Proposition 6.1.5 *Let $\Omega = \{x \in \mathbb{R}^N : \rho(x) < 0\}$ have C^2 boundary. If Ω is analytically convex, then we may write Ω as an increasing union of strongly analytically convex domains.*

Proof: To simplify the proof we shall assume that Ω has at least C^3 boundary.

Assume without loss of generality that $N \geq 2$ and $0 \in \Omega$. For $\epsilon > 0$ and $M \in \mathbb{N}$ to be selected, let

$$\rho_\epsilon(x) = \rho(x) + \epsilon \frac{|x|^{2M}}{M} \quad \text{and} \quad \Omega_\epsilon = \{x : \rho_\epsilon(x) < 0\}.$$

Then $\Omega_\epsilon \subset \Omega_{\epsilon'}$ if $\epsilon' < \epsilon$ and $\cup_{\epsilon > 0} \Omega_\epsilon = \Omega$. If $M \in \mathbb{N}$ is large and ϵ is small, then Ω_ϵ is strongly analytically convex. This explicitly exhibits the required exhaustion. ∎

Proposition 6.1.6 *If Ω is connected and strongly analytically convex, then Ω is geometrically convex.*

Proof: We use a connectedness argument.

Clearly $\Omega \times \Omega$ is connected. Set $S = \{(P_1, P_2) \in \Omega \times \Omega : (1-\lambda)P_1 + \lambda P_2 \in \Omega, \text{ all } 0 \leq \lambda \leq 1\}$. Then S is plainly open and non-empty.

To show S is closed, we fix a defining function $\tilde{\rho}$ for Ω as in Lemma 6.1.4 and argue by contradiction. If S is not closed in $\Omega \times \Omega$, then there exist $P_1, P_2 \in \Omega$ such that the function

$$t \mapsto \tilde{\rho}((1-t)P_1 + tP_2)$$

assumes an interior maximum value of 0 on $[0, 1]$. But the positive definiteness of the real Hessian of $\tilde{\rho}$ contradicts that assertion. ∎

Proposition 6.1.7 *If Ω is connected and weakly analytically convex, then Ω is geometrically convex.*

Proof: By Proposition 6.1.5, Ω is exhausted by strongly analytically convex domains Ω_j. By Proposition 6.1.6, each Ω_j is geometrically convex, so Ω is geometrically convex. ∎

Proposition 6.1.8 *Let $\Omega \subset\subset \mathbb{R}^N$ have C^2 boundary and be geometrically convex. Then Ω is weakly analytically convex.*

Proof: Seeking a contradiction, we suppose that for some $P \in \partial\Omega$ and some $w \in T_P(\partial\Omega)$ we have

$$\sum_{j,k} \frac{\partial^2 \rho}{\partial x_j \partial x_k}(P) w_j w_k = -2K < 0. \tag{6.6}$$

Suppose without loss of generality that coordinates have been selected in $\mathbb{R}^N$ so that $P = 0$ and $(0, 0, \ldots, 0, 1)$ is the unit outward normal vector to $\partial\Omega$ at P. We may further normalize the defining function ρ so that $\partial\rho/\partial x_N(0) = 1$. Let $Q = Q^t = tw + \epsilon \cdot (0, 0, \ldots, 0, 1)$, where $\epsilon > 0$ and $t \in \mathbb{R}$. Then, by Taylor's expansion, we have

$$\begin{aligned} \rho(Q) &= \rho(0) + \sum_{j=1}^N \frac{\partial \rho}{\partial x_j}(0) Q_j + \frac{1}{2} \sum_{j,k=1}^N \frac{\partial^2 \rho}{\partial x_j \partial x_k}(0) Q_j Q_k + o(|Q|^2) \\ &= \epsilon \frac{\partial \rho}{\partial x_N}(0) + \frac{t^2}{2} \sum_{j,k=1}^N \frac{\partial^2 \rho}{\partial x_j \partial x_k}(0) w_j w_k + \mathcal{O}(\epsilon^2) + o(t^2) \\ &= \epsilon - Kt^2 + \mathcal{O}(\epsilon^2) + o(t^2). \end{aligned}$$

Thus if $t = 0$ and $\epsilon > 0$ is small enough then $\rho(Q) > 0$. However, for that same value of ϵ, if $|t| > \sqrt{2\epsilon/K}$ then $\rho(Q) < 0$. This contradicts the definition of geometric convexity. ∎

Remark 6.1.9 The reader can already see in the proof of Proposition 6.1.8 how useful the quantitative version of convexity can be.

The assumption that $\partial\Omega$ be C^2 is not very restrictive, for convex functions of one variable are twice differentiable almost everywhere (see Zygmund [1] as well as the discussion later in this section). On the other hand, C^2 smoothness of the boundary is essential for our approach to the subject.

> **Exercise:** Prove that if Ω_0 is *any* bounded domain with C^2 boundary, then there is a relatively open subset U of $\partial\Omega_0$ such that every point of U is strongly analytically convex. (Hint: Fix $x_0 \in \Omega_0$ and choose $P \in \partial\Omega_0$ that is as far as possible from x_0).

Convex Functions

We next turn our attention to convex functions. We begin our discussion by restricting our attention to functions of one real variable.

Definition 6.1.10 *Let $I = (s,t)$ be an open interval in $\mathbb{R}$. A function $f : I \to \mathbb{R}$ is called* **convex** *if, for all $a, b \in I$ and all $0 \leq \lambda \leq 1$, we have that*

$$f(\lambda a + (1-\lambda)b) \leq \lambda f(a) + (1-\lambda) f(b). \tag{6.7}$$

6.1. THE CLASSICAL NOTION OF CONVEXITY

This definition is plainly equivalent to requiring the set

$$\{(x,y) \in \mathbb{R} \times \mathbb{R} : s < x < t,\ f(x) \leq y\}$$

to be geometrically convex.

Convex functions are remarkably regular, as we shall now demonstrate. Let us begin by making a small change in notation. If $a < c < b$ are in the domain of the convex function f, then by rewriting (6.7) with $c = \lambda a + (1 - \lambda)b$ so that $\lambda = (b - c)/(b - a)$, we have

$$f(c) \leq \frac{b-c}{b-a}f(a) + \frac{c-a}{b-a}f(b). \tag{6.8}$$

In turn, (6.8) may be rewritten as

$$\frac{b-c}{b-a}[f(c) - f(a)] \leq \frac{c-a}{b-a}[f(b) - f(c)]$$

or

$$\frac{f(c) - f(a)}{c - a} \leq \frac{f(b) - f(c)}{b - c}. \tag{6.9}$$

This says, quite simply, that the slopes of secants to the graph of f increase as we move from left to right along the graph.

In particular, if $[\alpha, \beta]$ is a compact subinterval of the domain $I = (s, t)$ of f then every quotient

$$\frac{f(x+h) - f(x)}{h}$$

for $x \in [\alpha, \beta]$ and $0 < h < \beta - x$, is bounded below by $[f(q) - f(p)]/(q - p)$ for any $s < p < q \leq \alpha$ and bounded above by $[f(v) - f(u)]/(v - u)$ for any $\beta \leq u < v < t$. In particular, it follows that f is absolutely continuous.

Thus f is differentiable off a set of Lebesgue measure zero, and it follows from (6.9) that the derivative function is monotone increasing. We may take this first derivative function to be corrected so that it is monotone increasing everywhere. Being a monotone function, the derivative is in turn differentiable except on a set of Lebesgue measure zero (see *e.g.* Folland [1]). We conclude that a convex function of a single variable is twice differentiable except on a set of measure zero and that second derivative is non-negative wherever it exists.

Now a convex function of several real variables may be defined in a manner similar to that in the first paragraph:

Definition 6.1.11 *Let $\Omega \subset \mathbb{R}^N$ be geometrically convex; that is, assume that whenever $a, b \in \Omega$ and $0 \leq \lambda \leq 1$ then $\lambda a + (1 - \lambda)b \in \Omega$. A function*

$$f : \Omega \to \mathbb{R}$$

is termed **convex** *if, for all $a, b \in \Omega$ and all $0 \leq \lambda \leq 1$, Eq. (6.7) is satisfied, that is,*

$$f(\lambda a + (1 - \lambda)b) \leq \lambda f(a) + (1 - \lambda)f(b).$$

By the discussion of one variable convex functions, we see immediately that if f is, in addition, C^2 then f will have non-negative second derivative in every direction $\xi \in \mathbb{R}^N$ at each point of its domain. An elementary calculation, similar to the one we performed when deriving the condition for analytic convexity of a domain in space, then shows that this observation entails the explicit inequality:

$$\sum_{j,k=1}^{N} \frac{\partial^2 f}{\partial x_j \partial x_k}(P) \xi_j \xi_k \geq 0$$

for all points $P \in \Omega$ and for all direction vectors $\xi \in \mathbb{R}^N$. In other words, the Hessian matrix of f must be positive semi-definite.

Definition 6.1.12 *If the Hessian matrix of the convex function f is positive definite at a point $P \in \Omega$, then we say that f is* **strongly convex** *at P.*

In the converse direction, a C^2 function f of several real variables that has positive semi-definite Hessian matrix at each point of its (convex) domain will satisfy condition (6.7). This is true because, once points a, b in the domain of f and a number $0 \leq \lambda \leq 1$ have been fixed then we may restrict attention to points in the domain lying on the line passing through a and b. Then the positive semi-definiteness of the Hessian translates to non-negativity of the second derivative of this restricted function along this line segment. Thus, by elementary calculus, (6.7) will hold.

The consideration of smoothness of an arbitrary convex function of several real variables is rather more subtle than that for functions of a single variable. A proof that a convex function of several variables is twice differentiable at almost all points of its domain (the Aleksandrov-Buseman-Feller theorem) may be found in Evans and Gariepy [1]; see also Bianchi et al. [1]. Its full explication would require the development of considerable auxiliary machinery. We shall forgo the details at this time.

A detailed discussion of convex functions of a single variable, from a traditional point of view, appears in Zygmund [1], pp. 21–26. Of particular interest in that context are the applications to integration theory. We shall sketch some of those applications now.

Suppose that ϕ, ψ are two non-negative functions defined on an interval $[0, D]$, each continuous, strictly increasing, and vanishing at the origin. Further assume that $\phi = \psi^{-1}$. Define

$$\Phi(x) = \int_0^x \phi(t)\, dt \quad \text{and} \quad \Psi(x) = \int_0^x \psi(t)\, dt.$$

(It is not difficult to see, in light of the preceding discussions, that Φ and Ψ are convex functions.) Such a pair of functions is said to be **complementary in the sense of Young**. An examination of Figure 6.1 shows that

6.1. THE CLASSICAL NOTION OF CONVEXITY

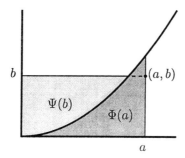

Figure 6.1: **Complementary Functions**

if $0 \leq a, b \leq D$, then
$$ab \leq \Phi(a) + \Psi(b). \tag{6.10}$$

To derive an interesting and familiar example, we fix $\lambda > 0$. Set $\phi(t) = t^\lambda$ and $\psi(t) = t^{1/\lambda}$. Now set $p = 1 + \lambda$ and $p' = 1 + 1/\lambda$. Then calculation of Φ, Ψ and application of (6.10) yields the inequality

$$ab \leq \frac{a^p}{p} + \frac{b^{p'}}{p'}. \tag{6.11}$$

It is important to note that p and p' are connected by the equation

$$\frac{1}{p} + \frac{1}{p'} = 1.$$

If f, g are measurable functions on a measure space (X, μ), then we may apply (6.10) to $a = |f(x)|$ and $b = |g(x)|$. Thus

$$|f \cdot g| \leq \Phi(|f|) + \Psi(|g|).$$

In particular, by (6.11),

$$|f(x) \cdot g(x)| \leq \frac{1}{p}|f(x)|^p + \frac{1}{p'}|f(x)|^{p'}.$$

Integrating both sides gives

$$\int_X |f(x) \cdot g(x)| \, d\mu(x) \leq \frac{1}{p} \int_X |f(x)|^p \, d\mu(x) + \frac{1}{p'} \int_X |g(x)|^{p'} \, d\mu(x). \tag{6.12}$$

This inequality tells us that if $f \in L^p$ and $g \in L^{p'}$ then $f \cdot g \in L^1$. However the inequality is not in the most standard and useful form that comes up in measure theory. In particular, the inequality (6.12) would not stand up to a dimensional analysis. We remedy this matter by normalizing f, g so that

$$\int_X |f(x)|^p \, d\mu(x) = 1 \quad \text{and} \quad \int_X |g(x)|^{p'} \, d\mu(x) = 1.$$

Then inequality (6.12) becomes

$$\int_X |f(x) \cdot g(x)|\, d\mu(x) \leq 1.$$

This may be rewritten as

$$\int_X |f(x) \cdot g(x)|\, d\mu(x) \leq \int_X |f(x)|^p\, d\mu(x)^{1/p} \cdot \int_X |g(x)|^{p'}\, d\mu(x)^{1/p'}. \qquad (6.13)$$

This is the familiar inequality of Hölder.

The theory of Orlicz spaces is inspired by the treatment of Hölder's inequality that we have just given. We now outline its principal features.

We let ϕ, ψ, Φ, Ψ be as above and, following Zygmund [1], p. 170 ff., we declare a measurable function f on the measure space (X, μ) to be in L_Φ if $\Phi \circ |f|$ is integrable.

Now suppose that the function f on (X, μ) has the property that $f \cdot g$ is integrable for every function $g \in L_\Psi$. We then say that $f \in \tilde{L}_\Phi$. We set

$$\|f\|_{L_\Phi} \equiv \|f\|_\Phi = \sup\left\{ \left| \int_X f(t) \cdot g(t)\, d\mu(t) \right| : \int_X \Psi(|g|)(t)\, d\mu(t) \leq 1 \right\}.$$

Certainly the inequality (6.12) or (6.13) implies that $L_\Phi \subset \tilde{L}_\Phi$. Of course $\tilde{\Phi}$ is a normed linear space, and it is called an **Orlicz space**. One can prove that if $f \in \tilde{L}_\Phi$ then $\|f\|_\Phi$ is finite, but this assertion is not entirely obvious.

It turns out that Orlicz spaces are very natural generalizations of L^p spaces, and they often arise in questions of harmonic analysis and partial differential equations. As an example, the Hardy-Littlewood maximal function is bounded on L^p for $p > 1$ but is definitely unbounded on L^1. It is reasonable to seek function spaces X such that $L^p \subset X \subset L^1$ for all $p > 1$ and for which the Hardy-Littlewood maximal function is bounded (in some strong sense) on X. The sharp result is that if f satisfies

$$\int_X |f(t)| \cdot \log^+ |f(t)|\, d\mu(t) < \infty, \qquad (6.14)$$

where

$$\log^+ |f(t)| \equiv \begin{cases} |f(t)| & \text{if } |f(t)| \geq 1 \\ 0 & \text{if } |f(t)| < 1, \end{cases}$$

then Mf is L^1 (here Mf is the Hardy-Littlewood maximal function—see Section 3.5). The class of functions that satisfy (6.14) is commonly denoted by $L(\log^+ L)$.

In the study of convergence of Fourier series, one studies function spaces of the form $L(\log^+ L)^a (\log^+ \log^+ L)^b$ in an effort to find the sharp function space on which pointwise convergence obtains.

Obversely, in the study of partial differential equations it is common to encounter spaces of functions that satisfy an integrability condition of the form

$$\int_X \exp(c \cdot |f(t)|^a) \, d\mu(t).$$

Because

$$e^t = \sum_{j=0}^{\infty} \frac{t^j}{j!},$$

it is clear that a function satisfying the exponential integrability condition will be in every L^p class, and that the L^p norms blow up at a certain rate. It turns out that functions of bounded mean oscillation are exponentially integrable in a certain sense. The reference Krantz [6] discusses these ideas in some detail.

In a certain sense, which we shall not specify in any detail (but which fits naturally into the Orlicz duality of L_Φ and L_Ψ), functions in the exponential integrability classes are dual to functions in the logarithmic integrability classes. The source Zygmund [1] contains further information on this topic.

6.2 Other Characterizations of Convexity

Supporting Hyperplanes

Definition 6.2.1 *Let $S \subset \mathbb{R}^N$ be any set and let $\Pi \subset \mathbb{R}^N$ be a hyperplane. We say that Π is a **supporting hyperplane** for S at $P \in S$ if $P \in \Pi$ and S lies in one or the other of the two closed half-spaces determined by Π.*

One can use induction on N to show that every boundary point of a geometrically convex set S has a supporting hyperplane (see Theorem 6.2.5). The geometrically convex sets in $\mathbb{R}^1$ are simply the intervals, so the first non-trivial case is $\mathbb{R}^2$, and interestingly this case is the key to the entire argument.

Lemma 6.2.2 *Suppose $\Omega \subset \mathbb{R}^2$ is open and geometrically convex. For each θ, set*

$$R(\theta) = \Omega \cap \{(t\cos\theta, t\sin\theta) : 0 < t < \infty\}.$$

(i) *If*

$$R(\theta_1) \neq \emptyset, \quad R(\theta_2) \neq \emptyset, \quad \text{and} \quad \theta_1 < \theta_2 < \theta_1 + \pi,$$

then

$$R(\theta) \neq \emptyset \quad \text{for} \quad \theta \in [\theta_1, \theta_2].$$

(ii) *If*
$$R(\theta) \neq \emptyset, \quad R(\theta + \pi) \neq \emptyset,$$
then the origin is a point of Ω.

Proof: Obvious. ∎

Lemma 6.2.3 *If Ω is a geometrically convex, open subset of $\mathbb{R}^2$ and the origin is a boundary point of Ω, then there exists θ^* such that*
$$R(\theta^*) = R(\theta^* + \pi) = \emptyset,$$
where $R(\theta)$ is as in Lemma 6.2.2. That is,
$$L(\theta^*) \equiv \{ (t\cos\theta^*, t\sin\theta^*) : -\infty < t < \infty \}$$
is a supporting line for Ω at the origin.

Proof: Since Ω is non-empty, there exists θ_0 so that $R(\theta_0) \neq \emptyset$. Set
$$\theta_1 = \inf\{\theta : R(\theta) \neq \emptyset, \quad \theta_0 - \pi < \theta < \theta_0\}$$
and
$$\theta_2 = \sup\{\theta : R(\theta) \neq \emptyset, \quad \theta_0 < \theta < \theta_0 + \pi\}.$$
By Lemma 6.2.2(i), we have $R(\theta) \neq \emptyset$ for $\theta \in (\theta_1, \theta_0] \cup [\theta_0, \theta_2) = (\theta_1, \theta_2)$.

Arguing by contradiction, we can see that we must have $\theta_2 - \theta_1 \leq \pi$. If not, then there would exist θ such that $\theta_1 < \theta < \theta + \pi < \theta_2$. But then, by Lemma 6.2.2(ii), the origin would be a point of Ω, contradicting the assumption that the origin is a boundary point of Ω.

Because Ω is open, we have $\theta_1 < \theta_0 < \theta_2$, and thus
$$\theta_0 - \pi < \theta_1 < \theta_0 < \theta_2 < \theta_0 + \pi. \tag{6.15}$$
Also, because Ω is open and by the inequalities in (6.15), we have $R(\theta_1) = R(\theta_2) = \emptyset$. Thus, we may take $\theta^* = \theta_1$. ∎

Remark 6.2.4 *In the proof of Lemma 6.2.3, we could equally well have chosen $\theta^* = \theta_2 - \pi$. Of course, it may happen, and does when the boundary is smooth, that $\theta_1 = \theta_2 - \pi$.*

Theorem 6.2.5 *If $\Omega \subset \mathbb{R}^N$ is open and geometrically convex, then every boundary point of Ω has a supporting hyperplane.*

Proof: We argue by induction on N. The result is trivial in case $N = 1$ and is shown by Lemma 6.2.3 in case $N = 2$. Suppose now that $N \geq 3$, P is a boundary point of $\Omega \subset \mathbb{R}^N$, and the result holds in $\mathbb{R}^{N-1}$.

6.2. OTHER CHARACTERIZATIONS OF CONVEXITY

We may suppose an orthonormal coordinate system $(x_1, x_2, \ldots, x_N)$ has been chosen with P at the origin. We will now show that there is a line L passing through P that does not intersect Ω. Set

$$\Omega^* = \Omega \cap \{ (x_1, x_2, 0, 0, \ldots, 0) : x_1, x_2 \in \mathbb{R} \}.$$

If Ω^* is empty, then the x_1-axis can be chosen as L. If Ω^* is non-empty, then, by Lemma 6.2.3, there is θ^* so that

$$L(\theta^*) = \{ (t \cos \theta^*, t \sin \theta^*, 0, 0, \ldots, 0) : -\infty < t < \infty \}$$

is a supporting line for Ω^*. In this case, we take $L = L(\theta^*)$.

Rotating coordinates if necessary, we may assume that L is the x_1-axis. Let $\Omega^\dagger$ be the orthogonal projection of Ω on the hyperplane

$$\Pi_1 = \{ (0, x_2, x_3, \ldots, x_N) : x_2, x_3, \ldots, x_N \in \mathbb{R} \}.$$

Identifying the hyperplane Π_1 with $\mathbb{R}^{N-1}$, we may apply the induction hypothesis to obtain an $(N-2)$-dimensional plane

$$\Pi^\dagger \subset \{ (0, x_2, x_3, \ldots, x_N) : x_2, x_3, \ldots, x_N \in \mathbb{R} \}$$

that contains the origin and is such that $\Omega^\dagger \cap \Pi^\dagger = \emptyset$. Then

$$\Pi = \{ (x_1, x_2, x_3, \ldots, x_N) : (0, x_2, x_3, \ldots, x_N) \in \Pi^\dagger \}$$

is a supporting hyperplane to Ω at P. ∎

Corollary 6.2.6 *If S is a geometrically convex set, then every boundary point of S has a supporting hyperplane.*

Proof: It is easy to see that if the interior of S is non-empty, then

$$\partial S = \partial \left(\overset{\circ}{S} \right).$$

Thus if the interior of S is non-empty, then the corollary follows from Theorem 6.2.5. The only way that the convex set S can have no interior point is if the entire set S is contained in a hyperplane. In the case that S is contained in a hyperplane, then that hyperplane is a supporting hyperplane for every point of S, and the conclusion holds. ∎

Corollary 6.2.7 *If S is a geometrically convex set, then the Lebesgue measure of the boundary of S is zero.*

Proof: By Corollary 3.5.6, almost every point of ∂S has density 1. But by Corollary 6.2.6, every point of ∂S has density less than or equal to $1/2$. Thus ∂S must have Lebesgue measure zero. ∎

It turns out that, for non-empty open sets, the existence of a supporting hyperplane at each boundary point is a characterization of geometric convexity.

Theorem 6.2.8 *If $\Omega \subset \mathbb{R}^N$ is open and every boundary point of Ω has a supporting hyperplane, then Ω is geometrically convex.*

Proof: Set
$$S = \bigcap_{P \in \partial\Omega} H(P),$$
where $H(P)$ is the closed half-space that is determined by the supporting hyperplane to Ω at P and that contains Ω. Clearly, $\Omega \subset \overset{\circ}{S}$.

To see that $\overset{\circ}{S} \subset \Omega$, we argue by contradiction. If $Q \in \overset{\circ}{S} \setminus \Omega$, then let $P_0 \in \partial\Omega$ be such that
$$|P_0 - Q| = \operatorname{dist}(Q, \Omega)..$$
Since $\operatorname{dist}(Q, \Omega) > 0$ and, by definition, $\Omega \subset H(P_0)$, we have $Q \notin H(P_0)$. Since $S \subset H(P_0)$, we have $Q \notin S$, contradicting the assumption $Q \in \overset{\circ}{S} \setminus \Omega$. ∎

Theorems 6.2.5 and 6.2.8 give us another method for recognizing convexity analytically, a method that applies when the defining function for a domain is only C^1.

Corollary 6.2.9 *Suppose $\Omega = \{x \in \mathbb{R}^N : \rho(x) < 0\}$, where ρ is a C^1 defining function for the domain Ω. Then Ω is geometrically convex if and only if, for each $P \in \partial\Omega$,*
$$\nabla \rho(P) \cdot (Q - P) < 0 \tag{6.16}$$
holds for all $Q \in \Omega$.

Families of Functions

Definition 6.2.10 *Let $\mathcal{L}$ be the collection of affine functions on $\mathbb{R}^N$. Thus $f \in \mathcal{L}$ means that*
$$f(x_1, \ldots, x_N) = b + a_1 x_1 + \cdots + a_N x_N.$$

(i) *If $K \subset \mathbb{R}^N$ then we define*
$$\widehat{K} \equiv \{x \in \mathbb{R}^N : f(x) \leq \max_{y \in K} f(y) \text{ for every } f \in \mathcal{L}\}.$$

*The set $\widehat{K}$ is termed the **hull of K with respect to $\mathcal{L}$**.*

(ii) *We say that a set S is **convex with respect to the family $\mathcal{L}$** if, whenever $K \subset\subset S$ then $\widehat{K} \subset\subset S$.*

Proposition 6.2.11 *Let $\Omega \subset \mathbb{R}^N$ be a connected, open set. Then Ω is geometrically convex if and only if it is convex with respect to the family $\mathcal{L}$.*

6.2. OTHER CHARACTERIZATIONS OF CONVEXITY

Proof: It follows easily from the existence of supporting hyperplanes, Corollary 6.2.6, that if Ω is geometrically convex, then it is convex with respect to $\mathcal{L}$.

On the other hand, if Ω is convex with respect to $\mathcal{L}$ and $P, Q \in \Omega$, then we can take $K = \{P, Q\}$. We note that $\widehat{K}$ equals the segment $\overline{PQ}$, so $\overline{PQ} \subset \Omega$. Thus Ω is geometrically convex. ∎

The notion of convexity with respect to a family of functions is a useful one, especially when that family is either a linear space or forms a cone. The function theory of several complex variables is another subject in which this circle of ideas arises (see Krantz [3] and Hörmander [2]).

Nearest Point Property

Theorem 6.2.12 *If $S \subset \mathbb{R}^N$ is closed and geometrically convex, then for each $P \in \mathbb{R}^N$ there exists a unique $Q \in S$ with*

$$|Q - P| = \text{dist}(P, S), \tag{6.17}$$

and the mapping $P \mapsto Q$ is a Lipschitz function with Lipschitz constant 1.

Proof: It is easy to see that if there were two distinct points Q and Q' that both satisfied (6.17), then the midpoint $(Q + Q')/2$ would be strictly closer to P and would be a point of S, which is impossible.

In showing that the Lipschitz constant of the mapping is 1, we may assume that S is not contained in an $(N-1)$-dimensional hyperplane. Suppose P_1 and P_2 are distinct points of $\mathbb{R}^N \setminus S$, and Q_i is the point of S satisfying (6.17) with P replaced by P_i. Let Π be the 2-dimensional plane determined by P_1, Q_1, and Q_2. Let L_i be the line of intersection of Π with the supporting hyperplane for S at Q_i. (We leave it to the reader to consider the special case in which Π is contained in the supporting hyperplane at Q_2). We also let P_2' be the orthogonal projection of P_2 onto Π. Note that the line from P_2' to Q_2 must be orthogonal to L_2. Because of the convexity of S, the situation must be as illustrated in Figure 6.2. It is now easy to verify that $|P_2' - P_1| \geq |Q_2 - Q_1|$. Since $|P_2 - P_1| \geq |P_2' - P_1|$, the result follows. ∎

Interestingly, the following converse of Theorem 6.2.12 is also true.

Theorem 6.2.13 *If S is a non-empty closed set such that for each $P \in \mathbb{R}^N$ there is a unique $Q \in S$ satisfying*

$$|Q - P| = \text{dist}(P, S), \tag{6.17}$$

then S is geometrically convex.

Proof: Arguing by contradiction, suppose there are points $P_1, P_2 \in S$ such that there is a point Q on the line segment between them with $Q \notin S$. We

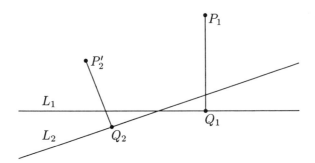

Figure 6.2: **The Nearest Point Configuration**

can then choose $r > 0$ so that

$$\overline{\mathbb{B}}(Q,r) \cap S = \emptyset,$$

and can then let $\overline{\mathbb{B}}(Q',r')$ be the ball with largest possible radius such that

$$\overline{\mathbb{B}}(Q,r) \subset \overline{\mathbb{B}}(Q',r') \quad \text{and} \quad \mathbb{B}(Q',r') \cap S = \emptyset. \tag{6.18}$$

Because there is a unique point P' of S nearest to Q', we must have

$$\overline{\mathbb{B}}(Q',r') \cap S = \{P'\}.$$

Heuristically, the argument proceeds as follows: While keeping the ball $\overline{\mathbb{B}}(Q,r)$ on the inside, expand a ball so that its radius is as large as possible with only the surface of the ball making contact with S. Presumably the center of the ball might shift as the radius increases, but finally a center Q' and radius r' will have been chosen so that r' is as large as possible subject to the condition (6.18). We note that the nearest point property tells us that the ball $\overline{\mathbb{B}}(Q',r')$ can only be touching S at one point. Physical experience indicates that contact with a ball at only one point is not enough to hold the ball fixed; there is a lot of freedom to move the ball $\overline{\mathbb{B}}(Q',r')$ away from contact with S by moving its center. The difficulty is in keeping $\overline{\mathbb{B}}(Q,r)$ inside while moving the ball. Nonetheless, if we succeed in moving the ball away from contact with S while keeping $\overline{\mathbb{B}}(Q,r)$ on the inside, then we can increase the radius of the ball. This contradicts the maximality of the radius r'. The rest of the formal argument makes this precise.

Notice that if $\mathbf{v}$ is any unit vector with

$$\mathbf{v} \cdot (Q' - P') > 0,$$

6.2. OTHER CHARACTERIZATIONS OF CONVEXITY

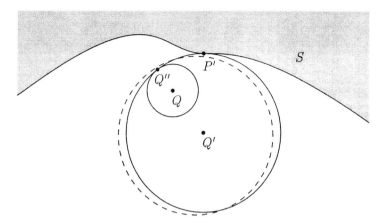

Figure 6.3: **Moving the Largest Ball**

then, for all sufficiently small $\epsilon > 0$,

$$\overline{\mathbb{B}}(Q' + \epsilon \mathbf{v}, r') \cap S = \emptyset.$$

If we were to also have

$$\overline{\mathbb{B}}(Q, r) \subset \overline{\mathbb{B}}(Q' + \epsilon \mathbf{v}, r'),$$

then a larger value of r' could be chosen and still satisfy (6.18), which is the desired contradiction.

In fact, one can choose $\mathbf{v} = Q' - P'$ if $\overline{\mathbb{B}}(Q, r) \subset \overline{\mathbb{B}}(Q', r')$, and otherwise choose $\mathbf{v} = Q'' - P'$ where Q'' is the unique point of $\overline{\mathbb{B}}(Q, r) \setminus \mathbb{B}(Q', r')$ (see Figure 6.3). ■

The concept of a set of positive reach was introduced in Federer [3]:

Definition 6.2.14 *Let $E \subset \mathbb{R}^N$ be a closed set. We call E a **set of positive reach** if there is an open neighborhood U of E so that each element $x \in U$ has a unique nearest point in E.*

The **reach** of a set E is the supremum of the numbers r so that every point in

$$\{P : \text{dist}(P, E) < r\}$$

has a unique nearest point in E.

Theorems 6.2.12 and 6.2.13 show us that the only sets of infinite reach are the convex sets. In Section 1.2 we showed that a domain with C^2 boundary is a set of positive reach. The following example shows that a domain that is less smooth than C^2 need not be a set of positive reach.

Example 6.2.15 For $\delta > 0$ consider the domain

$$\Omega = \Omega_\delta = \{(x,y) \in \mathbb{R}^2 : y > |x|^{2-\delta},\ x^2 + y^2 < 1\}.$$

The second inequality in the definition of Ω is provided only to make the domain bounded; it is really superfluous. Consider points $P = P_\epsilon = (0, \epsilon) \in \Omega$, $\epsilon > 0$. We claim that if ϵ is small enough then P has *two* nearest points in $\partial \Omega$. In particular, $(0,0)$ is *not* the nearest boundary point to P.

One can see this by direct computation; this is a tedious approach, but it is educational and we sketch it. We assume here that $0 < \delta, \epsilon << 1$. Now let $\phi(t) = (t, |t|^{2-\delta})$ be a point of the boundary curve. Then the distance squared of P to $\phi(t)$ is

$$f(t) = t^2 + (|t|^{2-\delta} - \epsilon)^2.$$

Then

$$f'(t) = 2t + 2(|t|^{2-\delta} - \epsilon) \cdot (2 - \delta)(\operatorname{sgn} t)|t|^{1-\delta}.$$

Notice the following about f :

(i) $f' < 0$ when $t > 0$ is so small that $(t^{2-\delta} - \epsilon) < -\epsilon/2$ and $t^{-\delta} > 8/\epsilon$;

(ii) f' is positive when t is large enough that $(t^{2-\delta} - \epsilon)$ is positive.

This discussion already shows that f does not have a local minimum at 0 (recall that f is even). It does have a minimum at a unique positive value t_0 (the point where f' vanishes) and a corresponding local minimum at $-t_0$ by evenness. These are the abscissas of the two nearest points on $\partial \Omega$ to P.

Extreme Points and the Krein-Milman Theorem

Definition 6.2.16 *A point x in a set $E \subset \mathbb{R}^N$ is called* **extreme** *if whenever $a, b \in E$ and $\lambda a + (1 - \lambda)b = x$ for some $0 < \lambda < 1$, then $a = b = x$.*

If E is the closed unit ball in $\mathbb{R}^N$, then each boundary point is extreme (exercise). Interior points are never extreme. The example of the closed unit ball is potentially misleading, for one might infer that an extreme point is one where all curvatures are positive. While the converse of this statement is true, the domains $E_m = \{(x_1, x_2) \in \mathbb{R}^2 : |x_1|^2 + |x_2|^{2m} < 1\}$ for $m = 2, 3, \ldots$ have all boundary points extreme while the curvature fails to be positive at the points $(\pm 1, 0)$.

A domain for which every boundary point is extreme is called **strictly convex**. Compare this concept with the notion of *strong convexity* of a domain. A strongly convex domain is surely strictly convex, but the converse is not true as the example at the end of the last paragraph shows.

All of our various characterizations of convexity may be applied to detect extreme points. Consider the point of view of affine, or linear functions. A

6.2. OTHER CHARACTERIZATIONS OF CONVEXITY

point x in the boundary of a convex domain $\Omega \subset \mathbb{R}^N$ is extreme if and only if there exists a linear function α on $\mathbb{R}^N$ such that α assumes its maximum value on $\bar{\Omega}$ at x and only at x.

Now consider the point of view of defining functions. A boundary point x of a C^2, bounded, convex domain $\Omega = \{x \in \mathbb{R}^N : \rho(x) < 0\}$ is extreme provided that for each tangent vector ξ to $\partial\Omega$ at x there is a positive integer $m = m(\xi)$ such that the mth directional derivative of ρ in the direction ξ is non-zero. Unfortunately, this condition is not quite necessary, as the example

$$\Omega = \{(x_1, x_2) : |x_1|^2 + 2e^{-1/|x_2|^2} < 1\}$$

and the boundary point $(1,0)$ shows.

We shall not provide detailed verification of these various descriptions of extreme points, as we shall not be using them in what follows. A characterization of extreme points that *is* germane to our discussions is as follows: Let $\Omega \subset \mathbb{R}^N$ be a bounded, convex domain. A point $x \in \partial\Omega$ is extreme if there exists a hyperplane $\mathcal{P}$ such that $\mathcal{P} \cap \bar{\Omega} = \{x\}$. In the case that Ω has C^2 boundary, then this hyperplane is unique and will coincide with the tangent plane at x. If the boundary is not smooth, then this plane need not be unique (consider, for instance, the case that Ω is a truncated half cone and the point x is the vertex). Note that the converse of this assertion is *not* true.

In functional analysis it is useful to characterize convex sets in terms of linear functionals. For example, a closed set E in a Banach space is convex if and only if, for every point y that is external to E, there is a linear functional φ such that $\varphi(y) > \varphi(e)$ for all $e \in E$. Of course a linear functional is completely determined by its null space, which is a codimension-one subspace of the total space so that the functional analysis description is just a reformulation of the standard hyperplane separation condition. For our purposes in Euclidean space, it is more convenient (and more geometric) to work directly with the hyperplanes, and to formulate statements in terms of separating hyperplanes rather than functionals.

There are several useful ways to think about the concept of convex hull. The simplest to state, and least useful to work with, is the following:

Definition 6.2.17 *Let $E \subset \mathbb{R}^N$ be any set. The* **closed convex hull** *of E is defined to be the intersection of all closed geometrically convex sets that contain E.*

Let $E \subset \mathbb{R}^N$ be any set and let $x, y \in E$. A **convex combination** of the points a, b is any point of the form $\lambda a + (1 - \lambda)b$, $0 \leq \lambda \leq 1$. More generally, if $a_1, a_2, \ldots, a_k$ are elements of E and $\lambda_j \geq 0$, $\sum_j \lambda_j = 1$, then $\sum_{j=1}^{k} \lambda_j a_j$ is called a convex combination of points of E. (Notice that this more general definition of convex combination is obtained by iterating the first.) Obviously a convex set contains any convex combination of its points.

We have already encountered the notion of convex combination in our discussion of extreme points: a point $x \in E$ is extreme if it cannot be written as a convex combination of points $a, b \in E$ with at least one of a or b different from x.

Notice that if $a, b \in E$ and if c is a convex combination of a and b, then any convex combination of a and c, for instance, is also a convex combination of a and b. It is plausible that if E is a bounded set, then its convex hull can be generated by adjoining to E all possible convex combinations of points of E. In fact this is the case:

Proposition 6.2.18 *Let $E \subset \mathbb{R}^N$ be a bounded set. The closed convex hull $\hat{E}$ of E coincides with the smallest closed set containing all points of the form $\sum_{j=1}^{k} \lambda_j e_j$ with $e_j \in E$, $\lambda_j \geq 0$, and $\sum_j \lambda_j = 1$.*

Proof: Clearly any geometrically convex set containing E will contain any convex combination of points of E. Therefore the closed convex hull will contain all these convex combinations. That proves one direction.

Conversely, suppose that x is a point that is not in the closure of the set S of all convex combinations of elements of E. Then there is a hyperplane $\mathcal{P}$ that separates x from S. But then x cannot be an element of the convex hull of E, for the half space determined by $\mathcal{P}$ and containing E is a convex set that excludes x. That completes the other half of the proof. ∎

Informally stated, the Krein-Milman Theorem asserts that a compact, convex set is the closed convex hull of its extreme points. In point of fact the heart of the matter is the following more technical result, also due to Krein and Milman [1]:

Proposition 6.2.19 *Let $E \subset \mathbb{R}^N$ be a non-empty, compact, convex set. Then E possesses an extreme point.*

Proof: Let P be a point not in the set. Let $\bar{B}(P, r)$ be the smallest closed ball that is centered at P and contains E (just take the intersection of all such balls). Then any point in $\partial \bar{B}(P, r) \cap E$ must be extreme for E since it is extreme for $\bar{B}(P, r)$. ∎

In fact this proof may be used to show that E must have more than one extreme point. As an exercise, give a lower bound for the number of extreme points this proof guarantees for a compact, convex set in $\mathbb{R}^N$ (your answer should be a function of N). The example of the unit simplex will be useful to consider.

Theorem 6.2.20 (Krein-Milman) *Let $E \subset \mathbb{R}^N$ be a non-empty, compact, convex set. Let S be the set of extreme points of E, which we know by the proposition is non-void. Then E is the closed convex hull of S.*

Proof: Clearly the closed convex hull of S will be contained in E.

6.3 Exhaustion Functions

For the converse direction, let x be a point of E that is not the closed convex hull of S. Then, just as in the proof of Proposition 6.2.18, there is a hyperplane separating x from the convex hull of S. Therefore surely there is an extreme point of E that lies on the same side of the hyperplane as does x. That is a contradiction. ∎

6.3 Exhaustion Functions

Definition 6.3.1 *Let $\Omega \subset \mathbb{R}^N$ be a domain. A function $f : \Omega \to \mathbb{R}$ is called an* **exhaustion function** *for Ω if, whenever c is a real constant, $\{x \in \Omega : f(x) < c\}$ is relatively compact in Ω.*

The purpose of this section is to examine the following result:

Proposition 6.3.2 *Let $\Omega \subset \mathbb{R}^N$ be a domain. Then Ω is geometrically convex if and only if it possesses an exhaustion function that is convex.*

In fact there is a stronger result which is extremely useful:

Theorem 6.3.3 *Let $\Omega \subset \mathbb{R}^N$ be a domain. Then Ω is geometrically convex if and only if it possesses a C^∞ exhaustion function that is strongly convex.*

Before proceeding to proofs, we record some additional results that are in the spirit of exhaustion functions. The functions that we describe in these next two results are sometimes called **bounded exhaustion functions**.

Theorem 6.3.4 *Let $\Omega \subset \mathbb{R}^N$ be any bounded geometrically convex set with C^k boundary, $k \geq 2$. Then there is a C^k function u on $\overline{\Omega}$ such that*

(i) *u is convex on $\overset{\circ}{\Omega}$;*

(ii) *$u < 0$ on Ω;*

(iii) *$u \equiv 0$ on $\partial\Omega$.*

Finally, we have:

Theorem 6.3.5 *Let $\Omega \subset \mathbb{R}^N$ be a bounded geometrically convex domain with C^k boundary, $k \geq 2$. Then there is a C^k convex function u defined on all of $\mathbb{R}^N$ such that $\nabla u \neq 0$ on $\partial\Omega$ and*

$$\Omega = \{x \in \mathbb{R}^N : u(x) < 0\}.$$

Now we shall establish the first result of this section. Theorem 6.3.3 will be a consequence of that one. We begin by constructing a function that behaves like an exhaustion *near the boundary*, and then we modify it elsewhere to suit.

Lemma 6.3.6 *Let $\Omega \subset \mathbb{R}^N$ be a geometrically convex domain (bounded or unbounded). For $x \in \Omega$, let $\delta(x)$ denote the Euclidean distance from x to the complement of Ω. Then the function*

$$\mu(x) \equiv -\log \delta(x)$$

is a convex exhaustion function for Ω.

Proof: That the function μ is an exhaustion function is obvious. The claim of convexity amounts to checking that if $a, b \in \Omega$ and if $0 \leq \lambda \leq 1$, then

$$-\log(\lambda \delta(a) + (1-\lambda)\delta(b)) \leq -\lambda \log \delta(a) - (1-\lambda) \log \delta(b).$$

Some elementary calculations show that this is equivalent to

$$[\delta(a)]^\lambda \cdot [\delta(b)]^{(1-\lambda)} \leq \delta(\lambda a + (1-\lambda)b).$$

The concavity of the logarithm function shows that

$$[\delta(a)]^\lambda \cdot [\delta(b)]^{(1-\lambda)} \leq \lambda \delta(a) + (1-\lambda)\delta(b).$$

So it suffices for us to check that

$$\lambda \delta(a) + (1-\lambda)\delta(b) \leq \delta(\lambda a + (1-\lambda)b).$$

But this inequality is obvious from the definition of geometric convexity. ∎

Lemma 6.3.7 *If Φ is a convex exhaustion function for the domain Ω (bounded or unbounded), then there exists a C^∞ exhaustion function that is strongly convex.*

We defer the proof of this lemma for a moment and turn to a result that may be of separate interest:

Proposition 6.3.8 *Let f be convex on an open set $U \subset \mathbb{R}^N$. Let $P \in U$ and let $r > 0$ be sufficiently small. Then*

$$f(P) \leq \frac{1}{\omega_{N-1}} \int_{\mathbb{S}(0,1)} f(P + r\xi) \, d\mathcal{H}^{N-1}\xi. \tag{6.19}$$

Proof: We certainly know that if ξ is any direction vector of unit length, then

$$f(P) \leq \frac{f(P + r\xi) + f(P - r\xi)}{2}. \tag{6.20}$$

Now integrate both sides of (6.20) over ξ in the unit sphere with respect to the $(N-1)$-dimensional Hausdorff measure. The result follows. ∎

6.3. EXHAUSTION FUNCTIONS

It should be noted that an inequality such as (6.19) can be used to characterize subharmonic functions. There is no such characterization for convex functions. All we are observing here is that a convex function is subharmonic. While this assertion may at first be surprising, we may dispel the surprise by noting that if u is convex on $U \subset \mathbb{R}^N$, then $\partial^2 u/\partial x_j^2 \geq 0$, $j = 1, \ldots N$, hence $\Delta u = \sum_j \partial^2 u/\partial x_j^2 \geq 0$. The reader should see Hörmander [2] for more on these matters.

A corollary of the Proposition 6.3.8 is the following:

Corollary 6.3.9 *Let f be convex on an open set $U \subset \mathbb{R}^N$. If ϕ is a radial function on $\mathbb{R}^N$ (this means that $\phi(x) = \phi(x')$ if $|x| = |x'|$) and if, furthermore, $\phi \in C_c$ and $\int \phi = 1$, then*

$$f(x) \leq r^{-N} \int f(t)\phi((x-t)/r)\, d\mathcal{L}^N(t).$$

Proof: The corollary is proved by introducing polar coordinates and invoking the already-proved sub-averaging property over spheres. ∎

Proof of Lemma 6.3.7: Let $\Omega_c \equiv \{x \in \Omega : \Phi(x) + |x|^2 < c\}, c \in \mathbb{R}$. Then each $\Omega_c \subset\subset \Omega$, by the definition of exhaustion function, and $c' > c$ implies that $\Omega_c \subset\subset \Omega_{c'}$. Let $0 \leq \phi \in C_c^\infty(\mathbb{R}^N)$, $\int \phi = 1$, ϕ radial. We may assume that ϕ is supported in $\mathbb{B}(0,1)$. Pick $\epsilon_j > 0$ such that $\epsilon_j < \text{dist}(\Omega_{j+1}, \partial\Omega)$. For $x \in \Omega_{j+1}$, set

$$\Phi_j(x) = \int_\Omega [\Phi(t) + |t|^2]\epsilon_j^{-N}\phi((x-t)/\epsilon_j)\, dV(t) + |x|^2 + 1.$$

Then Φ_j is C^∞ and strongly convex on Ω_{j+1}. We know that $\Phi_j(t) > \Phi(t) + |t|^2$ on $\overline{\Omega}_j$. Let $\chi \in C^\infty(\mathbb{R})$ be a convex function with $\chi(t) = 0$ when $t \leq 0$ and $\chi'(t), \chi''(t) > 0$ when $t > 0$. Observe that $\Psi_j(x) \equiv \chi(\Phi_j(x) - (j-1))$ is positive and convex on $\Omega_j \setminus \overline{\Omega}_{j-1}$ and is, of course, C^∞. Now we inductively construct the desired function Φ'. First, $\Phi_0 > \Phi$ on Ω_0. If a_1 is large and positive, then $\Phi_1' = \Phi_0 + a_1\Psi_1 > \Phi$ on Ω_1. Inductively, if $a_1, \ldots, a_{\ell-1}$ have been chosen, select $a_\ell > 0$ such that $\Phi_\ell' = \Phi_0 + \sum_{j=1}^\ell a_j\Psi_j > \Phi$ on Ω_ℓ. Since $\Psi_{\ell+k} = 0$ on Ω_ℓ, when $k > 0$, we see that $\Phi_{\ell+k}' = \Phi_{\ell+k'}'$ on Ω_ℓ for any $k, k' > 0$. So the sequence Φ_ℓ' stabilizes on compacta and $\Phi' \equiv \lim_{\ell \to \infty} \Phi_\ell'$ is a C^∞ strongly convex function that majorizes Φ. Hence Φ' is the smooth, strongly convex exhaustion function that we seek. ∎

A nice consequence of Lemma 6.3.7 is the following corollary:

Corollary 6.3.10 *Any geometrically convex domain is the increasing union of smooth, strongly analytically convex domains.*

Proof: Fix a geometrically convex domain Ω and let Ψ be a smooth, strongly convex exhaustion function. By Sard's theorem (see Section 5.1), almost every sublevel set $\Omega_c \equiv \{x \in \Omega : \Psi(x) < c\}$ has smooth boundary. Of course it will also be strongly analytically convex. ∎

A useful tool in treating bounded exhaustion functions is the Minkowski functional. We describe that functional now; the reader is referred to Rudin [3] and Yoshida [1] for further reading on this topic.

Definition 6.3.11 *Let $\Omega \subset \mathbb{R}^N$ be a domain. Assume that 0 is a point of Ω. If $x \in \mathbb{R}^N$ then we define the* **Minkowski functional**, $\mathcal{M}_\Omega$, *by setting*

$$\mathcal{M}(x) = \mathcal{M}_\Omega(x) = \inf\{\lambda \in \mathbb{R} : \lambda > 0 \text{ and } (1/\lambda)x \in \Omega\}.$$

In case Ω is the unit ball then $\mathcal{M}(x)$ is just $\|x\|$. For more general Ω, the Minkowski functional is a device for putting a quasi-norm on $\mathbb{R}^N$. We are particularly interested in the case when Ω is geometrically convex.

Proposition 6.3.12 *Let Ω be a geometrically convex, bounded subset of $\mathbb{R}^N$ containing the origin. Then the Minkowski functional for Ω is a convex function.*

Proof: Because the convex set can be intersected with the plane determined by any two points and the origin, it will suffice to prove the result in case $N = 2$.

First suppose that $\partial \Omega$ is smooth—at least C^2. Let ρ be a defining function for Ω. We define a function f by setting

$$f(x, y, z) = \rho\left(\tfrac{x}{z}, \tfrac{y}{z}\right).$$

We will denote the partial derivatives of ρ with respect to its first and second arguments by $D_1 \rho$ and $D_2 \rho$, respectively. Similar notation also will be used for the second partial derivatives of ρ. We have

$$\frac{\partial f}{\partial x} = \frac{1}{z} D_1 \rho, \quad \frac{\partial f}{\partial y} = \frac{1}{z} D_2 \rho, \quad \frac{\partial f}{\partial z} = -\left(\frac{x}{z^2} D_1 \rho + \frac{y}{z^2} D_2 \rho\right), \quad (6.21)$$

$$\frac{\partial^2 f}{\partial x^2} = \frac{1}{z^2} D_{11} \rho, \quad \frac{\partial^2 f}{\partial y^2} = \frac{1}{z^2} D_{22} \rho, \quad \frac{\partial^2 f}{\partial x \partial y} = \frac{1}{z^2} D_{12} \rho, \quad (6.22)$$

$$\frac{\partial^2 f}{\partial z^2} = \frac{2x}{z^3} D_1 \rho + \frac{2y}{z^3} D_2 \rho + \frac{x^2}{z^4} D_{11} \rho + \frac{2xy}{z^4} D_{12} \rho + \frac{y^2}{z^4} D_{22} \rho, \quad (6.23)$$

$$\frac{\partial^2 f}{\partial x \partial z} = -\frac{1}{z^2} D_1 \rho - \frac{x}{z^3} D_{11} \rho - \frac{y}{z^3} D_{12} \rho, \quad (6.24)$$

$$\frac{\partial^2 f}{\partial y \partial z} = -\frac{1}{z^2} D_2 \rho - \frac{x}{z^3} D_{12} \rho - \frac{y}{z^3} D_{22} \rho. \quad (6.25)$$

6.3. EXHAUSTION FUNCTIONS

Since the Minkowski functional satisfies $f(x, y, \mathcal{M}(x,y)) = 0$, we can differentiate implicitly to find

$$\frac{\partial f}{\partial x} + \frac{\partial f}{\partial z}\frac{\partial \mathcal{M}}{\partial x} = 0, \qquad \frac{\partial f}{\partial y} + \frac{\partial f}{\partial z}\frac{\partial \mathcal{M}}{\partial y} = 0, \qquad (6.26)$$

$$\frac{\partial^2 f}{\partial x^2} + 2\frac{\partial^2 f}{\partial x \partial z}\frac{\partial \mathcal{M}}{\partial x} + \frac{\partial^2 f}{\partial z^2}\left(\frac{\partial \mathcal{M}}{\partial x}\right)^2 + \frac{\partial f}{\partial z}\frac{\partial^2 \mathcal{M}}{\partial x^2} = 0, \qquad (6.27)$$

$$\frac{\partial^2 f}{\partial x \partial y} + \frac{\partial^2 f}{\partial x \partial z}\frac{\partial \mathcal{M}}{\partial y} + \frac{\partial^2 f}{\partial y \partial z}\frac{\partial \mathcal{M}}{\partial x}$$
$$+ \frac{\partial^2 f}{\partial z^2}\frac{\partial \mathcal{M}}{\partial x}\frac{\partial \mathcal{M}}{\partial y} + \frac{\partial f}{\partial z}\frac{\partial^2 \mathcal{M}}{\partial x \partial y} = 0, \quad (6.28)$$

$$\frac{\partial^2 f}{\partial y^2} + 2\frac{\partial^2 f}{\partial y \partial z}\frac{\partial \mathcal{M}}{\partial y} + \frac{\partial^2 f}{\partial z^2}\left(\frac{\partial \mathcal{M}}{\partial y}\right)^2 + \frac{\partial f}{\partial z}\frac{\partial^2 \mathcal{M}}{\partial y^2} = 0, \qquad (6.29)$$

Multiplying equations (6.27), (6.28), and (6.29) by $\left(\frac{\partial f}{\partial z}\right)^2 \xi^2$, $\left(\frac{\partial f}{\partial z}\right)^2 2\xi\eta$, and $\left(\frac{\partial f}{\partial z}\right)^2 \eta^2$, respectively, using (6.26), and collecting all terms involving second derivatives of $\mathcal{M}$ on one side of the equation, we have

$$-\left(\frac{\partial f}{\partial z}\right)^3 \left(\frac{\partial^2 \mathcal{M}}{\partial x^2}\xi^2 + 2\frac{\partial^2 \mathcal{M}}{\partial x \partial y}\xi\eta + \frac{\partial^2 \mathcal{M}}{\partial y^2}\eta^2\right) = \mathbf{v} \cdot H_f \cdot \mathbf{v} \qquad (6.30)$$

where H_f is the Hessian of f and $\mathbf{v}$ is the vector

$$\mathbf{v} = \begin{pmatrix} \xi \frac{\partial f}{\partial z} \\ \eta \frac{\partial f}{\partial z} \\ \xi \frac{\partial f}{\partial x} + \eta \frac{\partial f}{\partial y} \end{pmatrix}. \qquad (6.31)$$

We note that, because the gradient of ρ is an outward pointing normal vector, we have

$$x D_1 \rho + y D_2 \rho > 0 \qquad (6.32)$$

and hence

$$\frac{\partial f}{\partial z} < 0,$$

when we evaluate at $z = \mathcal{M}(x, y)$. Thus the sign of

$$\frac{\partial^2 \mathcal{M}}{\partial x^2}\xi^2 + 2\frac{\partial^2 \mathcal{M}}{\partial x \partial y}\xi\eta + \frac{\partial^2 \mathcal{M}}{\partial y^2}\eta^2$$

will agree with the sign of $\mathbf{v} \cdot H_f \cdot \mathbf{v}$.

Using (6.22)–(6.25), we have

$$\mathbf{v} \cdot H_f \cdot \mathbf{v} = \frac{2}{z}\left(\xi\frac{\partial f}{\partial x} + \eta\frac{\partial f}{\partial y}\right)\nabla\rho \cdot \mathbf{w} + \mathbf{w} \cdot H_\rho \cdot \mathbf{w} \qquad (6.33)$$

where

$$\mathbf{w} = \begin{pmatrix} \frac{x}{z^2}\left(\xi\frac{\partial f}{\partial x} + \eta\frac{\partial f}{\partial y}\right) - \frac{\xi}{z}\frac{\partial f}{\partial z} \\ \frac{y}{z^2}\left(\xi\frac{\partial f}{\partial x} + \eta\frac{\partial f}{\partial y}\right) - \frac{\eta}{z}\frac{\partial f}{\partial z} \end{pmatrix}.$$

We have $\mathbf{w} \cdot H_\rho \cdot \mathbf{w} \geq 0$ by the convexity of Ω. Finally, using (6.21), we see that

$$\frac{2}{z}\left(\xi\frac{\partial f}{\partial x} + \eta\frac{\partial f}{\partial y}\right)\nabla\rho \cdot \mathbf{w} = \frac{2}{z^5}(xD_1\rho + yD_2\rho)(\xi D_1\rho + \eta D_2\rho)^2. \quad (6.34)$$

By (6.32), the right-hand-side of (6.34) is non-negative.

The proof in full generality follows by approximating Ω by an increasing family of smoothly bounded analytically convex domains and passing to the limit. ■

Corollary 6.3.13 *Let Ω be a convex, bounded subset of $\mathbb{R}^N$ containing the origin. If $\partial\Omega$ is C^k, $k \geq 1$, then the Minkowski functional for Ω is C^k, except at the origin.*

Proof: Let ρ be a C^k defining function for Ω. For $u \in \mathbb{R}^N$ and $0 < v$, set

$$f(u,v) = \rho(\tfrac{u}{v}).$$

As in the proof of Proposition 6.3.12, we note that the Minkowski functional satisfies

$$f(u, \mathcal{M}(u)) = 0,$$

so the result follows from the Implicit Function Theorem. ■

Notice that if Ω is a convex domain that contains the origin and if $\mathcal{M}$ is its Minkowski functional, then the function $u(x) \equiv \mathcal{M}(x) - 1$ is a bounded convex exhaustion function. By Corollary 6.3.13, it will be as smooth as the boundary of Ω (except at the origin). Notice further that this u is defined and convex on all of $\mathbb{R}^N$. Thus we have established Theorems 6.3.4 and 6.3.5.

Now the Minkowski functional $\mathcal{M}(x)$ can be used to establish a result on approximation from the outside when Ω is a bounded, convex domain. If $c > 1$, then $\Omega'_c \equiv \{x \in \mathbb{R}^N : \mathcal{M}(x) < c\}$ is a convex domain that contains Ω and $\cap_c \Omega_c = \Omega$ exhibits Ω as a decreasing intersection of convex domains. Further note that Ω is relatively compact in Ω_c for each $c > 1$. Since Ω_c can be exhausted by smooth, strongly convex domains, we may find a strongly convex domain U such that $\Omega \subset\subset U \subset\subset \Omega_c$. It follows that Ω is the decreasing intersection of smooth, strongly analytically convex domains.

6.4 Convexity of Order k

Let Ω be a smoothly bounded domain (C^{k+1} will be sufficiently smooth) and $P \in \partial\Omega$ a point of analytic convexity. We say that P is a point of **convexity of order** k if the tangent hyperplane to $\partial\Omega$ at P has order of contact precisely k with the boundary. Let us flesh out this definition.

Definition 6.4.1 *The following are equivalent definitions of the* **order of contact** *of a hyperplane Π with a hypersurface $M \subset \mathbb{R}^N$ at a point $P \in M$:*

(i) *We say that Π has order of contact k with $M \subset \mathbb{R}^N$ at P if there are orthonormal coordinates $(t_1, \ldots, t_{N-1}, u) = (t, u)$ on $\mathbb{R}^N$ and a smooth function $f(t) = f(t_1, t_2, \ldots, t_{N-1})$ on $\mathbb{R}^{N-1}$ such that, in these coordinates, $P = (0,0)$, $\Pi = \{(t,u) : u = 0\}$, and there are neighborhoods U of P in $\mathbb{R}^N$ and U' of 0 in $\mathbb{R}^{N-1}$ and a constant $0 < C < \infty$ for which*

$$U \cap M = \left\{(t, f(t)) : t \in U'\right\},$$

and

$$|f(t) - f(0)| \leq C|t|^k \quad \text{for } t \in U', \tag{6.35}$$

and further, for no choice of C does (6.35) hold with k replaced by $k+1$.

(ii) *We say that Π has order of contact k with $M \subset \mathbb{R}^N$ at P if, whenever M is expressed as $M = \{x \in \mathbb{R}^N : \rho(x) = 0\}$ with ρ a smooth function satisfying $\nabla \rho \neq 0$ on M, then there is a constant $0 < C < \infty$ such that*

$$|\rho(x)| \leq C|x - P|^k \quad \text{for } x \in \Pi, \tag{6.36}$$

but for no choice of C does the inequality (6.36) hold with k replaced by $k+1$.

This definition tells us that all tangential derivatives of ρ of order not exceeding $(k-1)$ vanish; but some tangential derivative of ρ of order precisely k does not vanish.

(iii) *We say that Π has order of contact k with $M \subset \mathbb{R}^N$ at P if, whenever M is expressed as $M = \{x \in \mathbb{R}^N : \rho(x) = 0\}$ with ρ a smooth function satisfying $\nabla \rho \neq 0$ on M, there are local curvilinear coordinates $t_1, t_2, \ldots, t_N$ on $\mathbb{R}^N$ such that, in these coordinates, $P = 0$, $\Pi = \{(t_1, t_2, \ldots, t_N) : t_N = 0\}$ and*

$$\rho(t_1, t_2, \ldots, t_N) = t_N + R(t_1, t_2, \ldots, t_{N-1}),$$

where, for some constant $0 < C < \infty$, R satisfies

$$|R(t_1, \ldots, t_{N-1})| \leq C\bigl(|t_1|^k + \ldots + |t_{N-1}|^k\bigr), \tag{6.37}$$

and for no choice of C does the inequality (6.37) hold with k replaced by $k+1$.

The equivalence of these definitions is easily proved: That (i)⟹(iii) follows by setting $\rho(t,u) = u - f(t)$. That (iii)⟹(ii) is immediate by changing coordinates. Finally, the implication (ii)⟹(i) is a consequence of the Implicit Function Theorem.

Notice that a strongly analytically convex point is a point that is convex of order 2—the canonical example is a boundary point of the unit ball. The point $(1,0)$ in the boundary of

$$E_m = \{(x_1, x_2) \in \mathbb{R}^2 : |x_1|^2 + |x_2|^{2m} < 1\}$$

is a convex point of order $k = 2m$.

The last example raises the question of whether an analytically convex point can have odd order. The answer is "no", for the simple reason that, if the lead term (after the linear term) in the Taylor expansion of ρ has odd order, then the boundary near P locally looks like the graph of an odd degree polynomial, therefore cannot be geometrically convex, and hence cannot be analytically convex (by Proposition 6.1.8).

A special feature of strongly convex points is that they are stable. This assertion can be formulated more precisely as follows.

Proposition 6.4.2 *Let $\Omega \subset \mathbb{R}^N$ be a domain with C^2 boundary. Let $P \in \partial\Omega$ be a point of strong convexity. Then any $x \in \partial\Omega$ that is sufficiently near P is also a point of strong convexity.*

Proof: By Lemma 6.1.4, we can suppose the defining function for Ω has a Hessian that is positive definite on all vectors, not just the tangent vectors at P. But the second derivatives of the defining function are continuous. Thus the Hessian at nearby points (whether in the boundary or not) will also be positive definite. ∎

The same proof establishes the following:

Proposition 6.4.3 *Let $\Omega = \{x \in \mathbb{R}^N : \rho(x) < 0\}$ be a domain with C^2 boundary. Let $P \in \partial\Omega$ be a point of strong analytic convexity. If $\phi \in C_c^2(\mathbb{R}^N)$ has support in a small neighborhood of P and has sufficiently small C^2 norm and if we set $\tilde{\rho}(x) = \rho(x) + \phi(x)$ then*

$$\tilde{\Omega} \equiv \{x \in \mathbb{R}^N : \tilde{\rho}(x) < 0\}$$

will also be strongly analytically convex at points that are near to P.

The analogue of Proposition 6.4.2 is the following result.

Theorem 6.4.4 *Let $\Omega \subset \mathbb{R}^N$ be a smoothly bounded analytically convex domain. Assume that $P \in \partial\Omega$ is a convex point of order k. Then boundary points that are sufficiently near to P are convex of order at most k.*

6.4. CONVEXITY OF ORDER k

Proof: In fact the assertion has nothing to do with convexity. Our assertion amounts to proving that if the defining function for a domain Ω vanishes at $P \in \partial\Omega$ to order precisely k, then it vanishes at nearby points to order *at most* k. This is really just an assertion about Taylor expansions.

To make the preceding paragraph precise, note that Definition 6.4.1(ii) tells us that if Ω has the smooth defining function ρ then, at P, which is a point of convexity of order k, there are vectors $\mathbf{v}_i = (v_{i,1}, v_{i,2}, \ldots, v_{i,N})$, $i = 1, 2, \ldots, k$, which are tangent to $\partial\Omega$ at P and are such that

$$0 \neq \sum_{j_1=1}^{N} \sum_{j_2=1}^{N} \cdots \sum_{i_k=1}^{N} v_{1,j_1} v_{2,j_2} \cdots v_{k,j_k} \frac{\partial^k \rho}{\partial x_1 \partial x_2 \ldots \partial x_k}(P). \quad (6.38)$$

But the inequality (6.38) is an open condition, so it will hold for all points P' sufficiently near to P and all vectors $\mathbf{v}'_1, \mathbf{v}'_2, \ldots, \mathbf{v}'_k$ sufficiently near to the vectors $\mathbf{v}_1, \mathbf{v}_2, \ldots, \mathbf{v}_k$, respectively. Thus for P' sufficiently near to P and with the $\mathbf{v}'_i$ defined to be the orthogonal projections of the $\mathbf{v}_i$ onto the tangent plane to $\partial\Omega$ at P', we have

$$0 \neq \sum_{j_1=1}^{N} \sum_{j_2=1}^{N} \cdots \sum_{i_k=1}^{N} v'_{1,j_1} v'_{2,j_2} \cdots v'_{k,j_k} \frac{\partial^k \rho}{\partial x_1 \partial x_2 \ldots \partial x_k}(P'),$$

proving the result. ∎

Definition 6.4.5 *Let $\Omega = \{x \in \mathbb{R}^N : \rho(x) < 0\}$ be a smooth, analytically convex domain. If $U \subset \mathbb{R}^N$ is open, $U \cap \partial\Omega$ will be said to be **convex of order k** if every point of $U \cap \partial\Omega$ is a point of convexity of order $\ell \leq k$.*

Theorem 6.4.4 tells us that any point of $\partial\Omega$ that is a point of convexity of order k has a neighborhood in $\partial\Omega$ that is convex of order k.

Finding an analogue of Proposition 6.4.3 for convex points of order k with $k > 2$ is somewhat subtle. Such points are only weakly analytically convex, and an arbitrarily small perturbation in *any* smooth topology can turn the point into a concave point. Thus we need to restrict attention to perturbations of a certain kind. Figure 6.4 illustrates the sort of perturbation we wish to consider.

Theorem 6.4.6 *Let $\Omega = \{x \in \mathbb{R}^N : \rho(x) < 0\}$ be a domain. Let $P \in \partial\Omega$ be a point of convexity of order k. Then there is a neighborhood U of P and a function $\phi \in C_c^{k+1}$ such that*

(i) *ϕ is supported in a small neighborhood of P*

(ii) *$\phi \leq 0$*

(iii) *$\phi < 0$ near P*

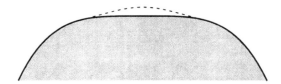

Figure 6.4: **Perturbing at a Point Convex of Order k**

and if $\widetilde{\Omega}$ is given by
$$\widetilde{\Omega} = \{x \in \mathbb{R}^N : \tilde{\rho} < 0\},$$
with $\tilde{\rho} = \rho + \phi$, then $U \cap \partial\widetilde{\Omega}$ is non-empty and is convex of order k. In short, points of order k can be "bumped" outward.

Remark 6.4.7 Notice that this theorem has real content. Consider for example, the unit box
$$B = \{(x,y) \in \mathbb{R}^2 : |x| < 1, |y| < 1\}.$$
Then $P = (1,0)$ is a point in ∂B. But there is no outward perturbation of ∂B near P that will still be analytically convex.

The following lemma will be crucial.

Lemma 6.4.8 *Suppose $U \cap \partial\Omega$ is convex of order k. The set W of points in $U \cap \partial\Omega$ that are not strongly analytically convex form a (relatively) closed, nowhere dense set in $\partial\Omega$.*

Proof: Since the condition for strong analytic convexity is open, it is clear that the set of strongly analytically convex points is open. Therefore its complement is closed.

If W has interior, then there is a relatively open set U in $\partial\Omega$ on which the Hessian vanishes. But that means that, on U, the defining function is linear. But then, because it is infinitely flat, that piece of the boundary is not convex of type k for any k. That is a contradiction. ∎

Proof of Theorem 6.4.6: We shall present the proof only in dimension 2, where it is quick and easy. The higher dimensional case involves interesting but nasty geometric details.

We may as well suppose that the defining function for Ω has the form $\rho(x,y) = -y + \mu(x)$ near P. Let $P = (p, \mu(p))$. Then $\mu'' \geq 0$ near $x = p$. By Lemma 6.4.8, choose points $a < p < b$ such that $\partial\Omega$ is strongly analytically convex at $(a, \mu(a))$ and at $(b, \mu(b))$. In other words, $\mu''(a) > c > 0$ and $\mu''(b) > c > 0$. Select $\delta > 0$ so that these inequalities persist on $(-\delta + a, a + \delta)$ and on $(-\delta + b, b + \delta)$.

6.4. CONVEXITY OF ORDER k

Let ϕ be a C_c^∞ function supported in $[a-\delta, b+\delta]$. Assume further that

$$\{x : \phi'(x) \neq 0\} \subset [a-\delta/2, a] \cup [b+\delta/2]$$

and that $\phi(x) \equiv -1$ for $a+\delta/2 \leq x \leq b-\delta/2$.

Now consider the function

$$\tilde{\rho} = \rho + \epsilon\phi.$$

If $\epsilon > 0$ is chosen small enough then $\sup |\epsilon\phi''| < c/2$. And of course ϕ'' will be supported in the set where $|\rho''| > c$. It follows that the domain $\tilde{\Omega} \equiv \{(x,y) : \tilde{\rho}(x,y) < 0\}$ is still convex of order k. And it is an outward perturbation of the original domain Ω as required. ∎

Chapter 7

Steiner Symmetrization

7.1 Basic Properties

The Swiss geometer Jakob Steiner (1796 – 1863) devoted much of his energy to seeking simple principles from which theorems in geometry could be derived in a natural way. The Isoperimetric Theorem provided one venue for this process and symmetrization was the simple principle he discovered. Mathematical mythology attributes the classical Isoperimetric Theorem (stating that the circle is the plane figure enclosing the greatest area for a given perimeter or equivalently the figure of least perimeter enclosing a given area) to Queen Dido, founder of Carthage. More reliable authority (see Knorr [1] and [2]) credits Zenodorus with a proof in the second or third century B.C. Steiner himself refers to another Swiss mathematician Simon-Antoine-Jean L'Huilier (1750 – 1840) as the one who provided the first definitive exposition of the Isoperimetric Theorem. L'Huilier published a treatise on the subject (in Latin) in 1782. Though an admirer of L'Huilier, Steiner was not satisfied with the status of the Isoperimetric Theorem, for Steiner felt that while L'Huilier's work was often cited, L'Huilier's methods were not rightfully appreciated or understood. Steiner thus sought a deeper and clearer understanding of the classical Isoperimetric Theorem and its variants.

Steiner's first results using symmetrization to prove the Isoperimetric Theorem were presented to the Berlin Academy of Science in 1836 (see Steiner [1]). Ultimately, Steiner gave five proofs showing that *if* there is a figure of least perimeter enclosing a given area, then it must be a circle. That Steiner did not successfully address the problem of existence of such a figure is not surprising, especially in view of the well-known history of Dirichlet's Principle (see the introduction of Courant [1]).[1]

Definition 7.1.1 *Let V be an $(N-1)$-dimensional vector subspace of $\mathbb{R}^N$.*

---
[1] Dirichlet and Steiner were contemporaries at Berlin University.

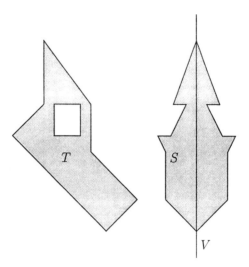

Figure 7.1: **Steiner Symmetrization**

Steiner symmetrization *with respect to V is the operation that associates with each bounded subset T of $\mathbb{R}^N$ the subset S of $\mathbb{R}^N$ having the property that, for each straight line ℓ perpendicular to V, $\ell \cap S$ is a closed line segment with center in V or is empty and the conditions*

$$\mathcal{H}^1(\ell \cap S) = \mathcal{H}^1(\ell \cap T)$$

and

$$\ell \cap S = \emptyset \quad \text{if and only if} \quad \ell \cap T = \emptyset$$

hold.

Figure 7.1 illustrates the application of Steiner symmetrization to an ellipse T in the plane. Symmetrization is performed about the vertical line V. The result of symmetrization is another ellipse S that is, of course, symmetric about V and is such that any horizontal line, that is, any line perpendicular to V, intersects T and S in line segments of the same length.

For Steiner's work the significant facts were that the area of a plane domain is unchanged by symmetrization about a line (by Cavalieri's Principle or, in modern terms, Fubini's Theorem) and that the perimeter is decreased (unless the figure was already symmetric about some line parallel to that with respect to which the Steiner symmetrization is performed). We will need to know more about Steiner symmetrization than just those two facts. In particular, we will need to know the many ways in which Steiner symmetrization is well-behaved. The sequence of propositions below will document this behavior.

7.1. BASIC PROPERTIES

Proposition 7.1.2 *If T is a bounded $\mathcal{L}^N$-measurable subset of $\mathbb{R}^N$ and if S is obtained from T by Steiner symmetrization, then S is $\mathcal{L}^N$-measurable and*
$$\mathcal{L}^N(T) = \mathcal{L}^N(S).$$

Proof: This is a consequence of Fubini's Theorem. ∎

Lemma 7.1.3 *Fix $0 < M < \infty$. If A and $A_1, A_2, \ldots$ are closed subsets of $\mathbb{R}^N \cap \overline{\mathbb{B}}(0, M)$ such that*
$$\bigcap_{i_0=1}^{\infty} \overline{\left[\bigcup_{i=i_0}^{\infty} A_i\right]} \subset A,$$
then
$$\limsup_i \mathcal{H}^N(A_i) \leq \mathcal{H}^N(A).$$

Proof: Let $\epsilon > 0$ be arbitrary. Then there exists an open set U with $A \subset U$ and
$$\mathcal{H}^N(U) \leq \mathcal{H}^N(A) + \epsilon.$$
A routine argument shows that, for all sufficiently large i, $A_i \subset U$. It follows that
$$\limsup_i \mathcal{H}^N(A_i) \leq \mathcal{H}^N(U),$$
and the fact that ϵ was arbitrary implies the lemma. ∎

Proposition 7.1.4 *If T is a compact subset of $\mathbb{R}^N$ and if S is obtained from T by Steiner symmetrization, then S is compact.*

Proof: Let V be an $(N-1)$-dimensional vector subspace of $\mathbb{R}^N$, and suppose that S is the result of Steiner symmetrization of T with respect to V. It is clear that the boundedness of T implies the boundedness of S. To see that S is closed, consider any sequence of points $p_1, p_2, \ldots$ in S that converges to some point p. Each p_i lies in a line ℓ_i perpendicular to V, and we know that
$$\operatorname{dist}(p_i, V) \leq \frac{1}{2}\mathcal{H}^1(\ell_i \cap S) = \frac{1}{2}\mathcal{H}^1(\ell_i \cap T).$$
We also know that the line perpendicular to V and containing p must be the limit of the sequence of lines $\ell_1, \ell_2, \ldots$. Further, we know that
$$\operatorname{dist}(p, V) = \lim_{i \to \infty} \operatorname{dist}(p_i, V).$$
The inequality
$$\limsup_i \mathcal{H}^1(\ell_i \cap T) \leq \mathcal{H}^1(\ell \cap T) \qquad (7.1)$$

would allow us to conclude that

$$\text{dist}(p, V) = \lim_{i \to \infty} \text{dist}(p_i, V) \leq \frac{1}{2} \limsup_{i \to \infty} \mathcal{H}^1(\ell_i \cap T) \leq \frac{1}{2}\mathcal{H}^1(\ell \cap T),$$

and thus that $p \in S$.

To obtain the inequality (7.1), we let q_i be the vector parallel to V that translates ℓ_i to ℓ, and apply Lemma 7.1.3, with N replaced by 1 and with ℓ identified with $\mathbb{R}$, to the sets $A_i = \tau_{q_i}(\ell_i \cap T)$, which are the translates of the sets $\ell_i \cap T$. We can take $A = \ell \cap T$, because T is closed. ∎

Because the Steiner symmetrization is constructed using closed line segments, Steiner symmetrization does not send open sets to open sets. For example, the Steiner symmetrization of the open unit square in $\mathbb{R}^2$

$$(0,1) \times (0,1) = \{(x,y) : 0 < x < 1,\ 0 < y < 1\}$$

about the x-axis is the set

$$(0,1) \times [-\tfrac{1}{2}, \tfrac{1}{2}] = \{(x,y) : 0 < x < 1,\ -\tfrac{1}{2} \leq y \leq \tfrac{1}{2}\}.$$

The following proposition tells us that the Steiner symmetrization of an open set is a G_δ set.

Proposition 7.1.5 *If T is a bounded open subset of $\mathbb{R}^N$ and if S is obtained from T by Steiner symmetrization, then S is a G_δ set.*

Proof: Let V be an $(N-1)$-dimensional vector subspace of $\mathbb{R}^N$, and suppose that S is the result of Steiner symmetrization of T with respect to V. For each positive integer i, let S_i be the subset of $\mathbb{R}^N$ having the property that, for each straight line ℓ perpendicular to V, the set $\ell \cap S_i$ is an open line segment with center in V or is empty and the conditions

$$\mathcal{H}^1(\ell \cap S_i) = \frac{1}{i} + \mathcal{H}^1(\ell \cap T) \quad \text{if } \ell \cap T \neq \emptyset$$

and

$$\ell \cap S_i = \emptyset \quad \text{if } \ell \cap T = \emptyset$$

hold.

First, we claim that each S_i is open. Consider i fixed. If a point p is an element of S_i and if ℓ is the line perpendicular to V passing through p, then

$$\text{dist}(p, V) < \frac{1}{2}\left(\frac{1}{i} + \mathcal{H}^1(\ell \cap T)\right)$$

must hold. Define $r_1 > 0$ by

$$2r_1 = \frac{1}{2}\left(\frac{1}{i} + \mathcal{H}^1(\ell \cap T)\right) - \text{dist}(p, V).$$

7.1. BASIC PROPERTIES

By general measure theory, $\ell \cap T$ must contain a compact subset C with

$$r_1 < \frac{1}{2}\left(\frac{1}{i} + \mathcal{H}^1(C)\right) - \text{dist}(p, V).$$

Since T is open, there is $r_2 > 0$ such that

$$\{x \in \mathbb{R}^N : \text{dist}(x, C) < r_2\} \subset T.$$

Now, if ℓ' is any line perpendicular to V and at a distance less than r_2 from ℓ, then

$$\mathcal{H}^1(\ell' \cap T) \geq \mathcal{H}^1(C)$$

holds, which implies

$$\text{dist}(p, V) + r_1 < \frac{1}{2}\left(\frac{1}{i} + \mathcal{H}^1(\ell' \cap T)\right).$$

We conclude that, for any point q, if $q - p$ resolves into components in the ℓ direction and in the hyperplane V of magnitude less than r_1 and r_2, respectively, then q is in S_i. This condition is satisfied by all points in the open ball centered at p and having radius $\min\{r_1, r_2\}$.

It is clear that $S \subset \cap_i S_i$. To complete the proof, we must show that $\cap_i S_i \subset S$. If $p \in \cap_i S_i$ and if ℓ is the line perpendicular to V passing through p, then for each i we must have

$$\text{dist}(p, V) < \frac{1}{2}\left(\frac{1}{i} + \mathcal{H}^1(\ell \cap T)\right),$$

so passing to the limit as i approaches ∞, we conclude that

$$\text{dist}(p, V) \leq \frac{1}{2}\mathcal{H}^1(\ell \cap T),$$

and thus that $p \in S$. ∎

Remark 7.1.6 It is natural to ask whether or not the Steiner symmetrization of a Borel subset of $\mathbb{R}^N$ is also a Borel set. The answer is "no", but we will only sketch the argument, since a complete, self-contained discussion would take us rather far afield. Let $\mathcal{N}$ be the cartesian product of infinitely many copies of the discretely topologized positive integers, as in Federer [4], §2.2.6, and let $\pi_1 : \mathbb{R}^{(N-1)} \times \mathcal{N} \to \mathbb{R}^{(N-1)}$ be projection on the first factor. By definition, a **Suslin set**[2] is any image $\pi_1(F)$ where F is a closed subset of $\mathbb{R}^{(N-1)} \times \mathcal{N}$. The classical fact (see Federer [4], §2.2.11) discovered by Suslin is that there exists a Suslin set that is not a Borel set.

Applying Federer [4], §2.2.9, we can find a Borel subset Γ of the standard, middle thirds removed, Cantor set $C \subset [0,1]$ such that $\mathcal{N}$ is homeomorphic to Γ. Let γ be the homeomorphism. If F is a closed subset of

---
[2] Suslin sets are also discussed in Section 1.3.

$\mathbb{R}^{(N-1)} \times \mathcal{N}$, then $(\mathrm{id} \times \gamma)(F)$ is a closed subset of $\mathbb{R}^{(N-1)} \times \Gamma$, where id is the identity map on $\mathbb{R}^{(N-1)}$. Thus $(\mathrm{id} \times \gamma)(F)$ is a Borel subset of $\mathbb{R}^{(N-1)} \times \mathbb{R}$. As stated above, we may and shall choose the closed set F so that $\pi_1(F)$ is not a Borel set.

Let $\pi_o : \mathbb{R}^{(N-1)} \times \Gamma \to \mathbb{R}^{(N-1)} \times \{0\}$ be orthogonal projection, and let inj $: \mathbb{R}^{(N-1)} \to \mathbb{R}^{(N-1)} \times \{0\}$ be the homeomorphism sending x to $(x, 0)$. Then the diagram

$$\begin{array}{ccc} \mathbb{R}^{(N-1)} \times \mathcal{N} & \xrightarrow{\mathrm{id} \times \gamma} & \mathbb{R}^{(N-1)} \times \Gamma \\ \downarrow \pi_1 & & \downarrow \pi_o \\ \mathbb{R}^{(N-1)} & \xrightarrow{\mathrm{inj}} & \mathbb{R}^{(N-1)} \times \{0\} \end{array}$$

is commutative.

Because C has 1-dimensional measure zero, the Steiner symmetrization with respect to $\mathbb{R}^{(N-1)} \times \{0\}$ of any subset T of $\mathbb{R}^{(N-1)} \times C$ is the same as the orthogonal projection of T on $\mathbb{R}^{(N-1)} \times \{0\}$. In particular, we set $T = (\mathrm{id} \times \gamma)(F)$ and conclude that if S is the Steiner symmetrization of T with respect to $\mathbb{R}^{(N-1)} \times \{0\}$, then

$$S = \pi_o \circ (\mathrm{id} \times \gamma)(F) = \mathrm{inj} \circ \pi_1(F).$$

Since $\pi_1(F)$ is not a Borel set and inj is a homeomorphism, we see that S is not a Borel set. ∎

Proposition 7.1.7 *If T is a bounded geometrically convex subset of $\mathbb{R}^N$ and S is obtained from T by Steiner symmetrization, then S is also a geometrically convex set.*

Proof: Let V be an $(N-1)$-dimensional vector subspace of $\mathbb{R}^N$, and suppose that S is the result of Steiner symmetrization of T with respect to V. Let x and y be two points of S. We let x' and y' denote the points obtained from x and y by reflection through the hyperplane V. Also, let ℓ_x and ℓ_y denote the lines perpendicular to V and passing through the points x and y, respectively. By the definition of the Steiner symmetrization and the convexity of T, we see that $\ell_x \cap T$ must contain a line segment, say from p_x to q_x, of length at least $\mathrm{dist}(x, x')$. Likewise, $\ell_y \cap T$ contains a line segment from p_y to q_y of length at least $\mathrm{dist}(y, y')$. The convex hull of the four points p_x, q_x, p_y, q_y is a trapezoid, Q, which is a subset of T.

We claim that the trapezoid, Q', which is the convex hull of x, x', y, y' must be contained in S. Let x'' be the point of intersection of ℓ_x and V. Similarly, define y'' to be the intersection of ℓ_y and V. For any $0 \leq \tau \leq 1$, the line ℓ'' perpendicular to V and passing through

$$(1 - \tau)x'' + \tau y''$$

intersects the trapezoid $Q \subset T$ in a line segment of length

$$(1 - \tau) \mathrm{dist}(p_x, q_x) + \tau \, \mathrm{dist}(p_y, q_y) \tag{7.2}$$

7.1. BASIC PROPERTIES

and it intersects the trapezoid Q' in a line segment, centered about V, of length
$$(1-\tau)\operatorname{dist}(x,x') + \tau \operatorname{dist}(y,y'). \tag{7.3}$$
But S must contain a closed line segment of ℓ'', centered about V, of length at least (7.2). Since (7.2) is at least as large as (7.3), we see that
$$\ell'' \cap Q' \subset \ell'' \cap S.$$
Since the choice of $0 \leq \tau \leq 1$ was arbitrary we conclude that $Q' \subset S$. In particular, the line segment from x to y is contained in Q' and thus in S. ∎

Proposition 7.1.8 *If T is a bounded geometrically convex subset of $\mathbb{R}^N$ and S is obtained from T by Steiner symmetrization, then* $\operatorname{diam}(S) \leq \operatorname{diam}(T)$.

Proof: The argument proceeds similarly to the proof of Proposition 7.1.7. Let V, x, y, x', y', ℓ_x, and ℓ_y be as in that proof. Since $\ell_x \cap T$ has $\mathcal{H}^1$-measure at least $\operatorname{dist}(x,x')$, it follows that $\ell_x \cap T$ contains two points p_x and q_x with $\operatorname{dist}(p_x,q_x) \geq \operatorname{dist}(x,x')$. Similarly, $\ell_y \cap T$ contains two points p_y and q_y with $\operatorname{dist}(p_y,q_y) \geq \operatorname{dist}(y,y')$. It is elementary to check that the longest diagonal of the trapezoid that has vertices p_x, q_x, p_y, q_y is at least as long as either of the diagonals of the trapezoid that has vertices x, x', y, y'. The basic fact needed is that the diagonals of the symmetric trapezoid with vertices x, x', y, y' are shorter than the longest diagonal of an unsymmetric trapezoid with the same height and base lengths. This is readily checked if one sets notation as in Figure 7.2, where X and Y are signed quantities satisfying $X + Y = A + B$, and considers the problem of minimizing $P + Q$, the sum of the lengths of the diagonals, as a function of X (here A and B are fixed and Y is constrained by the equation). ∎

We will be using the Hausdorff distance topology on the set of non-empty compact sets. The definition and relevant facts can be found in the Appendix, Section A.1.

One key to the utility of Steiner symmetrization is the fact that any collection of compact subsets of $\mathbb{R}^N$ that is closed under Steiner symmetrization with respect to all $(N-1)$-dimensional vector subspaces of $\mathbb{R}^N$ and is closed in the Hausdorff distance topology must contain a ball. Toward proving that fact we first present the following lemma:

Lemma 7.1.9 *Consider a sequence (finite or infinite) $T_1, T_2, \ldots$ of compact subsets of $\mathbb{R}^N$ and a sequence $V_1, V_2, \ldots$ of $(N-1)$-dimensional vector subspaces of $\mathbb{R}^N$ such that T_{i+1} is the result of Steiner symmetrization with respect to V_i applied to T_i. If $T_i \subset \overline{\mathbb{B}}(0,r)$ and $p \in \partial \overline{\mathbb{B}}(0,r)$, $\epsilon > 0$ are such that*
$$\mathbb{B}(p,\epsilon) \cap \partial \overline{\mathbb{B}}(0,r) \cap T_i = \emptyset,$$

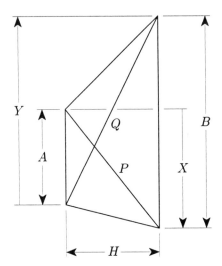

Figure 7.2: **Diagonals of a Trapezoid**

then
$$\mathbb{B}(p,\epsilon) \cap \partial\overline{\mathbb{B}}(0,r) \cap T_j = \emptyset$$

and
$$\mathbb{B}(p',\epsilon) \cap \partial\overline{\mathbb{B}}(0,r) \cap T_j = \emptyset$$

hold for $j \geq i$, where p' is the reflection of p through V_i.

Proof: It suffices to prove the conclusion for $j = i+1$. Let ℓ be the line perpendicular to V passing through p and p'. For any line ℓ' parallel to ℓ and at distance less than ϵ from ℓ, the Hausdorff measure of the intersection of ℓ' with T_i must be strictly less than the length of the intersection of ℓ' with $\overline{\mathbb{B}}(0,r)$, so the intersection of ℓ' with $\partial\overline{\mathbb{B}}(0,r)$ is not in T_{i+1}. ∎

Theorem 7.1.10 *If $\mathcal{C}$ is a non-empty family of non-empty compact subsets of $\mathbb{R}^N$ that is closed in the Hausdorff distance topology and that is closed under the operation of Steiner symmetrization with respect to any $(N-1)$-dimensional vector subspace of $\mathbb{R}^N$, then $\mathcal{C}$ contains a closed ball (possibly of radius 0) centered at the origin.*

Proof: Let $\mathcal{C}$ be such a family of compact subsets of $\mathbb{R}^N$ and set

$$r = \inf\{s : \text{ there exists } T \in \mathcal{C} \text{ with } T \subset \overline{\mathbb{B}}(0,s)\}.$$

If $r = 0$, we are done, so we may assume $r > 0$. By Theorem A.1.5, any uniformly bounded family of non-empty compact sets is compact in the

Hausdorff distance topology, so we can suppose there exists a $T \in \mathcal{C}$ with $T \subset \overline{\mathbb{B}}(0,r)$.

We claim that $T = \overline{\mathbb{B}}(0,r)$. If not, there exists $p \in \overline{\mathbb{B}}(0,r)$ and $\epsilon > 0$, such that $T \subset \overline{\mathbb{B}}(0,r) \setminus \mathbb{B}(p,\epsilon)$. Suppose T_1 is the result of Steiner symmetrization of T with respect to any arbitrarily chosen $(N-1)$-dimensional vector subspace V. Let ℓ be the line perpendicular to V and passing through p. For any line ℓ' parallel to ℓ and at distance less than ϵ from ℓ, the Hausdorff measure of the intersection of ℓ' with T must be strictly less than the length of the intersection of ℓ' with $\overline{\mathbb{B}}(0,r)$, so the intersection of ℓ' with $\partial \overline{\mathbb{B}}(0,r)$ is not in T_1. We conclude that if p_1 is either one of the points of intersection of the sphere of radius r about the origin with the line ℓ, then

$$\mathbb{B}(p_1, \epsilon) \cap \partial \overline{\mathbb{B}}(0,r) \cap T_1 = \emptyset.$$

Choose a finite set of distinct additional points $p_2, p_3, \ldots, p_k$ such that

$$\partial \overline{\mathbb{B}}(0,r) \subset \cup_{i=1}^k \mathbb{B}(p_i, \epsilon).$$

For $i = 1, 2, \ldots, k-1$, let T_{i+1} be the result of Steiner symmetrization of T_i with respect to the $(N-1)$-dimensional vector subspace perpendicular to the line through p_i and p_{i+1}. By the lemma it follows that

$$\mathbb{B}(p_i, \epsilon) \cap \partial \overline{\mathbb{B}}(0,r) \cap T_j = \emptyset$$

holds for $i \leq j \leq k$. Thus we have

$$T_k \cap \partial \overline{\mathbb{B}}(0,r) = \emptyset,$$

so

$$T_k \subset \overline{\mathbb{B}}(0,s)$$

holds for some $s < r$, a contradiction. ∎

7.2 The Isodiametric, Isoperimetric, and Brunn-Minkowski Inequalities

To obtain the main results of this section, we need to investigate the behavior of Lebesgue measure as a function on the collection of compact sets topologized by the Hausdorff distance. It would be convenient if the Lebesgue measure were a continuous function on the non-empty compact sets equipped with the Hausdorff distance topology, but this is just not true.

Example 7.2.1 If we set

$$A_n = \{i/n : i = 0, 1, \ldots, n\},$$

then we see that $\mathrm{HD}(A_n, [0,1]) = \frac{1}{2n} \to 0$ as $n \to \infty$, while $\mathcal{L}(A_n) = 0$ and $\mathcal{L}([0,1]) = 1$.

However, the Lebesgue measure *is* upper-semi-continuous, and if we restrict the domain to the non-empty compact, geometrically convex sets, then it *is* continuous. This assertion will follow from the next several propositions.

Proposition 7.2.2 *Let $A_n \subset \mathbb{R}^N$ be a non-empty compact set for each $n = 1, 2, \ldots$. Suppose $C \subset \mathbb{R}^N$ is such that $\mathrm{HD}(A_n, C) \to 0$ as $n \to \infty$. Then*
$$\mathcal{L}^N(C) \geq \limsup_{n \to \infty} \mathcal{L}^N(A_n)$$
holds.

Proof: This is a corollary of Lemma 7.1.3. ∎

Lemma 7.2.3 *Let $A_n \subset \mathbb{R}^N$ be a non-empty geometrically convex set for each $n = 1, 2, \ldots$. Suppose that $x_0, x_1, \ldots, x_N$ are affinely independent points (i.e. not all contained in a hyperplane) such that*
$$\sup\{\operatorname{dist}(x_i, A_n),\ i = 0, 1, \ldots N\} \to 0$$
as $n \to \infty$. If X is a compact set such that X is contained in the interior of the convex hull of $\{x_0, x_1, \ldots, x_N\}$, then, for all sufficiently large n,
$$X \subset \mathring{A}_n.$$

Proof: Let M be the matrix that has $x_1 - x_0, x_2 - x_0, \ldots, x_N - x_0$ as its rows. For $x \in X$, let $\lambda_1(x), \lambda_2(x), \ldots, \lambda_N(x)$ be the components of
$$M^{-1}(x - x_0). \tag{7.4}$$

In (7.4) we think of $x - x_0$ as a column vector. Setting
$$\lambda_0(x) = 1 - \sum_{i=1}^{N} \lambda_i(x),$$
we have
$$x = \sum_{i=0}^{N} \lambda_i(x) x_i,$$
so, by the hypothesis that X is contained in the interior of the convex hull of $\{x_0, x_1, \ldots, x_N\}$, we have
$$0 < \lambda_i(x) < 1, \quad \text{for } i = 1, 2, \ldots, N,$$
and
$$\sum_{i=1}^{N} \lambda_i(x) < 1$$

7.2. ISODIAMETRIC INEQUALITY ETC.

hold. Since X is compact and the $\lambda_i(x)$ are continuous functions of x, we conclude that there exists $\epsilon > 0$ so that

$$\epsilon \leq \lambda_i(x) \leq 1 - \epsilon, \quad \text{for } i = 1, 2, \ldots, N, \tag{7.5}$$

and

$$\sum_{i=1}^{N} \lambda_i(x) \leq 1 - \epsilon \tag{7.6}$$

hold for all $x \in X$.

For each $n = 1, 2, \ldots$, let a_i^n be a point of A_n nearest to x_i, for $i = 0, 1, \ldots, N$, and let M_n be the matrix such that M_n has $a_1^n - a_0^n, a_2^n - a_0^n, \ldots, a_N^n - a_0^n$ as its rows. For large enough n, M_n is non-singular and we can let $\lambda_1^n(x), \lambda_2^n(x), \ldots, \lambda_N^n(x)$ be the components of

$$M_n^{-1}(x - a_0^n).$$

Since matrix inversion is continuous on the set of non-singular matrices, and because of (7.2) and (7.6), we conclude that, for all sufficiently large n,

$$0 < \lambda_i^n(x) < 1, \quad \text{for } i = 1, 2, \ldots, N,$$

and

$$\sum_{i=1}^{N} \lambda_i^n(x) < 1$$

hold. Then the conclusion of the lemma follows from the fact that A_n is geometrically convex. ∎

Proposition 7.2.4 *Let $A_n \subset \mathbb{R}^N$ be a non-empty, compact, geometrically convex set for each $n = 1, 2, \ldots$. Suppose C is such that $\mathrm{HD}(A_n, C) \to 0$ as $n \to \infty$. Then C is also geometrically convex and*

$$\mathcal{L}^N(C) = \lim_{n \to \infty} \mathcal{L}^N(A_n)$$

holds.

Proof: It is easy to see that C must be geometrically convex. The more difficult question concerns the Lebesgue measure of C. Suppose without loss of generality that $\mathcal{L}^N(C) > 0$. Since C is geometrically convex, ∂C has $\mathcal{L}^N$-measure zero (by Corollary 6.2.7). Thus, in view of Proposition 7.2.2, it will suffice to show that for any compact $A \subset \overset{\circ}{C}$,

$$\liminf_{n \to \infty} \mathcal{L}^N(A_n) \geq \mathcal{L}^N(A)$$

holds.

Let A be any compact subset of $\mathring{C}$. Let $x \in A$ be arbitrary. We claim that there exist $x_0, x_1, \ldots, x_N$ in C which are affinely independent and such that x is an interior point of the convex hull of $\{x_0, x_1, \ldots, x_N\}$. This is actually obvious since we can suppose coordinates are chosen with x at the origin and then set $x_0 = (-r, -r, \ldots, -r)$ and, for $i = 1, 2, \ldots, N$, set x_i equal to r times the ith standard basis vector, where r is chosen sufficiently small but positive. Let U_x be an open neighborhood of x chosen so that $\overline{U}_x$ is contained in the interior of the convex hull of $\{x_0, x_1, \ldots, x_N\}$. By the preceding lemma, there exists an integer n_x such that $n \geq n_x$ implies

$$\overline{U}_x \subset A_n.$$

Since A is compact, finitely many of the sets U_x, $x \in A$, cover A, so we conclude that, for all sufficiently large n, $A \subset A_n$, from which the required inequality follows. ∎

We will also need the following general topological lemma.

Lemma 7.2.5 *Let $\mathcal{T}$ be a topological space, let $f : \mathcal{T} \to \mathbb{R}$ be continuous, and let $\{g_\gamma\}_{\gamma \in \Gamma}$ be a family of functions $g_\gamma : \mathcal{T} \to \mathcal{T}$ such that*

$$f[g_\gamma(x)] \leq f(x)$$

holds for all $x \in \mathcal{T}$ and all $\gamma \in \Gamma$. If $\mathcal{S} \subset \mathcal{T}$ and $\widetilde{\mathcal{S}}$ is the smallest closed subset of $\mathcal{T}$ that contains $\mathcal{S}$ and is closed under g_γ, for all $\gamma \in \Gamma$, then

$$\sup\{f(x) : x \in \widetilde{\mathcal{S}}\} \leq \sup\{f(x) : x \in \mathcal{S}\}$$

holds.

Proof: Set
$$\alpha = \sup\{f(x) : x \in \mathcal{S}\}.$$
Then
$$\mathcal{S}' = \{x \in \mathcal{T} : f(x) \leq \alpha\}$$
is closed in the topology of $\mathcal{T}$ and is closed under g_γ, for all $\gamma \in \Gamma$, so

$$\widetilde{\mathcal{S}} \subset \mathcal{S}'. \qquad \blacksquare$$

Now we can apply some of the tools we have collected. Consider any non-empty bounded $T \subset \mathbb{R}^N$. Since $T \subset \overline{\mathbb{B}}[p, \operatorname{diam}(T)]$ holds, for any choice of $p \in T$ we obtain the inequality

$$\mathcal{L}^N(T) \leq \Upsilon_N [\operatorname{diam}(T)]^N$$

(in case T is not $\mathcal{L}^N$-measurable we use the outer measure of T). Presumably a sharper inequality could be obtained by making a more careful

7.2. ISODIAMETRIC INEQUALITY ETC.

choice of p, with the choice of the center of a closed ball indicating the best for which one could hope. In fact, that best hope is fulfilled in the following general result:

Theorem 7.2.6 (Isodiametric Inequality) *For any non-empty bounded $T \subset \mathbb{R}^N$,*
$$\mathcal{L}^N(T) \leq \Upsilon_N \left[\tfrac{1}{2} \operatorname{diam}(T)\right]^N$$
holds.

Proof: It is no loss of generality to assume that T is compact and geometrically convex. Now let $\mathcal{T}$ be the space of non-empty compact, geometrically convex subsets of $\mathbb{R}^N$ with the Hausdorff distance topology. Let the function $f : \mathcal{T} \to \mathbb{R}$ be defined by
$$f(A) = \Upsilon_N \left[\tfrac{1}{2} \operatorname{diam}(A)\right]^N - \mathcal{L}^N(A).$$
By Proposition 7.2.4, f is continuous. For each $(N-1)$-dimensional vector subspace of $\mathbb{R}^N$, we let g_V be Steiner symmetrization with respect to V. It is a consequence of Proposition 7.1.8 that Lemma 7.2.5 can be applied to the set $\mathcal{S}$ containing the single element T. By Theorem 7.1.10 we know that the corresponding $\widetilde{\mathcal{S}}$ contains a ball. But f vanishes on any ball, so $f(T) \geq 0$ holds as required. ∎

Next, we will use Steiner symmetrization to give a proof of the Brunn-Minkowski Inequality. This inequality is concerned with the Lebesgue measure of the vector sum of two subsets of Euclidean space. The original result goes back to the Inaugural-Dissertation and Habilitationsschrift of Hermann Brunn (Brunn [1] and [2]) in the late 1880's. Minkowski's work on the topic, Minkowski [1], appeared in the late 1890's. Many proofs of the Brunn-Minkowski Inequality have been given, so we will not assert that our proof is new (but it is new to us).

Definition 7.2.7 *For subsets A and B of $\mathbb{R}^N$ the* **vector sum** *of A and B is the set*
$$A + B = \{a + b : a \in A, \ b \in B\}.$$

The line of argument ahead of us would have been far shorter but for the unfortunate fact that the Steiner symmetrization of the vector sum of two sets is not necessarily equal to the vector sum of their Steiner symmetrizations.

Example 7.2.8 Working in $\mathbb{R}^2$, we set
$$A = [0,1] \times \{0\} \quad \text{and} \quad B = \{0\} \times [0,1],$$
so
$$A + B = [0,1] \times [0,1].$$

Consider Steiner symmetrization with respect to any 1-dimensional vector subspace V other than either of the coordinate axes. The Steiner symmetrizations, A' and B', of A and B, respectively, are both closed line segments in V, so $A' + B'$ is also a closed line segment in V. But the Steiner symmetrization, $(A+B)'$, of $A+B$ must have area 1, so $(A+B)' \neq A'+B'$.

The very simple case dealt with in the following theorem is the basic fact on which our proof of the Brunn-Minkowski Inequality rests.

Lemma 7.2.9 *Suppose the non-empty sets $A, B \subset \mathbb{R}$ are both disjoint unions of finitely many open intervals. Then*

$$\mathcal{L}^1(A + B) \geq \mathcal{L}^1(A) + \mathcal{L}^1(B)$$

holds.

Proof: We write

$$A = \bigcup_{i=1}^{m} (\alpha_i, \beta_i)$$

and

$$B = \bigcup_{j=1}^{n} (\gamma_j, \delta_j),$$

where

$$\alpha_1 < \beta_1 \leq \alpha_2 < \beta_2 \leq \ldots \leq \alpha_m < \beta_m$$

and

$$\gamma_1 < \delta_1 \leq \gamma_2 < \delta_2 \leq \ldots \leq \gamma_n < \delta_n$$

hold, and argue by induction on $m + n$. The conclusion of the lemma is obviously true when $m + n = 2$.

Assume now $m + n \geq 3$. Without loss of generality, we may assume that $m \geq 2$. Set

$$A' = \bigcup_{i=1}^{m-1} (\alpha_i, \beta_i).$$

By the induction hypothesis, we have

$$\mathcal{L}^1(A' + B) \geq \mathcal{L}^1(A') + \mathcal{L}^1(B).$$

Noting that

$$(\alpha_m + \gamma_n, \beta_m + \delta_n) \subset (A + B) \setminus (A' + B),$$

we conclude that

$$\begin{aligned}
\mathcal{L}^1(A + B) &\geq (\beta_m - \alpha_m) + \mathcal{L}^1(A' + B) \\
&\geq (\beta_m - \alpha_m) + \mathcal{L}^1(A') + \mathcal{L}^1(B) \\
&= \mathcal{L}^1(A) + \mathcal{L}^1(B)
\end{aligned}$$

as needed. ∎

7.2. ISODIAMETRIC INEQUALITY ETC.

Lemma 7.2.10 *Let A and B be non-empty subsets of $\mathbb{R}$. Then*

$$\mathcal{L}^1(A+B) \geq \mathcal{L}^1(A) + \mathcal{L}^1(B)$$

holds.

Proof: We may and shall assume that both A and B are sets of finite measure. Let $\epsilon > 0$ be given. We can find compact sets $A' \subset A$ and $B' \subset B$ such that

$$\mathcal{L}^1(A') + \mathcal{L}^1(B') \geq \mathcal{L}^1(A) + \mathcal{L}^1(B) - \epsilon.$$

Let

$$U_1 \supset U_2 \supset \ldots \supset A'$$

and

$$V_1 \supset V_2 \supset \ldots \supset B'$$

be decreasing sequences of open sets such that

$$\bigcap_{i=1}^{\infty} U_i = A' \quad \text{and} \quad \bigcap_{i=1}^{\infty} V_i = B'.$$

Since A' and B' are compact, we may assume each U_i and each V_i is a finite union of bounded open intervals. We note that

$$\bigcap_{i=1}^{\infty}(U_i + V_i) = A' + B'$$

so, using the result of Lemma 7.2.9, we have

$$\begin{aligned}
\mathcal{L}^1(A+B) &\geq \mathcal{L}^1(A' + B') \\
&= \lim_{i \to \infty} \left[\mathcal{L}^1(U_i + V_i) \right] \\
&\geq \lim_{i \to \infty} \left[\mathcal{L}^1(U_i) + \mathcal{L}^1(V_i) \right] \\
&= \mathcal{L}^1(A') + \mathcal{L}^1(B') \\
&\geq \mathcal{L}^1(A) + \mathcal{L}^1(B) - \epsilon.
\end{aligned}$$

Since $\epsilon > 0$ was arbitrary, the result follows. ∎

Lemma 7.2.11 *Let A and B be non-empty subsets of $\mathbb{R}^N$. Let V be an $(N-1)$-dimensional vector subspace of $\mathbb{R}^N$. If X, Y, and Z are the result of Steiner symmetrization with respect to V applied to A, B, and $A+B$, respectively, then*

$$X + Y \subset Z$$

and

$$[\mathcal{L}^N(X+Y)]^{1/N} - [\mathcal{L}^N(X)]^{1/N} - [\mathcal{L}^N(Y)]^{1/N}$$
$$\leq [\mathcal{L}^N(A+B)]^{1/N} - [\mathcal{L}^N(A)]^{1/N} - [\mathcal{L}^N(B)]^{1/N} \quad (7.7)$$

hold.

Proof: First we show that $X + Y \subset Z$. Let $x + y \in X + Y$ be arbitrary. Set $z = x + y$. Let ℓ_x be the line perpendicular to V, passing through x. Similarly define ℓ_y and ℓ_z. By the definition of Steiner symmetrization, $\ell_x \cap A$ is non-empty and

$$\mathcal{H}^1(\ell_x \cap A) \geq 2\operatorname{dist}(x, V)$$

holds. Likewise, $\ell_y \cap B$ is non-empty and

$$\mathcal{H}^1(\ell_y \cap B) \geq 2\operatorname{dist}(y, V)$$

holds. If $a \in \ell_x \cap A$ and $b \in \ell_y \cap B$, then $a + b \in \ell_z \cap (A + B)$. It follows that $\ell_z \cap (A + B)$ is non-empty, and by Lemma 7.2.10, isometrically identifying ℓ_x, ℓ_y, and ℓ_z with $\mathbb{R}$, we have

$$\mathcal{H}^1[\ell_z \cap (A+B)] \geq \mathcal{H}^1(\ell_x \cap A) + \mathcal{H}^1(\ell_y \cap B)$$
$$\geq 2\operatorname{dist}(x,V) + 2\operatorname{dist}(y,V) = 2\operatorname{dist}(z,V).$$

Thus we have $z \in Z$, as required.

The inequality (7.7) follows, since $\mathcal{L}^N(X) = \mathcal{L}^N(A)$, $\mathcal{L}^N(Y) = \mathcal{L}^N(B)$, and $\mathcal{L}^N(X+Y) \leq \mathcal{L}^N(Z) = \mathcal{L}^N(A+B)$. ∎

We could now prove the Brunn-Minkowski Inequality in the limited context of compact, geometrically convex sets, but the result is true more generally, so we will introduce another family of compact sets on which the Lebesgue measure is a continuous function.

Definition 7.2.12 *For $\epsilon > 0$, let e_ϵ and c_ϵ be the maps sending the compact set C to the compact sets $e_\epsilon(C)$ and $c_\epsilon(C)$ defined by*

$$e_\epsilon(C) = \{x : \operatorname{dist}(x, C) \leq \epsilon\},$$
$$c_\epsilon(C) = \{x : \operatorname{dist}(x, \mathbb{R}^N \setminus C) \geq \epsilon\}.$$

We will say that a compact set, C, is ϵ-thick if

$$C = e_\epsilon \circ c_\epsilon(C).$$

The collection of non-empty ϵ-thick sets will be denoted by $\mathcal{K}_\epsilon$ and will be topologized by the Hausdorff distance.

7.2. ISODIAMETRIC INEQUALITY ETC.

Remark 7.2.13 The ϵ-thick sets were used in Parks [1] to study the problem of area minimization in arbitrary dimensions and codimensions. We think of $e_\epsilon(C)$ as being the "expansion" of C and $c_\epsilon(C)$ as being the "core" of C. Acute corners are among the things that the hypothesis that a set is ϵ-thick rules out. Thus a triangle with interior in the plane cannot be ϵ-thick. More generally, an N-simplex in $\mathbb{R}^N$, $N \geq 2$, is not ϵ-thick.

The following lemma is an almost immediate consequence of the definition of ϵ-thick sets; its two corollaries are clear.

Lemma 7.2.14 *For $\epsilon > 0$, $A \in \mathcal{K}_\epsilon$ holds if and only if for each $x \in A$ there exists $y \in A$ with $x \in \overline{\mathbb{B}}(y, \epsilon) \subset A$.*

Corollary 7.2.15 *If A is a non-empty compact set and $\epsilon > 0$, then $e_\epsilon(A) \in \mathcal{K}_\epsilon$.*

Corollary 7.2.16 *If $\epsilon > 0$ and $A, B \in \mathcal{K}_\epsilon$, then $A + B \in \mathcal{K}_\epsilon$.*

Verifying the continuity of the Lebesgue measure with respect to the Hausdorff distance topology on $\mathcal{K}_\epsilon$ requires more work.

Proposition 7.2.17 *Fix $\epsilon > 0$. The collection of non-empty ϵ-thick sets is closed in the Hausdorff distance topology, and the Lebesgue measure is a continuous function from $\mathcal{K}_\epsilon$ to $\mathbb{R}$.*

Proof: Suppose that $\{A_i\}_{i=1}^\infty \subset \mathcal{K}_\epsilon$ and C is a non-empty compact set such that $\mathrm{HD}(A_i, C) \to 0$ as $i \to \infty$. Let $x \in C$ be arbitrary. Let $x_i \in A_i$ be such that
$$|x - x_i| = \mathrm{dist}(x, A_i).$$
By Lemma 7.2.14, for each i there exists y_i such that
$$x_i \in \overline{\mathbb{B}}(y_i, \epsilon) \subset A_i.$$
Passing to a subsequence if necessary, but without changing notation, we may assume that y_i converges to some y, and it follows that
$$x \in \overline{\mathbb{B}}(y, \epsilon) \subset C.$$
Since $x \in C$ was arbitrary, we see by Lemma 7.2.14 that $C \in \mathcal{K}_\epsilon$.

Set $\mu = \liminf \mathcal{L}^N(A_i)$. Pass to a subsequence, but without changing notation, so that
$$\mu = \lim \mathcal{L}^N(A_i).$$
Let $D \subset C$ be countable and dense in C. Applying the argument of the previous paragraph to each point in D, and using a diagonalization argument, we can pass to a subsequence, again without changing notation, such that for each $d \in D$ there exist $x_i(d) \in \overline{\mathbb{B}}(y_i(d), \epsilon) \subset A_i$ and $y(d) \in C$ with
$$x_i(d) \to d, \qquad y_i(d) \to y(d), \qquad d \in \overline{\mathbb{B}}(y(d), \epsilon) \subset C.$$

In the remainder of the proof we will use this subsequence.

For $I = 1, 2, \ldots$, set

$$B_I = \bigcap_{i \geq I} A_i, \quad \text{and} \quad E = \bigcup_{I=1}^{\infty} B_I.$$

We will show that $\mathcal{L}^N(C \setminus E) = 0$. Suppose $x \in C \setminus E$ is arbitrary. Choose any $0 < r < \frac{3}{4}\epsilon$. We can find a point $d \in D$ with $|d - x| < r/3$. There exists I_0 so that $i \geq I_0$ implies $|y_i(d) - y(d)| < r/3$. Then, for $i \geq I_0$, we have

$$\overline{\mathbb{B}}(y(d), \epsilon - r/3) \subset \overline{\mathbb{B}}(y_i(d), \epsilon) \subset A_i$$

and $|y(d) - x| \leq |y(d) - d| + |d - x| < \epsilon + r/3$. If $I \geq I_0$ we have

$$\overline{\mathbb{B}}(y(d), \epsilon - r/3) \subset B_I.$$

Assuming $y(d) - x \neq 0$, let $\mathbf{u}$ be the unit vector in that direction; then we have

$$\begin{aligned}
\overline{\mathbb{B}}(x, r) \cap B_I &\supset \overline{\mathbb{B}}(x, r) \cap \overline{\mathbb{B}}(y(d), \epsilon - r/3) \\
&\supset \overline{\mathbb{B}}(x, r) \cap \overline{\mathbb{B}}[x + (\epsilon + r/3)\mathbf{u}, \epsilon - r/3] \\
&\supset \overline{\mathbb{B}}[x + (5r/6)\mathbf{u}, r/6].
\end{aligned}$$

If on the other hand $y(d) = x$, then

$$\overline{\mathbb{B}}(x, r) \cap B_I \supset \overline{\mathbb{B}}(x, r) \cap \overline{\mathbb{B}}(y(d), \epsilon - r/3) = \overline{\mathbb{B}}(x, r).$$

In either case, we see that

$$\mathcal{L}^N(\overline{\mathbb{B}}(x, r) \setminus B_I) \leq \left(1 - \frac{1}{6^N}\right) \Upsilon_N r^N$$

holds for $I \geq I_0$. We conclude that $C \setminus E$ cannot have density 1 at x, but this implies that $\mathcal{L}^N(C \setminus E) = 0$.

Finally, $C \setminus E$ is the decreasing intersection of the sets $C \setminus B_I$, so $\mathcal{L}^N(C \setminus B_I) \to 0$ as $I \to \infty$. But B_I is a subset of A_I, so $\mathcal{L}^N(C \setminus A_I) \to 0$ as $I \to \infty$. It follows that

$$\mu = \lim \mathcal{L}^N(A_I) \geq \mathcal{L}^N(C),$$

and the proposition follows by Proposition 7.2.2. ∎

Proposition 7.2.18 *Let A and B be ϵ-thick subsets of $\mathbb{R}^N$. Then*

$$[\mathcal{L}^N(A + B)]^{1/N} \geq [\mathcal{L}^N(A)]^{1/N} + [\mathcal{L}^N(B)]^{1/N}$$

7.2. ISODIAMETRIC INEQUALITY ETC.

Proof: Consider the topological space $\mathcal{T} = \mathcal{K}_\epsilon \times \mathcal{K}_\epsilon$. The function $f : \mathcal{T} \to \mathbb{R}$ defined by

$$f(A, B) = \left[\mathcal{L}^N(A + B)\right]^{1/N} - \left[\mathcal{L}^N(A)\right]^{1/N} - \left[\mathcal{L}^N(B)\right]^{1/N}$$

is continuous. For each $(N-1)$-dimensional vector subspace V of $\mathbb{R}^N$, let the map $g_V : \mathcal{T} \to \mathcal{T}$ be simultaneous Steiner symmetrization with respect to V, by which we mean $g_V(A, B) = (A', B')$ where A' is the Steiner symmetrization of A and B' is the Steiner symmetrization of B. As a consequence of Lemma 7.2.11, the conditions of Lemma 7.2.5 are satisfied. Let $\mathcal{S}$ consist of the single pair (A, B). By Theorem 7.1.10, we know that if $\widetilde{\mathcal{S}}$ is as in Lemma 7.2.5, then among the elements of $\widetilde{\mathcal{S}}$ is a pair of balls. But f vanishes on any pair of balls, so the result follows by Lemma 7.2.5. ∎

We can now prove the general result:

Theorem 7.2.19 (The Brunn-Minkowski Inequality) *Let A and B be non-empty subsets of $\mathbb{R}^N$. Then*

$$\left[\mathcal{L}^N(A + B)\right]^{1/N} \geq \left[\mathcal{L}^N(A)\right]^{1/N} + \left[\mathcal{L}^N(B)\right]^{1/N}.$$

Remark 7.2.20 Following our usual convention, if T is a non-measurable set, then $\mathcal{L}(T)$ denotes the outer measure of the set.

Proof: Let A and B be given. Let λ_a and λ_b be any constants satisfying with

$$\mathcal{L}^N(A) > \lambda_a \geq 0 \quad \text{and} \quad \mathcal{L}^N(B) > \lambda_b \geq 0.$$

Find non-empty compact $A' \subset A$, $B' \subset B$ with

$$\mathcal{L}^N(A') \geq \lambda_a \quad \text{and} \quad \mathcal{L}^N(B') \geq \lambda_b.$$

For $\epsilon > 0$, by Proposition 7.2.18, we have

$$\left[\mathcal{L}^N(e_\epsilon(A') + e_\epsilon(B'))\right]^{1/N} \geq \left[\mathcal{L}^N(e_\epsilon(A'))\right]^{1/N} + \left[\mathcal{L}^N(e_\epsilon(B'))\right]^{1/N}$$
$$\geq [\lambda_a]^{1/N} + [\lambda_b]^{1/N}.$$

Taking the limit as $\epsilon \downarrow 0$ we obtain

$$\left[\mathcal{L}^N(A + B)\right]^{1/N} \geq \left[\mathcal{L}^N(A' + B')\right]^{1/N} \geq [\lambda_a]^{1/N} + [\lambda_b]^{1/N}.$$

Since λ_a and λ_b were arbitrary, the theorem follows. ∎

The classical application of the Brunn-Minkowski Inequality is a proof of the Isoperimetric Inequality in $\mathbb{R}^N$ comparing the N-dimensional measure of a set to the $(N-1)$-dimensional area of its boundary. This will bring us almost to the goal Steiner originally sought: We will have shown that no figure encloses greater N-dimensional volume, given the $(N-1)$-dimensional area of its boundary, than does a sphere.

The Brunn-Minkowski Inequality is the perfect tool for the proof of the Isoperimetric Inequality, *if* the $(N-1)$-dimensional boundary surface area is measured using the Minkowski content. Recall Definition 3.3.1 from Section 3.3.

Definition *Suppose $A \subset \mathbb{R}^N$ and $0 \leq K \leq N$. The K-dimensional upper Minkowski content of A, denoted by $\mathcal{M}^{*K}(A)$, is defined by*

$$\mathcal{M}^{*K}(A) = \limsup_{r \downarrow 0} \frac{\mathcal{L}^N\{x : \operatorname{dist}(x, A) < r\}}{\Upsilon_{N-K} r^{N-K}}.$$

Similarly, the K-dimensional lower Minkowski content of A is denoted by $\mathcal{M}_*^K(A)$ and defined by

$$\mathcal{M}_*^K(A) = \liminf_{r \downarrow 0} \frac{\mathcal{L}^N\{x : \operatorname{dist}(x, A) < r\}}{\Upsilon_{N-K} r^{N-K}}.$$

In case the K-dimensional upper Minkowski content and the K-dimensional lower Minkowski content of A are equal, then their common value is called the K-dimensional Minkowski content of A and is denoted by $\mathcal{M}^K(A)$.

Remark 7.2.21 The Minkowski content is traditionally used for integer values of K. By Federer [4], §3.2.29, for any classical K-dimensional smooth surface, the Minkowski content is equal to the standard value for the K-dimensional measure of the surface.

Theorem 7.2.22 (The Isoperimetric Inequality)
Suppose that $S \subset \mathbb{R}^N$ and $\mathcal{L}^N[S] < \infty$. Then

$$\mathcal{M}_*^{N-1}[\partial S] \geq N \Upsilon_N \left(\frac{\mathcal{L}^N[S]}{\Upsilon_N} \right)^{\frac{N-1}{N}}. \tag{7.8}$$

Proof: We may and shall assume that S is closed. If $\mathcal{L}^N(\partial S) > 0$ held, then the Minkowski content of ∂S would be infinite, so (7.8) would trivially hold. So we may also assume $\mathcal{L}^N(\partial S) = 0$.

Consider $r > 0$. Set $T = \{x : \operatorname{dist}(x, \mathbb{R}^N \setminus S) \geq r\}$. Observing that

$$\{x : \operatorname{dist}(x, S) < r\} \supset T + \mathbb{B}(0, 2r),$$

and applying Theorem 7.2.19 (the Brunn-Minkowski Inequality), we conclude that

$$\mathcal{L}^N(\{x : \operatorname{dist}(x, S) < r\}) \geq \left[(\mathcal{L}^N(T))^{1/N} + 2 \Upsilon_N^{1/N} r \right]^N$$
$$\geq \mathcal{L}^N(T) + 2N \left(\mathcal{L}^N(T) \right)^{(N-1)/N} \Upsilon_N^{1/N} r.$$

7.3. EQUALITY IN THE ISOPERIMETRIC INEQUALITY

Since
$$\{x : \mathrm{dist}(x, \partial S) < r\} \supset (\{x : \mathrm{dist}(x, S) < r\} \setminus T),$$
we have
$$\frac{\mathcal{L}^N\{x : \mathrm{dist}(x, \partial S) < r\}}{2r} \geq N\, \Upsilon_N{}^{1/N} (\mathcal{L}^N(T))^{(N-1)/N}.$$

The result follows since our assumption that $\mathcal{L}^N(\partial S) = 0$ implies
$$\lim_{r \downarrow 0} \mathcal{L}^N(T) = \mathcal{L}^N(S). \qquad \blacksquare$$

Remark 7.2.23 The results in this section have shown that the sphere minimizes area among surfaces enclosing a given volume, but we have not shown that the sphere is the unique such surface. We will give a proof of uniqueness in the next section. Steiner symmetrization also can be used to show that the sphere is the unique shape making the Isoperimetric Inequality an equality. An exposition of this important fact by G. Talenti, in the context of Caccioppoli sets, can be found in Gruber and Wills [1].

The Isoperimetric Inequality arises in an extraordinary number of mathematical contexts. In this book, we have already proved and used a form of it in Section 3.7. A survey about isoperimetric inequalities appears in Osserman [1], and we call the reader's attention to a proof using elementary complex variable theory that appears in Gamelin and Khavinson [1].

7.3 Equality in the Isoperimetric Inequality

In considering the uniqueness of the solution to the Isoperimetric Problem, we will change context to the category of C^2 surfaces. This will save us from considering some rather messy technical issues, and will allow us to present the beautiful method of Aleksandrov.

Theorem 7.3.1 *If $S \subset \mathbb{R}^N$ is a C^2 surface enclosing a region R such that any isotropic deformation of $\mathbb{R}^N$ that preserves the N-dimensional volume of R increases the $(N-1)$-dimensional area of S, then the mean curvature of S is constant.*

Proof: Consider any two distinct points p and q in S. Since S is a C^2 surface, we can find coordinate systems $(x_1, x_2, \ldots, x_N)$ and $(y_1, y_2, \ldots, y_N)$ with origin at p and q respectively, and such that the plane $x_N = 0$ is the tangent plane to S at p, while the plane $y_N = 0$ is the tangent plane to S at q. Let $f : \mathbb{R}^{N-1} \to \mathbb{R}$ be the C^2 function such that

$$\Big(x_1, x_2, \ldots, x_{N-1}, f(x_1, x_2, \ldots, x_{N-1})\Big)$$

gives S in a neighborhood of p. Let $g : \mathbb{R}^{N-1} \to \mathbb{R}$ be the C^2 function such that
$$\Big(y_1, y_2, \ldots, y_{N-1}, g(y_1, y_2, \ldots, y_{N-1})\Big)$$
gives S in a neighborhood of q. For simplicity also suppose that the x_N-axis and the y_N-axis point toward the exterior of R.

Let $\delta > 0$ be small enough that the open δ-ball about the origin of the x-coordinate system and the open δ-ball about the origin of the y-coordinate system do not intersect. Let $\eta(x_1, x_2, \ldots, x_{N-1})$ be smooth and compactly supported in the open $\delta/2$-ball, and bounded by $\delta/2$. Let $\zeta(y_1, y_2, \ldots, y_{N-1})$ have the same properties. For $(s, t) \in \mathbb{B}^2[(0,0), 1]$, define a comparison surface $S_{(s,t)}$ which agrees with S outside the δ-balls about the origin of the x and y coordinate systems, but in those balls is the graph of $f+s\eta$ and $g+t\zeta$, respectively. Let $\Delta V(s,t)$ denote the difference between the volume enclosed by $S_{(s,t)}$ and the volume enclosed by S. Likewise, let $\Delta A(s,t)$ denote the difference between the area of $S_{(s,t)}$ and the area of S. It is elementary to see that

$$\Delta V(s,t) = \int s\eta(x)\, d\mathcal{L}^{N-1}x + \int t\zeta(y)\, d\mathcal{L}^{N-1}y,$$

$$\Delta A(s,t) = \int \left\{\sqrt{1+|Df(x)+sD\eta(x)|^2} - \sqrt{1+|Df(x)|^2}\right\} d\mathcal{L}^{N-1}x$$
$$+ \int \left\{\sqrt{1+|Dg(y)+tD\zeta(y)|^2} - \sqrt{1+|Dg(y)|^2}\right\} d\mathcal{L}^{N-1}y,$$

where the integrals are over the open δ-balls.

To find comparison surfaces that fit the requirement of enclosing the same volume, we need to find choices of (s,t) for which $\Delta V(s,t) = 0$. In general one applies the Implicit Function Theorem, but in our case we can simplify matters by choosing η and ζ to be the same function and choosing $t = -s$. With these choices,

$$J(s) = \Delta A(s, -s) = \int \left\{\sqrt{1+|Df(x)+sD\eta(x)|^2} - \sqrt{1+|Df(x)|^2}\right\} d\mathcal{L}^{N-1}x$$
$$+ \int \left\{\sqrt{1+|Dg(x)-sD\eta(x)|^2} - \sqrt{1+|Dg(x)|^2}\right\} d\mathcal{L}^{N-1}x.$$

7.3. EQUALITY IN THE ISOPERIMETRIC INEQUALITY

Since $\Delta A(0,0)$ must be minimal, we can differentiate $J(s)$ with respect to s and conclude that $J'(0) = 0$. The result of this computation is

$$0 = J'(0) = \int \left\{ \frac{D\eta(x) \cdot Df(x)}{\sqrt{1+|Df(x)|^2}} - \frac{D\eta(x) \cdot Dg(x)}{\sqrt{1+|Dg(x)|^2}} \right\} d\mathcal{L}^{N-1}x.$$

The Gauss-Green Theorem reveals that

$$0 = \int \eta(x) \left\{ \operatorname{div}\left(\frac{Df(x)}{\sqrt{1+|Df(x)|^2}}\right) - \operatorname{div}\left(\frac{Dg(x)}{\sqrt{1+|Dg(x)|^2}}\right) \right\} d\mathcal{L}^{N-1}x.$$

As is usual, since the function η is essentially arbitrary, the other factor in the integrand must vanish identically, so we conclude that

$$\operatorname{div}\left(\frac{Df(x)}{\sqrt{1+|Df(x)|^2}}\right) - \operatorname{div}\left(\frac{Dg(x)}{\sqrt{1+|Dg(x)|^2}}\right) = 0.$$

Evaluating at the origin we find that

$$\Delta f(0) = \Delta g(0).$$

By Lemma 2.3.5, we see that $\frac{1}{N-1} \Delta f(0)$ and $\frac{1}{N-1} \Delta g(0)$ are the mean curvatures of S at p and q, respectively. ∎

In the context of C^2 surfaces, the question of the uniqueness of the solution to the Isoperimetric Problem has been reduced to the characterization of the surfaces of constant mean curvature in Euclidean space. The physical analogue of a 2-dimensional surface of constant mean curvature in $\mathbb{R}^3$ is a soap bubble, because the difference in air pressure between the inside and outside of the bubble is proportional to the mean curvature of the bubble. The question in those terms is whether or not a soap bubble enclosing just one region must be spherical. This was answered in the affirmative in 1958 by A. D. Aleksandrov. We present his elegant argument below.

Theorem 7.3.2 *If $S \subset \mathbb{R}^N$ is a connected C^2 surface of constant mean curvature enclosing a bounded region R, then S is a sphere.*

Proof: Choose a coordinate system $(x_1, x_2, \ldots, x_N)$ such that the origin is a point of S, the plane $x_N = 0$ is tangent to S at the origin, and S is contained in the half-space $x_N \geq 0$.

Consider a plane $V = \{x : x_N = d > 0\}$. We reflect the surface

$$S \cap \{(x_1, x_2, \ldots, x_N) : x_N \leq d\}$$

in the plane V to form the surface S_d. We claim there is a largest positive number d_0 such that S_d is contained in $S \cup R$ for all $0 < d < d_0$. To verify the claim it will suffice to show there is some $\delta > 0$ such that S_d is contained in $S \cup R$ for all $0 < d < \delta$. Because S is a C^2 surface, there is $\delta > 0$ such that orthogonal projection of

$$S \cap \{(x_1, x_2, \ldots, x_N) : x_N \leq 2\delta\}$$

into the plane $x_N = 0$ is injective. Let D be the image of this injection and for $(x_1, x_2, \ldots, x_{N-1}) \in D$ let $f(x_1, x_2, \ldots, x_{N-1})$ be the unique number less than or equal to δ such that

$$(x_1, x_2, \ldots, x_{N-1}, f(x_1, x_2, \ldots, x_{N-1})) \in S.$$

For $0 < d < \delta$ set

$$D' = \{(x_1, x_2, \ldots, x_{N-1}) : f(x_1, x_2, \ldots, x_{N-1}) \leq d\}$$

and let $g : D' \to \mathbb{R}$ be defined by

$$g(x_1, x_2, \ldots, x_{N-1}) = 2d - f(x_1, x_2, \ldots, x_{N-1}).$$

Then S_d is the graph of g and is clearly contained in $S \cup R$.

We know that S_{d_0} lies on one side of S and it either makes contact with S above the plane $x_N = d_0$ or it makes first order contact at some point of the intersection of S with the plane $x_N = d_0$. Either way the maximum principle of E. Hopf (see Gilbarg and Trudinger [1], Chapter 3) guarantees that

$$S_{d_0} = S \cap \{(x_1, x_2, \ldots, x_N) : x_N \geq d_0\}.$$

That is, S is symmetric in the plane.

Now the plane can be chosen orthogonal to any direction, so the family (with exactly one element) $\mathcal{C} = \{S \cup R\}$ is closed under Steiner symmetrization and thus is a closed ball by Theorem 7.1.10. ∎

Aleksandrov's Theorem answers a very natural question, but almost immediately another question arises: If the surface S is not the boundary of a region, that is, if it is not imbedded, [3] then can it have constant mean curvature and fail to be a sphere? A partial answer to this question was given even before Aleksandrov's work. H. Hopf (in Hopf [1]) showed that if a surface of constant mean curvature is topologically a sphere, then that surface is, in fact, a Euclidean sphere. Quite remarkably, H. Hopf's result cannot be extended to higher topological types. In 1984, H. Wente constructed a constant mean curvature torus in $\mathbb{R}^3$ (see Wente [1] and Walter [1]). Later, using perturbation techniques, N. Kapouleas produced constant mean curvature surfaces of any genus greater than or equal to 2.

---

[3] Imbedded surfaces with constant mean curvature in $\mathbb{R}^3$ were discussed earlier in Section 2.3.

Chapter 8

Topics Related to Complex Analysis

8.1 Quasiconformal Mappings

The subject of quasiconformal mappings exists for at least four reasons.

- First, there is the remarkable theorem of Liouville (which we shall prove later): except for isometries, dilations, and inversions (to be defined below), there are no smooth conformal mappings in dimension three or greater. Thus one seeks a natural generalization of conformal mappings to higher dimensions.

- Second, many properties that are commonly proved for conformal mappings (in dimension two) are really properties of quasiconformal mappings. Mathematical facts, particularly geometric ones, are often most clearly understood when the fewest hypotheses are used.

- Third, taking the first remark into account, any basic geometric property that distinguishes among dimensions—particularly among Euclidean spaces—is of *a priori* interest.

- Fourth, the theory is used in the study of differentials on, and moduli for, Riemann surfaces—work pioneered in Teichmüller [1].

Let $U, V \subset \mathbb{R}^N$ be open sets. Recall that a C^1 mapping $F : U \to \mathbb{R}^N$ is said to be **conformal** if, at each point $P \in U$, the Jacobian matrix is an orthogonal transformation or an orthogonal transformation composed with a dilation or contraction. (Using the language of Riemannian geometry, a coordinate-free rendition of conformality may be formulated. Also, the notion may be formulated for arbitrary Riemannian metrics. But the formulation we have given will suffice for our purposes.) Both the orthogonal translation and the dilation will, in general, depend on the point P.

Note that, in $\mathbb{R}^1$, any C^1 mapping is conformal. For the role of the orthogonal transformation is moot, and the dilation is simply given by the scalar value $|F'(P)|$. In $\mathbb{R}^2$, it is a standard result of complex function theory that the sense preserving conformal mappings are precisely the complex analytic mappings and conversely.

Of course it is clear from the definition we have given that any isometry of $\mathbb{R}^N$, or any dilation, or any composition of these, is a conformal mapping.

There is a third type of conformal mapping—in any dimension—which we shall now describe. By an inversion we shall mean a mapping of the form
$$x \longmapsto \frac{x - P}{\|x - P\|^2}.$$

Here P is a fixed point in space and $\| \ \|$ denotes the standard norm on $\mathbb{R}^N$. Such a mapping is undefined at the point P, but we may compactify $\mathbb{R}^N$ by adjoining a point at infinity that will serve as the image of P under this mapping. In practice, this topological detail is of no interest.

Let us check that inversion is a conformal mapping. We may assume that $P = 0$ (in other contexts, the resulting mapping is called the Kelvin inversion—see Krantz [3]). If we use y to denote the variable in the target space then it is straightforward to calculate that

$$\frac{\partial y_k}{\partial x_j} = \begin{cases} \dfrac{1}{\|x\|^2} - \dfrac{2x_j^2}{\|x\|^4} & \text{if } j = k \\ -\dfrac{2x_j x_k}{\|x\|^4} & \text{if } j \neq k \ . \end{cases}$$

One checks by hand that the matrix

$$\mathcal{J} = \left(\frac{\partial y_k}{\partial x_j}\right)$$

is a dilation of an orthogonal matrix. In fact, the dilation factor is $1/\|x\|^2$.

So we see that in one dimension the theory of conformal mappings is meaningless, and in two dimensions (by the Riemann Mapping Theorem) it is rich. We have given three examples of smooth conformal mappings in $\mathbb{R}^N$, and later we shall see that these are all that there are. Now we turn our attention to quasiconformal mappings in $\mathbb{R}^2$.

The theory of quasiconformal mappings is currently an active field of research. Results in the subject tend to become quite technical rather quickly. In the present treatment we shall content ourselves with a survey of the elementary ideas. It is both convenient and appropriate to first study quasiconformal mappings in dimension two. For here the notation and the statements are particularly elegant, and we may clearly see the connections with the more classical notion of conformality. At the end of the section, we shall make some remarks about the higher dimensional case.

8.1. QUASICONFORMAL MAPPINGS

The books Ahlfors [1] and Lehto and Virtanen [1] are excellent sources for an introduction to quasiconformal mappings in dimension two.

Equivalent Definitions of Conformal Mappings

Grötzsch, who invented quasiconformal mappings (in Grötzsch [1]) but did not call them by that name[1], considered the fact that a square S (together with its interior) in the plane cannot be mapped conformally to a rectangle R (together with its interior) that is not a square—in such a fashion that vertices go to vertices. He wished to study the most nearly conformal, vertex-preserving map of the square to the rectangle, and thus desired a measure of "almost conformality."

Alternatively, one can consider a mapping F of a complex variable to be simply a mapping of (a region U in) $\mathbb{R}^2$ into $\mathbb{R}^2$. Considered in this way, the Jacobian of F is just a 2×2 matrix—or a linear mapping—and it will take circles to ellipses. If we mandate a bound on the eccentricity of the ellipses that arise, then the result is a quasiconformal mapping.

Thus we may think of quasiconformal maps in terms of either distortion of squares or distortion of circles. We will also have a differential characterization of quasiconformal mappings.

We will develop the habit of identifying $\mathbb{R}^2$ with $\mathbb{C}$. When we are thinking of real analysis, we write our variables as (x, y); in complex analysis we write them as z and $\bar{z}$. Of course, these two notations are related by

$$\begin{aligned} z &= x + iy \\ \bar{z} &= x - iy. \end{aligned} \quad (8.1)$$

The vectors $\frac{\partial}{\partial x}$ and $\frac{\partial}{\partial y}$ are a basis for the tangent space of $\mathbb{R}^2$. In complex notation, it is customary to take

$$\begin{aligned} \frac{\partial}{\partial z} &\equiv \frac{1}{2}\left(\frac{\partial}{\partial x} - i\frac{\partial}{\partial y}\right) \\ \frac{\partial}{\partial \bar{z}} &\equiv \frac{1}{2}\left(\frac{\partial}{\partial x} + i\frac{\partial}{\partial y}\right) \end{aligned} \quad (8.2)$$

as basis vectors. Notice that

$$\frac{\partial}{\partial z} z = 1 \qquad \frac{\partial}{\partial z} \bar{z} = 0$$

$$\frac{\partial}{\partial \bar{z}} z = 0 \qquad \frac{\partial}{\partial \bar{z}} \bar{z} = 1 \, .$$

These four conditions uniquely determine $\frac{\partial}{\partial z}$ and $\frac{\partial}{\partial \bar{z}}$ as first order linear differential operators with constant coefficients.

It is convenient to write, for a C^1 diffeomorphism $F : U \to V$ of planar domains, $F(x, y) = u(x, y) + iv(x, y) \approx (u(x, y), v(x, y))$. Then the Jacobian

---
[1] Ahlfors is credited with the terminology

determinant of F is
$$\mathcal{J} = u_x v_y - v_x u_y.$$
(Here we use subscripts to denote derivatives.) An easy calculation shows that
$$\mathcal{J} = |F_z|^2 - |F_{\bar z}|^2 \neq 0.$$
Of course the Jacobian never vanishes (because F is a diffeomorphism). If we assume in advance that F is sense-preserving at each point, then $\mathcal{J} > 0$ at each point and therefore
$$|F_z| > |F_{\bar z}|. \tag{8.3}$$

We take (8.3) as a standing hypotheses in what follows.

In fact the columns of the Jacobian matrix are F_x and F_y. In complex notation, we may say that the columns are $(F_z + F_{\bar z})$ and $(F_z - F_{\bar z})$. Thus any directional derivative of f in the direction $\tau = (\mu, \lambda)$ is
$$\mathcal{D}_\tau F = \mu(F_z + F_{\bar z}) + \lambda(F_z - F_{\bar z}),$$
where $\mu, \lambda \in \mathbb{C}$ and $|\mu|^2 + |\lambda|^2 = 1$. Clearly
$$|F_z| - |F_{\bar z}| \leq |\mathcal{D}_\tau F| \leq |F_z| + |F_{\bar z}|.$$
Indeed the upper extremum is realized when
$$\mu = \frac{1}{2}\left(\frac{\overline{F_z}}{|F_z|} + \frac{\overline{F_{\bar z}}}{|F_{\bar z}|}\right), \qquad \lambda = \frac{1}{2}\left(\frac{\overline{F_z}}{|F_z|} - \frac{\overline{F_{\bar z}}}{|F_{\bar z}|}\right).$$
The lower extremum is realized when
$$\mu = \frac{1}{2}\left(\frac{\overline{F_z}}{|F_z|} - \frac{\overline{F_{\bar z}}}{|F_{\bar z}|}\right), \qquad \lambda = \frac{1}{2}\left(\frac{\overline{F_z}}{|F_z|} + \frac{\overline{F_{\bar z}}}{|F_{\bar z}|}\right).$$

We think of the Jacobian of F, frozen at a point, as a linear map; and we think of this matrix operating on a circle, centered at the origin, in the tangent plane. Then the image is an ellipse and the ratio of its major axis to its minor axis is, according to our calculations,
$$D_F \equiv \frac{|F_z| + |F_{\bar z}|}{|F_z| - |F_{\bar z}|}. \tag{8.4}$$

We call the number D_F the **dilation factor of the mapping** F (also called the **dilitation factor**).

Already we can see that F is conformal if and only if $D_F = 1$ at each point, and this in turn is true precisely when $F_{\bar z} = 0$ at each point. This last equation, $F_{\bar z} = 0$, is equivalent to the Cauchy-Riemann equations
$$\frac{\partial u}{\partial x} = \frac{\partial v}{\partial y} \qquad \frac{\partial u}{\partial y} = -\frac{\partial v}{\partial x}.$$

In conclusion, the following are equivalent:

8.1. QUASICONFORMAL MAPPINGS

- F is conformal
- circles go to circles
- $D_F \equiv 1$
- the Cauchy-Riemann equations are satisfied
- F is holomorphic

It is sometimes convenient to formulate our condition in terms of the quantity

$$d_F \equiv \frac{|F_{\bar{z}}|}{|F_z|}. \tag{8.5}$$

Clearly

$$D_F = \frac{1 + |F_{\bar{z}}|/|F_z|}{1 - |F_{\bar{z}}|/|F_z|} = \frac{1 + d_F}{1 - d_F}.$$

In particular, F is conformal at $P \in U$ if and only if $d_F(P) = 0$.

Definition 8.1.1 *Let $K \geq 1$ be a constant. The mapping F is said to be K-quasiconformal if $D_F \leq K$ at each point of its domain U. If K is not specified then we simply say that F is* **quasiconformal**.

Setting $k = (K - 1)/(K + 1)$ we see that F is K-quasiconformal if and only if $d_F \leq k$, that is, if and only if

$$|f_{\bar{z}}| \leq k|f_z|. \tag{8.6}$$

Thus, F is conformal precisely when $K \equiv 1$ or $k \equiv 0$.

Quasiconformal Mappings of Plane Domains

It is now instructive to return to Grötzsch's formulation of quasiconformality. Again, we follow the exposition in Ahlfors [1]. Let Q be a closed rectangle with sides α, β, and let Q' be a closed rectangle with sides α', β'. We assume that the sides α, α' are horizontal and that β, β' are vertical. We consider a continuously differentiable mapping F from Q to Q'. The mapping is assumed to take the side α to α' and the side β to β'; thus we assume that the origin and neighboring vertices go to vertices. If $\alpha/\beta \neq \alpha'/\beta'$ then the mapping F cannot be conformal; we wish to measure how far from conformality it may be, and we wish to determine that mapping F that is most nearly conformal (by whatever yardstick we choose to measure proximity).

We may assume that $\alpha/\beta < \alpha'/\beta'$ (otherwise rename the rectangles and vertices). Assume that each of Q, Q' has lower left corner at the origin. Now we perform an elementary calculation, using nothing more than calculus:

$$\alpha' \leq \int_0^\alpha |d(F(x + iy))| \leq \int_0^\alpha \left(|F_z| + |F_{\bar{z}}|\right) dx. \tag{8.7}$$

Multiplying both sides of (8.7) by β, we obtain

$$\begin{aligned} \alpha'\beta &\leq \beta \cdot \int_0^\alpha (|F_z| + |F_{\bar{z}}|)\, dx \\ &= \int_0^\beta \int_0^\alpha (|F_z| + |F_{\bar{z}}|)\, dx\, dy \\ &= \iint_Q \left[\frac{\sqrt{|F_z| + |F_{\bar{z}}|}}{\sqrt{|F_z| - |F_{\bar{z}}|}}\right] \cdot \left[\sqrt{|F_z|^2 - |F_{\bar{z}}|^2}\right] dx\, dy\,. \end{aligned}$$

As a result, we have

$$\begin{aligned} (\alpha')^2 \beta^2 &\overset{\text{(Schwarz)}}{\leq} \iint_Q \frac{|F_z| + |F_{\bar{z}}|}{|F_z| - |F_{\bar{z}}|} dx\, dy \cdot \iint_Q |F_z|^2 - |F_{\bar{z}}|^2\, dx\, dy \\ &= \iint_Q D_F\, dx\, dy \cdot \iint_Q J_F\, dx\, dy \\ &\leq [(\text{area of } Q) \cdot \sup_Q D_F] \cdot (\text{area of } Q'). \end{aligned}$$

Dividing through by $[(\text{area } Q) \cdot (\text{area } Q')] = \alpha\beta \cdot \alpha'\beta'$, we obtain

$$\frac{\alpha'/\beta'}{\alpha/\beta} \leq \sup_Q D_F.$$

In fact this minimum is actually attained by the linear mapping of Q to Q':

$$L(z) = \frac{1}{2}\left(\frac{\alpha'}{\alpha} + \frac{\beta'}{\beta}\right)z + \frac{1}{2}\left(\frac{\alpha'}{\alpha} - \frac{\beta'}{\beta}\right)\bar{z}.$$

We have proved the following proposition:

Proposition 8.1.2 *Let Q and Q' be rectangles as above, with side lengths α, β and α', β' respectively. There exists a K-quasiconformal mapping of Q to Q' if and only if*

$$\frac{1}{K} \leq \frac{\alpha'/\beta'}{\alpha/\beta} \leq K.$$

An important problem is to characterize when two domains in the plane are quasiconformally equivalent. A complete answer can be given to the more restrictive problem of characterizing which domains are quasiconformally equivalent to an open metric ball (or disc).

Theorem 8.1.3 *A domain Ω in the plane is quasiconformally equivalent to an open metric ball if and only if $\partial\Omega$ is a nondegenerate continuum.*

8.1. QUASICONFORMAL MAPPINGS

For details, we refer the reader to Gehring [2] and to the references therein.

Liouville's Theorem

Having set the stage, by placing conformal mappings in the more general context of quasiconformal mappings, we shall prove Liouville's theorem which says, in effect, that there are no interesting conformal mappings of Euclidean space in dimensions 3 and higher. We state the result for C^4 mappings, but weaker hypotheses suffice (see Gehring [1] and Reshetnyak [1]).

Theorem 8.1.4 (Liouville [1]) *Let $U \subset \mathbb{R}^N$ be an open set, and let $F: U \to \mathbb{R}^N$ be a C^4 mapping that is conformal. Then F is the composition of an isometry, a dilation, and an inversion.*

Proof: For the proof, we follow closely the argument in Dubrovin, Fomenko, and Novikov [1]. We restrict our attention to $N = 3$.

Let x be a point in the domain of F and y the corresponding image point. It follows from conformality that if two vectors ξ and η are orthogonal with respect to the standard Euclidean inner product in the x-coordinates, then their images will certainly be orthogonal in the standard Euclidean inner product in the y-coordinates. Let us denote the dilation factor at each point by $\lambda(x)$. We shall assume that λ never vanishes. Thus, if $\mathcal{J}$ denotes the Jacobian matrix of F,

$$\|\mathcal{J}\eta\|^2 = \lambda(x)\|\eta\|^2$$

at each point x.

Now fix three mutually orthogonal vectors η^1, η^2, η^3. Let $\{\eta_j^k\}_{j=1}^3$ be the coordinates of η^k in the x-coordinate system, $k = 1, 2, 3$. Of course the three vectors $\mathcal{J}\eta^1$, $\mathcal{J}\eta^2$, and $\mathcal{J}\eta^3$ are then mutually orthogonal in the y-coordinate system. We differentiate the equation $\langle \mathcal{J}\eta^1, \mathcal{J}\eta^2 \rangle = 0$ with respect to the coordinate x_j to obtain

$$\frac{\partial}{\partial x_j}\langle \mathcal{J}\eta^1, \mathcal{J}\eta^2 \rangle$$

$$= \left\langle \sum_\ell \frac{\partial^2 y}{\partial x_j \partial x_\ell}\eta_\ell^1, \mathcal{J}\eta^2 \right\rangle + \left\langle \mathcal{J}\eta^1, \sum_\ell \frac{\partial^2 y}{\partial x_j \partial x_\ell}\eta_\ell^2 \right\rangle = 0. \quad (8.8)$$

We multiply both sides of (8.8) by η_j^3 and sum over j to obtain

$$\left\langle \sum_{\ell,j} \frac{\partial^2 y}{\partial x_j \partial x_\ell}\eta_\ell^1 \eta_j^3, \mathcal{J}\eta^2 \right\rangle + \left\langle \mathcal{J}\eta^1, \sum_{\ell,j} \frac{\partial^2 y}{\partial x_j \partial x_\ell}\eta_\ell^2 \eta_j^3 \right\rangle. \quad (8.9)$$

Now plainly we may permute the indices in (8.9) to obtain three such equations. Adding two of these, and subtracting the third, yields the equation

$$\left\langle \sum_{j,\ell} \frac{\partial^2 y}{\partial x_j \partial x_\ell}\eta_j^1 \eta_\ell^2, \mathcal{J}\eta^3 \right\rangle = 0. \quad (8.10)$$

There are two other equations, similar to (8.10), that may be obtained by permuting indices.

Because we are in dimension 3, and since the $\mathcal{J}\eta^\ell$ are mutually perpendicular, equation (8.10) implies that

$$\sum_{j,\ell} \frac{\partial^2 y}{\partial x_j \partial x_\ell} \eta_j^1 \eta_\ell^2 = \mu(x)[\mathcal{J}\eta^1] + \nu(x)[\mathcal{J}\eta^2] \tag{8.11}$$

for some coefficients $\mu(x)$ and $\nu(x)$. Taking the inner product of this equation with $\mathcal{J}\eta^1$ or with $\mathcal{J}\eta^2$ enables us to solve for μ and ν. Thus we have

$$\mu = \frac{1}{\|\mathcal{J}\eta^1\|^2} \left\langle \sum_{j,\ell} \frac{\partial^2}{\partial x_j \partial x_\ell} \eta_j^1 \eta_\ell^2, \mathcal{J}\eta^1 \right\rangle, \tag{8.12}$$

$$\nu = \frac{1}{\|\mathcal{J}\eta^2\|^2} \left\langle \sum_{j,\ell} \frac{\partial^2}{\partial x_j \partial x_\ell} \eta_j^1 \eta_\ell^2, \mathcal{J}\eta^2 \right\rangle. \tag{8.13}$$

But we know that $\|\mathcal{J}\eta\|^2 = \lambda \|\eta\|^2$. It is convenient in this argument to replace λ with $\widetilde{\lambda} = 1/\lambda$. With this notation, the equations (8.12) and (8.13) may be rewritten as

$$\mu = \frac{\widetilde{\lambda}}{\|\eta^1\|^2} \left[\frac{1}{2} \sum_m \eta_m^2 \frac{\partial}{\partial x_m} \langle \mathcal{J}\eta^1, \mathcal{J}\eta^1 \rangle \right]$$

$$= \frac{\widetilde{\lambda}}{\|\eta^1\|^2} \left[\frac{1}{2} \sum_m \eta_m^2 \frac{\partial}{\partial x_m} \left(\frac{\|\eta^1\|^2}{\widetilde{\lambda}(x)} \right) \right]$$

$$= -\frac{1}{2\widetilde{\lambda}} \sum_m \eta_m^2 \frac{\partial \widetilde{\lambda}}{\partial x_m}$$

and, by a similar calculation,

$$\nu = -\frac{1}{2\widetilde{\lambda}} \sum_m \eta_m^1 \frac{\partial \widetilde{\lambda}}{\partial x_m}.$$

Substituting these last two expressions for μ and ν into (8.11), and performing some manipulations, we obtain that

$$\frac{\partial^2 y}{\partial x_j \partial x_\ell} \eta_j^1 \eta_\ell^2 = -\frac{1}{2\widetilde{\lambda}} \left(\sum_{j,\ell} \frac{\partial \widetilde{\lambda}}{\partial x_j} \frac{\partial y}{\partial x_\ell} \eta_\ell^2 \eta_j^1 + \sum_{j,\ell} \frac{\partial \widetilde{\lambda}}{\partial x_\ell} \frac{\partial y}{\partial x_j} \eta_\ell^1 \eta_j^2 \right). \tag{8.14}$$

If we set $\rho = \sqrt{\widetilde{\lambda}}$, then

$$\frac{\partial^2 (\rho y)}{\partial x_j \partial x_\ell} = \frac{\partial \rho}{\partial x_j} \frac{\partial y}{\partial x_\ell} + \frac{\partial \rho}{\partial x_\ell} \frac{\partial y}{\partial x_j} + \rho \frac{\partial^2 y}{\partial x_j \partial x_\ell} + y \frac{\partial^2 \rho}{\partial x_j \partial x_\ell}$$

$$= \frac{1}{2\sqrt{\widetilde{\lambda}}} \left(\frac{\partial \widetilde{\lambda}}{\partial x_j} \frac{\partial y}{\partial x_\ell} + \frac{\partial \widetilde{\lambda}}{\partial x_\ell} \frac{\partial y}{\partial x_j} + 2\widetilde{\lambda} \frac{\partial^2 y}{\partial x_j \partial x_\ell} \right) + y \frac{\partial^2 \rho}{\partial x_j \partial x_\ell}. \tag{8.15}$$

8.1. QUASICONFORMAL MAPPINGS

Equation (8.15), together with (8.14), gives us that

$$\sum_{j,\ell} \frac{\partial^2}{\partial x_j \partial x_\ell}(\rho y \eta_j^1 \eta_\ell^2) = \sum_{j,\ell}\left(\frac{\partial^2 \rho}{\partial x_j \partial x_\ell}\eta_j^1 \eta_\ell^2\right) y. \tag{8.16}$$

Differentiating (8.16) with respect to x_p and then multiplying through by η_p^3 and summing over p, we obtain

$$\sum_{j,\ell,p} \frac{\partial^3(\rho y)}{\partial x_j \partial x_\ell \partial x_p}\eta_j^1 \eta_\ell^2 \eta_p^3$$

$$= \sum_{j,\ell}\left(\frac{\partial^2 \rho}{\partial x_j \partial x_\ell}\eta_j^1 \eta_\ell^2\right)\mathcal{J}\eta^3 + \sum_{j,\ell,p}\left(\frac{\partial^3 \rho}{\partial x_j \partial x_\ell \partial x_p}\eta_j^1 \eta_\ell^2 \eta_p^3\right)y \tag{8.17}$$

Now notice that the first and last expressions in (8.17) are invariant under permutations of the indices $1, 2, 3$. It follows that the middle expression is likewise invariant. Therefore we have

$$\left(\frac{\partial^2 \rho}{\partial x_j \partial x_\ell}\eta_j^1 \eta_\ell^2\right)\mathcal{J}\eta^3 = \left(\frac{\partial^2 \rho}{\partial x_j \partial x_\ell}\eta_j^2 \eta_\ell^3\right)\mathcal{J}\eta^1.$$

Recall that none of $\mathcal{J}\eta^1, \mathcal{J}\eta^2, \mathcal{J}\eta^3$ is zero (since our transformation is supposed to have non-vanishing Jacobian at each point). Thus the mutual orthogonality of these expressions implies that

$$\sum_{\ell,p} \frac{\partial^2 \rho}{\partial x_\ell \partial x_p}\eta_\ell^1 \eta_p^2 \equiv 0. \tag{8.18}$$

In short, the bilinear form represented by the left side of (8.18) vanishes on any pair of orthogonal vectors. It follows that the coefficient matrix of such a form must be scalar, that is, a multiple of the identity. As a result,

$$\frac{\partial^2 \rho}{\partial x_j \partial x_p} = \begin{cases} \sigma(x) & \text{if } j = p \\ 0 & \text{if } j \neq p. \end{cases} \tag{8.19}$$

We shall next demonstrate that σ is identically constant.

Let ξ^1, ξ^2, ξ^3 be fixed vectors at the point x. We differentiate (8.19) three times, and then multiply by $\xi_j^1 \xi_\ell^2 \xi_p^3$; finally we sum over the repeated indices. The result is

$$\sum_{j,\ell,p} \frac{\partial^3 \rho}{\partial x_j \partial x_\ell \partial x_p}\xi_j^1 \xi_\ell^2 \xi_p^3 = \sum_{\ell}\left(\frac{\partial \sigma}{\partial x_\ell}\xi_\ell^1\right)\langle \xi^2, \xi^3\rangle. \tag{8.20}$$

We obtain a similar equation by interchanging ξ^1 and ξ^2, and subtracting that new equation from (8.20). The result is

$$\left\langle \left(\sum_m \frac{\partial \sigma}{\partial x_m}\xi_m^1\right)\xi^2 - \left(\sum_\ell \frac{\partial \sigma}{\partial x_\ell}\xi_\ell^2\right)\xi^1, \xi^3 \right\rangle = 0. \tag{8.21}$$

Since (8.21) holds for all vectors ξ^i, it follows that σ is identically constant. As a result,
$$\frac{\partial^2 \rho}{\partial x_\ell \partial x_j} = \sigma \delta_{\ell j} = \begin{cases} \sigma & \text{if } \ell = j \\ 0 & \text{if } \ell \neq j. \end{cases}$$

It follows that
$$\rho = \sqrt{\lambda} = a_1 \|x - P\|^2 + b_1,$$
with a_1, b_1 constant.

Of course the inverse mapping to F is also conformal, with conformality factor $1/\lambda$. It follows by similar reasoning that
$$\frac{1}{\sqrt{\widetilde{\lambda}}} = a_2 \|y - Q\|^2 + b_2,$$
with a_2, b_2 constants and Q the image of P under F.

In summary, we have
$$\left(a_1 \|x - P\|^2 + b_1 \right) \left(a_2 \|y - P\|^2 + b_2 \right) = 1. \tag{8.22}$$

Now we reason as follows. Our transformation maps any sufficiently small sphere centered at P to a sphere centered at Q. Take any line segment L that passes through P. Of course this line segment is perpendicular to all the spheres centered at P. Thus the curve that is the image of L under F must likewise (by conformality) be perpendicular to all the spheres centered at Q. As a result, this image curve is also a line segment—passing through Q.

Consider now a fixed line passing through P: let P' be another fixed point on that line and x a variable point. Let Q' and y be the corresponding image points under F. By (8.22), we see that
$$a_2 \|y - Q\|^2 + b_2 = a_2 \left[\|y - Q'\|^2 + 2\|y - Q'\| \, \|Q' - Q\| + \|Q' - Q\|^2 \right] + b_2$$
$$= \frac{1}{a_1 \|x - P\|^2 + b_1}. \tag{8.23}$$

So we see that $\|y - Q'\|$ is an algebraic function of $\|x - P\|$.

Now if we parametrize the segment $\overline{yQ'}$ by a parameter τ, $t_1 \leq \tau \leq t$, then the formula $\sum_m (dy_m)^2 = (1/\widetilde{\lambda}) \sum_m (dx_m)^2$ teaches us that

$$\|y - Q'\| = \int_{t_1}^{t} \sqrt{\sum_m \left(\frac{dy_m}{d\tau} \right)^2} \, d\tau = \int_{t_1}^{t} \sqrt{\frac{1}{\widetilde{\lambda}} \sum_m \left(\frac{dx_m}{d\tau} \right)^2} \, d\tau.$$

If we select τ to be arc length along the segment Px, with $t_1 = \|P' - P\|$ and $t = \|x - P\|$, then this last line becomes
$$\|y - Q'\| = \int_{\|P' - P\|}^{\|x - P\|} \frac{d\tau}{\sqrt{\widetilde{\lambda}}} = \int_{\|P' - P\|}^{\|x - P\|} \frac{d\tau}{(a_1 \tau^2 + b_1)}.$$

8.1. QUASICONFORMAL MAPPINGS

If neither a_1 nor b_1 is zero, then it follows from calculus that this last line is a transcendental function of $\|x - P\|$, contradicting what we learned from (8.23). Thus either $a_1 = 0$ or $b_1 = 0$.

If $a_1 = 0$ then $\widetilde{\lambda}$ is a constant function and our transformation is plainly an isometry followed by a dilation. If instead $b_1 = 0$ then $a_1 \neq 0$ and we may follow the transformation $y = F(x)$ with the inversion

$$x^* = \frac{x - P}{\|x - P\|^2}$$

to see that

$$a_2 \|y - P\|^2 = a_1 \|x^*\|^2.$$

This brings us back to the other case of an isometry followed by a dilation. Thus, in this second instance, our mapping is the composition of an isometry, a dilation, and an inversion. ∎

Concluding Remarks

We conclude with a brief discussion of a metric characterization of quasiconformality.

Let C be a Jordan curve in the plane. We call C **quasiconformal** if there is a quasiconformal mapping of a domain U that contains C which carries C into a circle.

Just as, in complex function theory, we wish to know which curves can be carried by a conformal mapping to a circle, so it is of interest to identify the quasiconformal curves. With that goal in mind, we define an auxiliary concept.

Definition 8.1.5 *Let C be a Jordan curve in the plane. Let z, w be two points on C and consider the smaller (in diameter) of the two arcs determined by z and w. If the quotient of this diameter by the distance $|z - w|$ is bounded over all pairs z, w then we say that C is of **bounded turning**.*

The following theorem (which we shall not prove, but see Lehto and Virtanen [1], p. 100) gives a complete metric characterization of quasiconformal curves:

Theorem 8.1.6 *A Jordan curve in the plane is quasiconformal if and only if it is of bounded turning.*

It is enlightening to look at a simple example that illustrates the theorem. Define

$$\gamma(t) = (t, \sqrt{|t|}), \quad -1 \leq t \leq 1.$$

Of course this is not a closed curve, but it is easily closed up; and we are only interested in behavior near the origin. Let $z_j = (-1/j, 1/\sqrt{j})$ and $w_j = (1/j, 1/\sqrt{j})$ for $j = 1, 2, \ldots$. Then $|z_j - w_j| = 2/j$ while the diameter of the smaller part of the curve between z_j and w_j is on the order of $1/\sqrt{j}$. Of course the ratio in the definition of "bounded turning" is unbounded, so the curve is not quasiconformal. It is somewhat less intuitive to see directly from the definition that the curve is not quasiconformal; we leave the details for the interested reader.

In dimensions higher than two and for functions that are not necessarily differentiable, we have the following definition:

Definition 8.1.7 *Suppose Ω and Ω' are bounded domains in $\mathbb{R}^N$ and $F : \Omega \to \Omega'$ is a homeomorphism. We say that F is **quasiconformal** if the following function has a finite essential supremum with respect to Lebesgue measure:*

$$\mathcal{F}(x) = \limsup_{r \downarrow 0} \left(\frac{\max\{|F(x) - F(y)| : |x - y| = r\}}{\min\{|F(x) - F(y)| : |x - y| = r\}} \right). \tag{8.24}$$

As we have already seen in the 2-dimensional case, there are several equivalent definitions of quasiconformal mapping. The reader may see Caraman [1] for the details of such definitions and proofs of their equivalence.

Theorem 8.1.6 is only valid for Jordan curves in the plane; it does not generalize to surfaces in $\mathbb{R}^3$ that are homeomorphic to the sphere. We illustrate this with the following interesting example in dimension three, for which we are grateful to A. Baernstein.

Example 8.1.8 Let $\phi : \mathbb{R} \to \mathbb{R}$ be given by

$$\phi(t) = \begin{cases} 0 & \text{if } t \leq 1/2 \\ (t - 1/2)^2 & \text{if } t > 1/2 \end{cases}.$$

Now set

$$\Omega_1 = \{x \in \mathbb{R}^3 : |x| < 1, |x - (|x|, 0, 0)| > \phi(|x|)\}$$

and

$$\Omega_2 = \{x \in \mathbb{R}^3 : |x| < 1\} \cup \{x \in \mathbb{R}^3 : |x - (|x|, 0, 0)| < \phi(-|x| + 7/4)\}.$$

Then Ω_1 is not quasiconformally equivalent to the unit ball, but Ω_2 *is* quasiconformally equivalent to the ball.

Intuitively these assertions are true because Ω_2 is easier to flatten than a ball with a right circular cone protruding while Ω_1 is harder to flatten than a domain with a right circular conical indentation (the point being that a ball with a right circular cone added or deleted has Lipschitz boundary and thus is obviously quasiconformal equivalent to a ball).

Of course, both the analogous curves Ω_1 and Ω_2 that would result if $\mathbb{R}^3$ were replaced by $\mathbb{R}^2$ are conformally (not just quasiconformally) equivalent to the disc—by the Riemann mapping theorem.

The theory of quasiconformal mappings is a lovely mixture of geometry and analysis. Quasiconformal mappings have proved to be natural tools in both pure and applied applications. It is a rich theory that is still under development.

8.2 Weyl's Theorem on Eigenvalue Asymptotics of a Domain in Space

If one considers the modes of vibration of a plane domain, one observes that the resonant frequencies for the wave equation[2] are precisely the eigenvalues of the Laplace operator (see Courant and Hilbert [1], Chapter V); thus the spectrum (meaning the set of eigenvalues) of the Laplace operator for a domain is the mathematical equivalent of the sound of a drum head in the shape of the domain. (Of course, a real drum is much more complicated, but we ignore those complications.) The article Kac [1] formulated the question "Can you hear the shape of a drum?" This query captures the spirit of the present section: Which geometric features of a domain (in the plane, for instance) determine the spectral properties for the Laplace operator on that domain and, conversely, if one has complete knowledge of the spectral properties, then can one give a geometric description of the domain?

It is now known that the answer to Kac's original question is "no:" In fact, there are two (isometrically) distinct 2-dimensional drums with the same eigenfrequencies. (It is not known whether there are two such drums that are also *convex*.) It would take us far afield to develop the machinery necessary for presenting the famous example of Gordon, Webb, and Wolpert [1]. But what we can, and shall, do is to acquaint the reader with some of the basic ideas of the spectral theory of a domain in space. Our discussion will culminate with a presentation of Weyl's theorem for the Dirichlet eigenvalues of the domain.

In fact Weyl's theorem gives an estimate on the behavior of eigenvalues of the Dirichlet problem in terms of the *volume* of the domain. Volume is perhaps the most elementary, and the coarsest, geometric property of a domain. More sophisticated results derive spectral properties of a domain in a manifold from (for instance) curvature considerations. We refer the reader to Chavel [1] for more information on such advanced results.

Basic Terminology

Fix a domain Ω in real Euclidean N-space. For convenience, we assume that the domain has smooth boundary; sometimes we will consider domains with *piecewise smooth* boundary, where the pieces have transversal crossings. The main fact that we shall need is that, on either of these two types of domains, Green's Theorem is valid.

We also assume the domain to be connected. Let Δ denote the usual Laplace operator in Euclidean space. We consider the partial differential equation

$$\Delta u + \lambda u = 0. \tag{8.25}$$

---

[2] The wave equation is the standard equation for a vibrating membrane.

Any value of λ such that (8.25) has a non-trivial solution is called an **eigenvalue** and any such non-trivial solution is called an **eigenfunction**. Because of the connection with the wave equation, eigenvalues are sometimes called **eigenfrequencies**.

Definition 8.2.1

(i) The **Dirichlet eigenvalue problem** *is the problem of finding all values of λ for which there exist non-trivial solutions of (8.25) with*

$$u\big|_{\partial\Omega} \equiv 0. \tag{8.26}$$

(ii) The **Neumann eigenvalue problem** *is the problem of finding all values of λ for which there exist non-trivial solutions of (8.25) with*

$$\partial u/\partial\nu\big|_{\partial\Omega} \equiv 0. \tag{8.27}$$

Here ν is the unit outward normal vector field to $\partial\Omega$.

(iii) *A* **mixed eigenvalue problem** *is the problem of finding all values of λ for which there exist non-trivial solutions of (8.25) with*

$$u\big|_{\partial\Omega\setminus U} = 0, \quad \partial u/\partial\nu\big|_{U\cap\partial\Omega} = 0. \tag{8.28}$$

Here U is an open set such that $\partial\Omega\setminus U \neq \emptyset$ and $U \cap \partial\Omega \neq \emptyset$.

Lemma 8.2.2 *Let ϕ and ψ be smooth functions. If either ϕ satisfies the Dirichlet condition (8.26), ψ satisfies the Neumann condition (8.27), or both satisfy the mixed condition (8.28), then*

$$\int_\Omega \phi \, \Delta \psi \, dV = -\int_\Omega \operatorname{grad} \phi \cdot \operatorname{grad} \psi \, dV \tag{8.29}$$

holds.

Proof: By Green's Theorem, we have

$$\begin{aligned}
0 &= \int_{\partial\Omega} \phi \operatorname{grad} \psi \cdot \nu \, d\sigma \\
&= \int_\Omega \operatorname{div}(\phi \operatorname{grad} \psi) \, dV \\
&= \int_\Omega \operatorname{grad} \phi \cdot \operatorname{grad} \psi \, dV + \int_\Omega \phi \, \Delta \psi \, dV. \quad \blacksquare
\end{aligned}$$

The following corollary is an immediate consequence of Lemma 8.2.2.

8.2. WEYL'S THEOREM

Corollary 8.2.3 *Let ϕ and ψ be smooth functions. If both ϕ and ψ satisfy either the Dirichlet condition (8.26), the Neumann condition (8.27), or the mixed condition (8.28), then*

$$\int_\Omega \phi \, \triangle \psi \, dV = \int_\Omega \psi \, \triangle \phi \, dV \qquad (8.30)$$

holds.

The content of Corollary 8.2.3 is summarized by saying that equation (8.25) is self-adjoint. Considering (8.29) with ϕ an eigenfunction corresponding to the eigenvalue λ and with $\phi = \psi$, we see that

$$\lambda = \frac{\int_\Omega |\text{grad}\,\phi|^2 \, dV}{\int_\Omega \phi^2 \, dV} > 0. \qquad (8.31)$$

Since the problem is elliptic, it is known (see Widom [1]) that the eigenvalues do not accumulate at the origin, nor in any finite part of the half line.

In conclusion, we typically write

$$0 < \lambda_1 \leq \lambda_2 \leq \cdots, \qquad (8.32)$$

where each eigenvalue is listed according to multiplicity.

We define the associated **summatory functions**: For the Dirichlet eigenvalue problem, we set

$$N(\lambda) = \text{the number of eigenvalues of (8.25), subject} \\ \text{to the Dirichlet boundary condition (8.26),} \qquad (8.33) \\ \text{which do not exceed } \lambda$$

and for the Neumann eigenvalue problem, we set

$$M(\lambda) = \text{the number of eigenvalues of (8.25), subject} \\ \text{to the Neumann boundary condition (8.27),} \qquad (8.34) \\ \text{which do not exceed } \lambda.$$

The standard notation $N(\lambda)$ for the Dirichlet summatory function must be distinguished by context from the dimension of $\mathbb{R}^N$.

Hermann Weyl's celebrated theorem says, with Υ_N denoting the volume of the unit ball in $\mathbb{R}^N$, that

$$\frac{N(\lambda)}{\lambda^{N/2}} \approx \frac{\Upsilon_N \text{vol}(\Omega)}{(2\pi)^N} \quad \text{as } \lambda \to +\infty. \qquad (8.35)$$

We shall derive separately the estimate on $N(\lambda)/\lambda^{N/2}$ from above and from below. Each of these will involve elegant geometric ideas related to measure, content, and capacity. The estimate from above will use, as an incidental tool, certain estimates on $M(\lambda)$.

In fact we shall proceed as follows:

A. We give some preliminary background about Hilbert space.

B. We begin by enunciating certain maximum-minimum and related monotonicity properties of the eigenvalues in question.

C. We perform certain explicit calculations for the case of Ω a rectangular parallelepiped in Euclidean space.

D. We apply the monotonicity properties in **B**, together with the calculations in **C**, to derive the estimates from above and below that lead to (8.35).

E. We provide the proofs of the results in **B**.

For some particulars of the arguments, we shall take the liberty of referring the reader to Chavel [1], and to the detailed literature cited therein.

A. Hilbert Spaces

In our discussion of **B**, we shall require knowledge of certain elementary properties of Hilbert space. Here we enunciate those properties, but we refer the reader to Rudin [2] for details of the proofs.

Consider a complex vector space $\mathcal{H}$ that is equipped with an inner product $\langle \cdot, \cdot \rangle$ that is conjugate symmetric in its entries. The inner product induces a **norm** on $\mathcal{H}$ by $\|x\| \equiv \langle x, x \rangle^{1/2}$. We say that two elements $x, y \in \mathcal{H}$ are **orthogonal** if $\langle x, y \rangle = 0$; in this circumstance we write $x \perp y$. The space $\mathcal{H}$ is called a **(complex) Hilbert space** if it is complete in the topology induced by the norm $\| \cdot \|$.

Note that it is also useful to consider inner product spaces over the ground field $\mathbb{R}$. In this case we of course do not require the inner product to be conjugate symmetric. When the space is complete, then it is called a **real Hilbert space**.

A finite or infinite collection $\{x_\alpha\}_{\alpha \in \mathcal{A}}$ of elements in $\mathcal{H}$, where $\mathcal{A}$ is some index set, is said to be **orthonormal** if $\langle x_\alpha, x_\beta \rangle = \delta_{\alpha,\beta}$, where $\delta_{\alpha,\beta}$ is the Kronecker delta. It is an exercise with Zorn's Lemma (or the Hilbert maximum principle) to see that any Hilbert space contains a **complete orthonormal system** $\{x_\alpha\}_{\alpha \in \mathcal{A}}$. This means that if $z \in \mathcal{H}$ and $z \perp x_\alpha$ for every $\alpha \in \mathcal{A}$ then $z = 0$, i.e., no nontrivial element may be added to the complete orthonormal system. Note that the index set $\mathcal{A}$ could, in principle, be uncountable. If the Hilbert space $\mathcal{H}$ happens to be **separable** then $\mathcal{H}$ possesses a complete orthonormal system that is *countable*; in fact, under these circumstances, any orthonormal system for $\mathcal{H}$ will be countable.

If $x_1, x_2, \ldots$ are orthonormal in $\mathcal{H}$ and if $x \in \mathcal{H}$ then we set

$$\alpha_j = \alpha_j(x) = \langle x, \alpha_j \rangle.$$

We sometimes call α_j the jth **Fourier coefficient** of x with respect to the

8.2. WEYL'S THEOREM

orthonormal system. **Bessel's inequality** then says that

$$\sum_j |\alpha_j|^2 \leq \|x\|^2$$

and **Parseval's identity** says that

$$\sum_{j=1}^{\infty} \alpha_j^2 = \|x\|^2.$$

For $\Omega \subset \mathbb{R}^N$ a domain, we can think of $L^2(\Omega)$ with the inner product

$$\langle \phi, \psi \rangle = \int_\Omega \phi \overline{\psi} \, dV$$

as a Hilbert space. We may consider eigenfunctions $\{\phi_j\}$ for any one of the eigenvalue problems in Definition 8.2.1, where ϕ_j is an eigenfunction corresponding to the eigenvalue λ_j, and multiplicities are represented by repetition. It follows from (8.30) that eigenfunctions ϕ_j, ϕ_k corresponding to distinct eigenvalues $\lambda_j \neq \lambda_k$ must be orthogonal; for we have

$$0 = \int_\Omega \phi_j \Delta \overline{\phi_k} - \overline{\phi_k} \Delta \phi_j \, dV = (\lambda_j - \lambda_k) \int_\Omega \phi_j \overline{\phi_k} \, dV$$

and the result is immediate. If K is the closed subspace of L^2 spanned by the $\{\phi_j\}$, then the eigenvalue problem makes sense on the orthogonal complement of K. Thus we may find another eigenvalue, and eigenfunction—thereby extending the list of eigenvalues and eigenfunctions. Thus an argument using Zorn's Lemma shows that (after normalization) the eigenfunctions for the Dirichlet eigenvalue problem (and likewise for the Neumann eigenvalue problem) form a complete orthonormal basis for $L^2(\Omega)$.

A consequence of (8.29) is that if the eigenfunctions are normalized so as to form an orthonormal system, then

$$\int_\Omega \mathbf{grad}\, \phi_j \cdot \mathbf{grad}\, \overline{\phi_k} \, dV = \lambda_j \, \delta_{j,k} \, . \tag{8.36}$$

B. Maximum-Minimum Methods

With the requisite Hilbert space background in place, we now begin to examine maximum-minimum methods. Suppose that X and Y are continuous vector fields on an open set $\Omega \subset \mathbb{R}^N$. For us the most important type of vector field will be the gradient of some C^1 function on Ω. We define the inner product

$$(X, Y) = \int_\Omega \langle X(s), Y(s) \rangle \, dVs,$$

where $\langle X(s), Y(s) \rangle$ is the usual inner product of vectors. The associated norm is

$$\|X\| = \sqrt{\int_\Omega |X(s)|^2 \, dVs} = \sqrt{\int_\Omega \langle X(s), X(s) \rangle \, dVs}.$$

Thus the space of (square integrable) vector fields on Ω is a linear space equipped with an inner product. Passing to the completion, we obtain an L^2 space. We denote the completed space, which is of course a Hilbert space, by $\mathcal{L}^2(\Omega)$.

As a special case of the discussion in the preceding paragraph, consider a C^1 function g on Ω and a C^1 vector field X on Ω. Then, as is easily checked,

$$(\mathbf{grad}\, g, X) = -(g, \mathbf{div}\, X). \tag{8.37}$$

This equation is just the integrated form of the assertion that the divergence operator is the (negative of the) adjoint of the gradient operator.

In analogy with standard distribution theory, we now define a weak form of (8.37). If $g \in L^2(\Omega)$ and $Z \in L^2(\Omega)$ then we say that Z is a **weak derivative** or **weak gradient** of g if

$$(Z, X) = -(g, \mathbf{div}\, X)$$

for any C^1 vector field X. Notice that, in case g is C^1, the new definition is justified by the formula (8.37). If the weak derivative exists then it is unique. The standard, and most important, justification for using the weak derivative is that it extends the operator **grad** to a larger domain so that it is now a **closed operator**.

For use in the present context, we recall from Section 4.1 some of the notation and terminology used in the study of Sobolev spaces. We let $H^1(\Omega)$ denote the subspace of $L^2(\Omega)$ consisting of those vector fields that have a weak derivative in $L^2(\Omega)$. We call $H^1(\Omega)$ the Sobolev space (of order 1). A useful inner product on $H^1(\Omega)$ is

$$(g, h)_1 = (g, h) + (\mathbf{grad}\, g, \mathbf{grad}\, h).$$

It is easy to check that, equipped with this inner product, $H^1(\Omega)$ is a Hilbert space. The associated norm is

$$\|g\|_1 = \sqrt{\|g\|^2 + \|\mathbf{grad}\, g\|^2}.$$

It is straightforward to check that, when Ω has smooth boundary, $C^\infty(\overline{\Omega})$ is dense in $H^1(\Omega)$.

The space $H^1(\Omega)$ is appropriate for the study of the Neumann eigenvalue problem, but for the Dirichlet eigenvalue problem we need to restrict ourselves to the completion of the C^∞ functions with compact support in Ω in the H^1 topology. This space is sometimes called $H^1_0(\Omega)$, and it is a *proper* subspace of H^1 (see Krantz [4] or Taylor [1] for details). For the

8.2. WEYL'S THEOREM

mixed eigenvalue problem, the appropriate function space is the completion of the C^∞ functions with compact support in $\Omega \cup U$, where we recall from Definition 8.2.1(iii) that the mixed eigenvalue problem also involves an open U such that $\partial\Omega \setminus U \neq \emptyset$ and $U \cap \partial\Omega \neq \emptyset$. To cover all three cases, we let $\mathcal{H}$ denote the appropriate function space, so

$$\mathcal{H} = \begin{cases} H_0^1(\Omega) & \text{for the Dirichlet problem,} \\ H^1(\Omega) & \text{for the Neumann problem,} \\ H^1 \text{ completion of } C_0^\infty(\Omega \cup U) & \text{for the mixed problem.} \end{cases} \quad (8.38)$$

One of the basic ideas in this subject is to extremize a certain quadratic form known as the **Dirichlet integral** or **energy integral**. This form is given by

$$D[f, g] = (\mathbf{grad}\, f, \mathbf{grad}\, g)$$

for $f, g \in H^1$. It is a classical observation that the Dirichlet integral is connected with the Laplace operator. For instance, the calculus of variations (*e.g.* see Courant and Hilbert [1], Chapter IV) tells us that a C^2 function f that minimizes the Dirichlet integral, subject only to a boundary condition

$$D[f, f],$$

must satisfy the equation

$$\triangle f = 0.$$

Our first main concern will be to determine the validity of the formula

$$(\triangle\phi, f) = -D[\phi, f], \quad (8.39)$$

where ϕ is an eigenfunction for some eigenvalue problem, and f lies in a subspace of H^1. The construction that we study is a variant of the standard Hilbert space construction of the **domain of the adjoint** of an unbounded Hilbert space operator.

To wit, by Lemma 8.2.2, equation (8.39) is certainly valid for f smooth and compactly supported in Ω. Now the identity (8.39) shows that, if we set $T_\phi(f) = (\triangle\phi, f)$, then

$$|T_\phi(f)| \leq |D[\phi, f]| \overset{\text{(Schwarz)}}{\leq} \|\phi\|_1 \|f\|_1.$$

Thus T_ϕ is a bounded linear functional, which may be extended to $\mathcal{H}$.

We now state our results. The proofs will be deferred until Subsection E.

Theorem 8.2.4 (Rayleigh) *Let $\Omega \subset \mathbb{R}^N$ be a domain with either C^∞ boundary or piecewise C^∞ boundary. Consider any one of the eigenvalue problems from Definition 8.2.1, and suppose that the eigenvalues are*

$$0 < \lambda_1 \leq \lambda_2 \leq \cdots,$$

where each eigenvalue is repeated according to its multiplicity. Then for any $f \in \mathcal{H}$, $f \neq 0$, we have

$$\lambda_1 \leq \frac{D[f,f]}{\|f\|^2}. \tag{8.40}$$

Equality holds in (8.40) if and only if f is actually an eigenfunction corresponding to the eigenvalue λ_1.

If $\{\phi_1, \phi_2, \ldots\}$ is a complete orthonormal basis of $L^2(\Omega)$ such that ϕ_j is an eigenfunction corresponding to the eigenvalue λ_j for each j, then any $f \in \mathcal{H}$ that satisfies the additional condition

$$(f, \phi_1) = (f, \phi_2) = \cdots = (f, \phi_{k-1}) = 0$$

will also satisfy the inequality

$$\lambda_k \leq \frac{D[f,f]}{\|f\|^2}. \tag{8.41}$$

Equality holds in (8.41) if and only if f is an eigenfunction corresponding to the eigenvalue λ_k.

Theorem 8.2.5 (The Maximum-Minimum Theorem) *Let there be given functions $u_1, u_2, \ldots, u_{k-1}$ in $L^2(\Omega)$, and set*

$$\mu = \inf \frac{D[f,f]}{\|f\|^2}, \tag{8.42}$$

where f varies over the subspace of functions in $\mathcal{H}$ that is orthogonal to $u_1, u_2, \ldots, u_{k-1}$ in the L^2 inner product. Then, with λ_k as in Rayleigh's Theorem 8.2.4, it holds that

$$\mu \leq \lambda_k.$$

Moreover, if we consider

$$\mu = \mu(u_1, u_2, \ldots, u_{k-1})$$

in (8.42) to be a function of $u_1, u_2, \ldots, u_{k-1}$, then the maximum of μ is λ_k and it is attained if and only if each u_ℓ is an eigenfunction corresponding to the eigenvalue λ_ℓ, $\ell = 1, 2, \ldots, k-1$.

Remark 8.2.6 Note that μ is defined to be the infimum as f varies over the appropriate subspace and that the conclusion of the second part of the theorem refers to the maximum that this infimum can obtain; hence the name "maximum-minimum theorem".

The next result will be fundamental for our analysis that leads to Weyl's theorem. Our strategy will be to compare the eigenvalue enumeration for a domain Ω with the enumeration for a system of rectangular parallelepipeds either sitting inside Ω or containing Ω (this technique continues to be one of the most important tools in the subject).

8.2. WEYL'S THEOREM

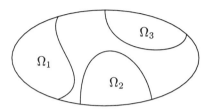

Figure 8.1: **Example Subdomains for the Monotonicity Theorem**

Theorem 8.2.7 (Domain Monotonicity) *Let $\Omega \subset \mathbb{R}^N$ be a domain with either C^∞ boundary or piecewise C^∞ boundary. Consider any one of the eigenvalue problems from Definition 8.2.1, and suppose that the eigenvalues are*

$$0 < \lambda_1 \leq \lambda_2 \leq \cdots,$$

where each eigenvalue is repeated according to its multiplicity. Suppose that $\Omega_1, \Omega_2, \ldots, \Omega_n$ are pairwise disjoint domains contained in Omega. Also suppose that the boundary of each Ω_j, $j = 1, 2, \ldots, n$, meets the boundary of Ω transversally (e.g. see Figure 8.1). For each $r = 1, 2, \ldots, m$, consider the eigenvalue problem on Ω_r which imposes vanishing Dirichlet data on $\partial\Omega_r \cap \Omega$ and imposes the original data on $\partial\Omega_r \cap \partial\Omega$, be it Dirichlet, Neumann or mixed; further, arrange all the eigenvalues of all the Ω_r as a single set in an increasing sequence:

$$0 \leq \tau_1 \leq \tau_2 \leq \cdots.$$

Then for all $j = 1, 2, \ldots$ we have

$$\lambda_j \leq \tau_j.$$

The theorem has the following immediate corollary.

Corollary 8.2.8 *Let Ω be an open domain as usual and $V \subset \Omega$ a subdomain. Consider the Dirichlet eigenvalue problem on V and any one of the eigenvalue problems from Definition 8.2.1 on Ω. Then we have*

$$\lambda_k(V) \geq \lambda_k(\Omega).$$

If in fact $\Omega \setminus \overline{V}$ is open in Ω, then the inequality is strict.

Theorem 8.2.9 (Monotonicity for Vanishing Neumann Data) *Let $\Omega \subset \mathbb{R}^N$ be a domain with either C^∞ boundary or piecewise C^∞ boundary. Consider any one of the eigenvalue problems from Definition 8.2.1, and suppose that the eigenvalues are*

$$0 < \lambda_1 \leq \lambda_2 \leq \cdots,$$

where each eigenvalue is repeated according to its multiplicity. Suppose that $\Omega_1, \Omega_2, \ldots, \Omega_n$ are pairwise disjoint domains that partition Ω, i.e. $\Omega = \overline{\Omega}_1 \cup \overline{\Omega}_2 \cup \ldots \cup \overline{\Omega}_r$, and are such that the boundaries of the Ω_j meet the boundary of Ω transversally. For each $r = 1, 2, \ldots, m$, consider the eigenvalue problem on Ω_r which imposes vanishing Neumann data on $\partial \Omega_r \cap \Omega$ and imposes the original data on $\partial \Omega_r \cap \partial \Omega$, be it Dirichlet, Neumann or mixed; further, arrange all the eigenvalues of all the Ω_r as a single set in an increasing sequence:
$$0 \leq \tau_1 \leq \tau_2 \leq \cdots.$$
Then for all $j = 1, 2, \ldots$ we have
$$\lambda_j \leq \tau_j.$$

C. Eigenvalue Problems on Rectangular Parallelepipeds

We now sketch some basic, and well-known, ideas about eigenvalue problems on **(i)** an interval in $\mathbb{R}$ and **(ii)** a rectangular parallelepiped in $\mathbb{R}^N$. These special cases will be important building blocks in the arguments that follow, but they also illustrate the key geometric insight in eigenvalue analysis.

Let us begin with an interval $[0, T]$. The differential equation analogous to (8.25) in dimension one is
$$\phi'' + \lambda \phi = 0. \tag{8.43}$$

The Dirichlet boundary condition is $\phi(0) = \phi(T) = 0$. Of course the general solution—which one can find by separation of variables—to (8.43) (for $\lambda > 0$) is
$$\phi_g(s) = A \cos \sqrt{\lambda} s + B \sin \sqrt{\lambda} s.$$
Here A and B are arbitrary real constants. However, only $A = 0$ and $\lambda = (\pi k/T)^2$, $k = 1, 2, \ldots$ give solutions that satisfy *both* boundary conditions. We thus devolve upon the sequence of eigenvalues
$$\lambda_k = \left(\frac{\pi k}{T}\right)^2$$
and corresponding eigenfunctions
$$\phi_k(s) = \sqrt{\frac{2}{T}} \sin\left(\frac{\pi k x}{T}\right).$$

Note that the constant preceding the sine function is chosen so that the eigenfunction will have norm 1. We see that the solution of the eigenvalue problem depends on the length of the interval (the volume of the domain) in a manner that is predictable from physical considerations. For example,

8.2. WEYL'S THEOREM

solving the wave equation by separation of variables leads precisely to this eigenvalue problem. And we know that, in the vibrating string problem, making the string longer forces the natural modes of vibration to be slower (i.e. produces lower notes, as on a guitar).

It is now a straightforward exercise to see that, on the domain

$$\Omega = (0, T_1) \times (0, T_2) \times \cdots \times (0, T_m),$$

the Dirichlet eigenvalue problem may be solved using separations of variables (try it for yourself using just two variables). We shall not replicate the analysis here. The resulting eigenvalues are

$$\left\{ \pi^2 \left(\frac{k_1^2}{\alpha_1^2} + \cdots \frac{k_m^2}{\alpha_m^2} \right) \right\}.$$

Now let us do a calculation to verify, by hand, that Weyl's theorem is true on the rectangular parallelepiped.

A useful way to organize our information about the eigenvalue asymptotics on a rectangular parallelepiped is by way of dualization. To this end, we think for a moment about tori. Let Γ be a lattice, that is a discrete additive subgroup of $\mathbb{R}^N$. We shall treat only Γ of rank N: this means that there is a linearly independent set of vectors $v_1, v_2, \ldots, v_n$ (called a **basis** for Γ) such that

$$\Gamma = \left\{ \sum_{j=1}^{N} \beta_j v_j : \beta_j \in \mathbb{Z} \; \forall j \right\}.$$

Then $\mathbb{R}^N / \Gamma$ is an N-dimensional torus, which we denote by $\mathcal{T}$.

To obtain a set of eigenfunctions for the Laplacian on $\mathcal{T}$, we consider the **dual lattice**:

$$\Gamma^* = \{ z \in \mathbb{R}^N : \langle x, z \rangle \in \mathbb{Z} \text{ for all } x \in \Gamma \}.$$

A natural dual basis $\{w_1, w_2, \ldots, w_N\}$ for Γ^* is determined by the equations

$$\langle w_j, v_k \rangle = \delta_{j,k},$$

where $\delta_{j,k}$ is the Kronecker delta.

Now for each $z \in \Gamma^*$ we may define the function

$$\phi_z(x) = e^{2\pi i \langle x, z \rangle}.$$

(Note that, in this discussion of tori, we shall be considering *complex valued solutions* of the equation $\Delta u + \lambda u = 0$ and complex valued eigenfunctions. This is for convenience only, and should lead to no confusion.) Note that the function ϕ_z is invariant under the action of Γ. This means that if $g \in \Gamma$ and τ_g denotes translation by g, i.e. $\tau_g(x) = x + g$, then $\phi_z \circ \tau_g = \phi_z$. So ϕ_z is a well-defined function on $\mathcal{T}$. Moreover,

$$\Delta \phi_z = -4\pi^2 |z|^2 \phi_z.$$

The corresponding eigenvalue is

$$\lambda = 4\pi^2 |z|^2. \tag{8.44}$$

In other words, ϕ_z is an eigenfunction of the Laplacian on the torus $\mathcal{T}$. It is not difficult, using some elementary Fourier analysis, to see that the functions ϕ_z span $L^2(\mathcal{T})$. Moreover, if $z_1, z_2, \ldots, z_p$ are distinct, then the corresponding functions $\phi_{z_1}, \phi_{z_2}, \ldots, \phi_{z_p}$ are linearly independent (exercise, or see Chavel [1]). Thus the functions ϕ_z account for all the eigenfunctions of the Laplacian on $\mathcal{T}$.

To summarize what we have just learned, for a given $\lambda > 0$, the eigenfunctions corresponding to λ are just the functions ϕ_z for values of z that satisfy (8.44). In other words, the value of the summatory function $N(\lambda)$ is just the number of elements of Γ^* that lie inside the ball $\overline{\mathbb{B}}(0, \sqrt{\lambda}/(2\pi))$. Thus, with $\mathcal{N}$ denoting the summatory function on $\mathcal{T}$,

$$N(\lambda) = \mathcal{N}(\sqrt{\lambda}/(2\pi)).$$

By the preceding calculations,

$$\begin{aligned} N(\lambda) = \mathcal{N}(\sqrt{\lambda}/(2\pi)) &= \mathcal{N}(\sqrt{\lambda}/\pi)/2^m \\ &\approx \frac{\Upsilon_m (\sqrt{\lambda}/\pi)^n (\operatorname{vol}\Omega)}{2^m} = \frac{\Upsilon_m \lambda^{m/2}(\operatorname{vol}\Omega)}{(2\pi)^m}. \end{aligned}$$

This is the Weyl formula in the special case when the domain Ω is a rectangular parallelepiped.

D. Eigenvalue Problems on Arbitrary Smooth Domains

Let us use these last two results, together with our monotonicity results, to derive the Weyl formula for an arbitrary smoothly bounded domain Ω in $\mathbb{R}^N$. To this end, let $R_1, R_2 \ldots, R_m$ be pairwise disjoint, open N-dimensional rectangles, each relatively compact in Ω. Fix a positive number λ. For $j = 1, 2, \ldots, m$, let $N_j(\lambda)$ be the number of Dirichlet eigenvalues for R_j that do not exceed λ. Recall that in the domain monotonicity theorem we numbered *all the eigenvalues of all the subdomains* of Ω, and then we compared that enumeration to the enumeration of eigenvalues for Ω itself. As a result, we can be sure that

$$N(\lambda) \geq \sum_{j=1}^{m} N_j(\lambda). \tag{8.45}$$

The inequality (8.45) implies

$$\liminf \frac{N(\lambda)}{\lambda^{N/2}} \geq \sum_{j=1}^{m} \liminf \frac{N_j(\lambda)}{\lambda^{N/2}} = \frac{\Upsilon_N}{(2\pi)^N} \sum_{j=1}^{N} \operatorname{vol}(R_j). \tag{8.46}$$

8.2. WEYL'S THEOREM

Notice that (8.46) holds for all possible choices of parallelepipeds lying in Ω. Taking the supremum on the right over all such choices, we find that

$$\liminf \frac{N(\lambda)}{\lambda^{N/2}} \geq \frac{\Upsilon_N}{(2\pi)^N} \text{vol}(\Omega).$$

This is half of the Weyl asymptotic formula.

For the other half, we must estimate the limit supremum from above. Let $R_1, R_2, \ldots, R_m$ be pairwise disjoint open rectangles with

$$\Omega \subset \text{int}\left(\overline{R}_1 \cup \overline{R}_2 \cup \cdots \cup \overline{R}_m\right).$$

Let $M(\lambda)$ denote the summatory function for the Neumann eigenvalues for the domain $\text{int}\left(\overline{R}_1 \cup \overline{R}_2 \cup \cdots \cup \overline{R}_m\right)$. Also, for each j, let $M_j(\lambda)$ be the summatory function for the Neumann eigenvalues on the individual domain R_j.

Now the Domain Monotonicity Theorem 8.2.9 for the Neumann problem tells us that

$$N(\lambda) \leq M(\lambda) \leq \sum_{j=1}^{m} M_j(\lambda).$$

Therefore

$$\limsup \frac{N(\lambda)}{\lambda^{N/2}} \leq \frac{\Upsilon_N}{(2\pi)^N} \sum_{j=1}^{m} \text{vol}(R_j).$$

Taking the infimum on the right over all possible choices of rectangular parallelepipeds, we find that

$$\limsup \frac{N(\lambda)}{\lambda^{N/2}} \leq \frac{\Upsilon_N}{(2\pi)^N} \text{vol}(\Omega).$$

Combining this inequality for the limit supremum with the earlier inequality for the limit infimum yields Weyl's asymptotic formula.

E. Proofs of the Results in B

Proof of Rayleigh's Theorem 8.2.4: By our earlier discussion, if ϕ is an eigenfunction and $f \in \mathcal{H}$ (recall $\mathcal{H}$ denotes the function space appropriate to the eigenvalue problem under consideration as defined in (8.38)), then $(\Delta \phi, f) = -D[\phi, f]$. For any such f, we set (as before) $\alpha_j = (f, \phi_j)$. We are assuming, in the second part of the theorem, that $\alpha_1 = \alpha_2 = \cdots = \alpha_{k-1} = 0$. Imitating the proof of Bessel's inequality and using (8.36), we have for $1 \leq k \leq r$ that

$$0 \leq D\left[f - \sum_{j=k}^{r} \alpha_j \phi_j, f - \sum_{j=k}^{r} \alpha_j \phi_j\right]$$

$$\begin{aligned}
&= D[f,f] - 2\sum_{j=k}^{r} \alpha_j D[f,\phi_j] + \sum_{j,\ell=k}^{r} \alpha_j \alpha_\ell D[\phi_j,\phi_\ell] \\
&= D[f,f] + 2\sum_{j=k}^{r} \alpha_j (f,\Delta\phi_j) + \sum_{j,\ell=k}^{r} \alpha_j \alpha_l D[\phi_j,\phi_\ell] \\
&= D[f,f] - 2\sum_{j=k}^{r} \alpha_j (f,\lambda_j\phi_j) + \sum_{j=k}^{r} \lambda_j \alpha_j^2 \\
&= D[f,f] - \sum_{j=k}^{r} \lambda_j \alpha_j^2.
\end{aligned}$$

We thus conclude that

$$\lambda_k \|f\|^2 = \lambda_k \sum_{j=1}^{\infty} \alpha_j^2 = \lambda_k \sum_{j=k}^{\infty} \alpha_j^2 \leq \sum_{j=k}^{\infty} \lambda_j \alpha_j^2 \leq D[f,f] < \infty.$$

This establishes the desired inequality.

In the case of equality, all the coefficients α_j must equal 0 for $\lambda_j > \lambda_k$ and then it follows that f is an eigenfunction corresponding to the eigenvalue λ_k. ∎

Proof of the Maximum-Minimum Theorem 8.2.5: Consider a function f that is orthogonal to the span of $\{u_1, u_2, \ldots, u_k\}$ and that, in addition, has the special form

$$f = \sum_{j=1}^{k} \alpha_j \phi_j,$$

with each ϕ_j an eigenfunction corresponding to the eigenvalue λ_j, $j = 1, 2, \ldots, k$. Thus

$$\sum_{j=1}^{k} \alpha_j (\phi_j, u_\ell) = 0 \tag{8.47}$$

for $\ell = 1, 2, \ldots, k-1$.

We think of the numbers (ϕ_j, u_ℓ) as given, for $j = 1, 2, \ldots, k$ and $\ell = 1, 2, \ldots, k-1$. And we think of the α_j as unknowns. Then the system in (8.47) has more unknowns (namely, k of them) than it has equations (namely, $k-1$ of them). So it certainly must have a non-trivial solution. But then, for these values of α_j, we see that

$$\mu \|f\|^2 \leq D[f,f] = \sum_{j=1}^{k} \lambda_j \alpha_j^2 \leq \lambda_k \|f\|^2.$$

This inequality gives the desired conclusion. ∎

8.2. WEYL'S THEOREM

Proof of the Domain Monotonicity Theorem 8.2.7: Of course we shall apply the maximum-minimum method from the last theorem.

Let $\phi_1, \phi_2, \ldots, \phi_k$ be the eigenfunctions that correspond to the eigenvalues $\lambda_1, \lambda_2, \ldots, \lambda_k$ on the full domain Ω. Corresponding to $j = 1, 2, \ldots, k$ let $\psi_j : \overline{\Omega} \to \mathbb{R}$ be an eigenfunction corresponding to the eigenvalue τ_j on the appropriate subdomain (recall that each τ_j comes from one of the subdomains Ω_j) and identically zero elsewhere. Note in particular that each ψ_j vanishes on the boundary of its particular subdomain. It follows that $\psi_j \in \mathcal{H}$. Moreover, it is straightforward to check that the ψ_j are orthogonal in $L^2(\Omega)$; after normalization, they may be taken to be orthonormal.

Thus we may select numbers $\alpha_1, \alpha_2, \ldots, \alpha_k$, not all zero, such that

$$\sum_{j=1}^{k} \alpha_j (\psi_j, \phi_\ell) = 0$$

for $\ell = 1, 2, \ldots, k-1$ (as in the proof of the maximum-minimum theorem). As a result, the function

$$f = \sum_{j=1}^{k} \alpha_j \psi_j$$

is orthogonal to $\phi_1, \phi_2, \ldots, \phi_{k-1}$ in $L^2(\Omega)$. Therefore

$$\lambda_k \|f\|^2 \leq D[f, f] = \sum_{j=1}^{k} \nu_j \alpha_j^2 \leq \tau_k \|f\|^2.$$

That is what we wished to prove. ∎

Proof of the Monotonicity Theorem for Neumann Data 8.2.9: Let $\psi_\ell : \overline{\Omega} \to \mathbb{R}$ be the eigenfunction corresponding to the eigenvalue τ_ℓ when ψ_ℓ is restricted to the corresponding subdomain; let ψ_ℓ be set identically equal to zero otherwise.

If f is any function in $\mathcal{H}$, then $f \in \mathcal{H}(\Omega_r)$, a fortiori for each r, where $\mathcal{H}(\Omega_r)$ is the function space for the eigenvalue problem on the subdomain Ω_r. Therefore if f is orthogonal to $\psi_1, \psi_2, \ldots, \psi_{k-1}$ in $L^2(\Omega)$ then

$$D[f, f] = \sum_{r=1}^{n} \int_{\Omega_r} |\mathbf{grad}\, f(s)|^2 \, ds \geq \sum_{r=1}^{n} \tau_k \iint_{\Omega_r} f^2(s) \, ds = \tau_k \|f\|^2.$$

But there is a non-trivial function $f = \sum_{j=1}^{k} \alpha_j \phi_j$ that is orthogonal to $\psi_1, \psi_2, \ldots, \psi_{k-1}$ in $L^2(\Omega)$; as a result,

$$D[f, f] \leq \lambda_k \|f\|^2.$$

This implies the theorem. ∎

Appendix

A.1 Metrics on the Collection of Subsets of Euclidean Space

Let $\mathcal{S}$ be the collection of all subsets of $\mathbb{R}^N$. It is often useful, especially in geometric applications, to have a metric on $\mathcal{S}$. In this section we address methods for achieving this end. In practice, $\mathcal{S}$ (the entire power set of $\mathbb{R}^N$) is probably too large a collection of objects to have a reasonable and useful metric topology (see Dugundji [1], Section IX.9 for several characterizations of metrizability). With these considerations in mind, we shall restrict attention to $\mathcal{B}$, the collection of *bounded* subsets of $\mathbb{R}^N$. We have:

Definition A.1.1 *Let S and T be elements of $\mathcal{B}$. We set*

$$\mathrm{HD}(S,T) = \max\{\sup_{s \in S} \mathrm{dist}(s,T)\,,\,\sup_{t \in T} \mathrm{dist}(t,S)\,\}. \qquad (A.1)$$

*This function is called the **Hausdorff distance**.*

To obtain the most satisfactory results, we further restrict attention to $\tilde{\mathcal{B}}$, the collection of both closed and bounded (*i.e.* compact) subsets of $\mathbb{R}^N$.

Lemma A.1.2 *Let $S, T \in \tilde{\mathcal{B}}$. Then there are points $s \in S$ and $t \in T$ such that $\mathrm{HD}(S,T) = |s - t|$.*

Proof: By the triangle inequality, the function $s \mapsto \mathrm{dist}(s,T)$ is a continuous function on S. So it achieves its maximum at some point $s' \in S$. Likewise, the function $t \mapsto \mathrm{dist}(t,S)$ achieves its maximum at some point $t' \in T$. Let $d = \max\{\mathrm{dist}(s',T), \mathrm{dist}(t',S)\}$, and suppose without loss of generality that this maximum is achieved by s'.

Since T is compact, there is a point $t^* \in T$ such that $d = |t^* - s'|$. Thus s', t^* are the points that we seek. ∎

Proposition A.1.3 *The function HD is a metric on $\tilde{\mathcal{B}}$.*

Proof: Clearly $\mathrm{HD} \geq 0$ and, if $S = T$, then $\mathrm{HD}(S,T) = 0$.

Conversely, if $\mathrm{HD}(S,T) = 0$ then let $s \in S$. By definition, there are points $t_j \in T$ such that $|s - t_j| \to 0$. Since T is compact, we may select a subsequence $\{t_{j_k}\}$ such that $t_{j_k} \to s$. But, since T is compact, we then conclude that $s \in T$. Hence $S \subset T$. Similar reasoning shows that $T \subset S$. Hence $S = T$.

Finally we come to the triangle inequality. Let $S, T, U \in \tilde{\mathcal{B}}$. Let $s \in S$, $t \in T$, $u \in U$. Then we have

$$|s - u| \leq |s - t| + |t - u|$$
$$\Downarrow$$
$$\mathrm{dist}(S, u) \leq |s - t| + |t - u|$$
$$\Downarrow$$
$$\mathrm{dist}(S, u) \leq \mathrm{dist}(S, t) + |t - u|$$
$$\Downarrow$$
$$\mathrm{dist}(S, u) \leq \mathrm{HD}(S, T) + |t - u|$$
$$\Downarrow$$
$$\mathrm{dist}(S, u) \leq \mathrm{HD}(S, T) + \mathrm{dist}(T, u)$$
$$\Downarrow$$
$$\mathrm{dist}(S, u) \leq \mathrm{HD}(S, T) + \sup_{u \in U} \mathrm{dist}(T, u)$$
$$\Downarrow$$
$$\sup_{u \in U} \mathrm{dist}(S, u) \leq \mathrm{HD}(S, T) + \sup_{u \in U} \mathrm{dist}(T, u).$$

By symmetry, we have

$$\sup_{s \in S} \mathrm{dist}(U, s) \leq \mathrm{HD}(U, T) + \sup_{s \in S} \mathrm{dist}(T, s)$$

and thus

$$\max\{\sup_{u \in U} \mathrm{dist}(S, u), \sup_{s \in S} \mathrm{dist}(U, s)\}$$
$$\leq \max\{\mathrm{HD}(S, T) + \sup_{u \in U} \mathrm{dist}(T, u), \mathrm{HD}(U, T) + \sup_{s \in S} \mathrm{dist}(T, s)\}.$$

We conclude that

$$\mathrm{HD}(U, S) \leq \mathrm{HD}(U, T) + \mathrm{HD}(T, S). \qquad \blacksquare$$

There are fundamental questions concerning completeness, compactness, etc. that we need to ask about any metric space.

Theorem A.1.4 *The metric space* $(\tilde{\mathcal{B}}, \mathrm{HD})$ *is complete.*

Proof: Let $\{S_j\}$ be a Cauchy sequence in the metric space $(\tilde{\mathcal{B}}, \mathrm{HD})$. We seek an element $S \in \tilde{\mathcal{B}}$ such that $S_j \to S$.

Elementary estimates, as in any metric space, show that the elements S_j are all contained in a common ball $B(0, R)$. We set S equal to the closure of

$$\bigcap_{j=1}^{\infty} \left(\bigcup_{\ell=j}^{\infty} S_\ell \right).$$

A.1. METRICS ON SUBSETS OF EUCLIDEAN SPACE 277

Then S is closed and bounded so it is an element of $\tilde{\mathcal{B}}$.

To see that $S_j \to S$, select $\epsilon > 0$. Choose J large enough so that if $j, k \geq J$ then $\mathrm{HD}(S_j, S_k) < \epsilon$. For $m > J$ set $T_m = \cup_{\ell=J}^{m} S_\ell$. Then it follows from the definition, and from Proposition A.1.3, that $\mathrm{HD}(S_J, T_m) < \epsilon$ for every $m > J$. Therefore, with $U_p = \cup_{\ell=p}^{\infty} S_\ell$ for every $p > J$, it follows that $\mathrm{HD}(S_J, U_p) \leq \epsilon$.

We conclude that that $\mathrm{HD}(S_J, \cap_{p=J+1}^{K} U_p) \leq \epsilon$. Hence, by the continuity of the distance, $\mathrm{HD}(S_J, S) \leq \epsilon$. That is what we wished to proved. ∎

The next proposition informs us of the seminal fact regarding the Hausdorff distance topology.

Theorem A.1.5 *The set of non-empty compact subsets of $\mathbb{R}^N$ with the Hausdorff distance topology is boundedly compact, i.e. any bounded sequence has a subsequence that converges to a compact set.*

Proof: Let $A_1, A_2, \ldots$ be a bounded sequence in the Hausdorff distance. We may assume without loss of generality that each A_i is a subset of the closed unit N-cube, C_0.

We will use an inductive construction and a diagonalization argument. Let $A_{0,i} = A_i$ for $i = 1, 2, \ldots$. For each $k \geq 1$, the sequence $A_{k,i}$, $i = 1, 2, \ldots$, will be a subsequence of the preceding sequence $A_{k-1,i}$, $i = 1, 2, \ldots$. Also, we will construct sets $C_0 \supset C_1 \supset \ldots$ inductively. Each C_k will be the union of a set of subcubes of the unit cube. The first set in this sequence is the unit cube itself C_0. For each $k = 0, 1, \ldots$, the sequence $A_{k,i}$, $i = 1, 2, \ldots$, and the set C_k are to have the properties that

$$C_k \cap A_{k,i} \neq \emptyset \quad \text{holds for } i = 1, 2, \ldots \tag{A.2}$$

and

$$A_{k,i} \subset C_k \quad \text{holds for all sufficiently large } i. \tag{A.3}$$

It is clear that (A.2) and (A.3) are satisfied when $k = 0$.

Assume $A_{k-1,i}$, $i = 1, 2, \ldots$ and C_{k-1} have been defined so that

$$C_{k-1} \cap A_{k-1,i} \neq \emptyset \quad \text{holds for } i = 1, 2, \ldots$$

and

$$A_{k-1,i} \subset C_{k-1} \quad \text{holds for all sufficiently large } i.$$

For each integer $k \geq 1$, subdivide the unit N-cube into 2^{kN} congruent subcubes of side-length 2^{-k}. We let $\mathcal{C}_k$ be the collection of subcubes of side-length 2^{-k} which are subsets of C_{k-1}. A subcollection, $\mathcal{C} \subset \mathcal{C}_k$, will be called **admissible** if there are infinitely many i for which

$$D \cap A_{k-1,i} \neq \emptyset \quad \text{holds for all } D \in \mathcal{C}. \tag{A.4}$$

Let C_k be the union of a maximal admissible collection of subcubes, which is immediately seen to exist, since $\mathcal{C}_k$ is finite. Let $A_{k,1}, A_{k,2}, \ldots$ be the subsequence of $A_{k-1,1}, A_{k-1,2}, \ldots$ consisting of those $A_{k-1,i}$ for which (A.4) is true. Observe that $A_{k,i} \subset C_k$ holds for sufficiently large i, else there is another subcube which could be added to the maximal collection while maintaining admissibility.

We set

$$C = \bigcap_{k=0}^{\infty} C_k$$

and claim that C is the limit in the Hausdorff distance of $A_{k,k}$ as $k \to \infty$. Of course C is non-empty by the finite intersection property. Let $\epsilon > 0$ be given. Clearly we can find an index k_0 such that

$$C_{k_0} \subset \{x : \text{dist}(x, C) < \epsilon\}.$$

There is a number i_0 such that for $i \geq i_0$ we have

$$A_{k_0, i} \subset C_{k_0} \subset \{x : \text{dist}(x, C) < \epsilon\}.$$

So, for $k \geq k_0 + i_0$, we know that

$$A_{k,k} \subset \{x : \text{dist}(x, C) < \epsilon\}$$

holds. We let $k_1 \geq k_0 + i_0$ be such that

$$\sqrt{N}\, 2^{-k_1} < \epsilon.$$

Let $c \in C$ be arbitrary. Then $c \in C_{k_1}$ so there is some cube, D, of sidelength 2^{-k_1} containing c and for which

$$D \cap A_{k_1, i} \neq \emptyset$$

holds for all i. But then if $k \geq k_1$, we have $D \cap A_{k,k} \neq \emptyset$, so

$$\text{dist}(c, A_{k,k}) \leq \sqrt{N}\, s^{-k} < \epsilon.$$

It follows that $\text{HD}(C, A_{k,k}) < \epsilon$ holds for all $k \geq k_1$. ∎

Next we give two more useful facts about the Hausdorff distance topology.

Proposition A.1.6 *Let $\mathcal{C}$ be the collection of all closed, bounded, convex sets in $\mathbb{R}^N$. Then $\mathcal{C}$ is a closed subset of $(\tilde{\mathcal{B}}, \text{HD})$.*

Proof: There are several amusing ways to prove this assertion. One is by contradiction. If $\{S_j\}$ is a convergent sequence in $\mathcal{C}$, then let $S \in \tilde{\mathcal{B}}$ be its limit. If S does not lie in $\mathcal{C}$ then S is not convex. Thus there is a segment ℓ with endpoints lying in S but with some interior point p not in S.

Let $\epsilon > 0$ be selected so that the open ball $U(p, \epsilon)$ does not lie in S. Let a, b be the endpoints of ℓ. Choose j so large that $\text{HD}(S_j, S) < \epsilon/2$. For such

A.1. METRICS ON SUBSETS OF EUCLIDEAN SPACE

j, there exist points $a_j, b_j \in S_j$ such that $|a_j - a| < \epsilon/3$ and $|b_j - b| < \epsilon/3$. But then each point $c_j(t) \equiv (1-t)a_j + tb_j$ has distance less than $\epsilon/3$ from $c(t) \equiv (1-t)a + tb$, $0 \leq t \leq 1$. In particular, there is a point p_j on the line segment ℓ_j connecting a_j to b_j such that $|p_j - p|\epsilon/3$. Noting that p_j must lie in S_j, we see that we have contradicted our statement about $U(p, \epsilon)$. Therefore S must be convex. ∎

Proposition A.1.7 *Let $\{S_j\}$ be a sequence of elements of $\tilde{\mathcal{B}}$, each of which is connected. Suppose that $S_j \to S$ in the Hausdorff metric. Then S must be connected.*

Proof: Suppose not. Then S is disconnected. So we may write $S = A \cup B$ with each of A and B closed and non-empty and $A \cap B = \emptyset$. Then there is a number $\eta > 0$ such that if $a \in A$ and $b \in B$ then $|a - b| > \eta$.

Choose j so large that $\mathrm{HD}(S_j, S) < \eta/3$. Define

$$A_j = \{s \in S_j : \mathrm{dist}(s, A) \leq \eta/3\} \quad \text{and} \quad B_j = \{s \in S_j : \mathrm{dist}(s, B) \leq \eta/3\}.$$

Clearly $A_j \cap B_j = \emptyset$ and A_j, B_j are closed and non-empty. Moreover, $A_j \cup B_j = S_j$. That contradicts the connectedness of S_j and completes the proof. ∎

Remark A.1.8 It is certainly possible to have totally disconnected sets E_j, $j = 1, 2, \ldots$, such that $E_j \to E$ as $j \to \infty$ and E is connected (exercise).

The proof of the last result may be modified to show that if $\{S_j\} \subset \tilde{\mathcal{B}}$, if $S_j \to S$ in the Hausdorff metric, and if each S_j has at most M connected components then S has at most M connected components.

It is not true that the collection of simply connected sets in $\tilde{\mathcal{B}}$ forms a closed subset. For instance, let $N = 2$ and let

$$S_j = \{(r \cos \theta, r \sin \theta) \in \mathbb{R}^2 : 1 \leq r \leq 2, 1/j \leq \theta \leq 2\pi\}.$$

Then each S_j is simply connected but the S_j converge to an annulus. Using elementary PL topology, one can construct counterexamples for every homotopy group or every Betti number.

Although the Hausdorff metric is the most far-reaching notion of distance among sets, it is by no means the only metric on sets. It is frequently useful to put a topology on the collection of smoothly bounded domains. We should like to briefly discuss some of the methods for doing that.

Definition A.1.9 *Fix an integer $k \geq 1$. Consider the collection $\mathcal{E}_k$ of all bounded domains in $\mathbb{R}^N$ with C^k boundary. Fix one such domain Ω_0. Associate to it the imbedding ϕ_0 into $\mathbb{R}^N$ given by $\phi_0(x) = x$. In other words, ϕ_0 is the identity mapping. Now let $\epsilon > 0$. Set*

$$\mathcal{U}_\epsilon = \{\Omega \in \mathcal{E}_k : \text{there is a surjective imbedding}$$
$$\phi \text{ of } \Omega_0 \text{ to } \Omega \text{ such that } \|\phi - \phi_0\|_{C^k} < \epsilon\}.$$

Here we use the standard C^k topology on functions and mappings.

Then the sets $\mathcal{U}_\epsilon$ form a sub-basis for a topology on $\mathcal{E}_k$. This is called the C^k **topology on domains**.

The C^k topology is equivalent to considering the boundary M of each element Ω of $\mathcal{E}_k$ as a regularly imbedded compact manifold and declaring $\Omega_j \to \Omega_0$ if the boundaries M_j of Ω_j converge to the boundary M_0 of Ω_0 in the sense that the imbedding maps converge.

It is convenient, but not necessary, to formulate a topology on smoothly bounded domains by considering an associated topology on imbeddings of the domains. Instead, one could look at a local boundary neighborhood and consider the boundary of Ω_0 as locally the graph of some function f_0 of $N-1$ variables. Then a sequence of domains Ω_j converges to Ω_0 in the C^k topology if each boundary is locally given as the graph of a function f_j and the f_j converge to f_0 in C^k. Properly formulated, we would have to cover $\partial\Omega$ with finitely many open neighborhoods U^ℓ and demand that f_j^ℓ converge to f_0^ℓ in a suitable sense.

When one is studying perturbation theory for partial differential equations, or other perturbation problems in geometric analysis, it is important to have a suitable topology on domains.

A.2 The Constants Associated to Euclidean Space

There are many special constants of mathematical analysis that arise from the consideration of the geometry of Euclidean space. A simple example is the constant π, which is the ratio of the circumference of any circle to its diameter. Similar considerations, in higher dimensions, give rise to the gamma function and the beta function. The elementary representation theory of the orthogonal group also gives rise to many of the special functions of mathematical physics; these, in turn, are the source of the constant e, the Bessel functions, and so forth.

In this section, we shall touch on just a few of these topics. The book Vilenkin [1] is an excellent source for some of the topics mentioned in the first paragraph. Another fine source is Erdelyi [1].

We begin by introducing two of the special functions of mathematical physics. Our point of view in the present section is that one "discovers" these functions by considering geometric questions in space; nonetheless, it is best to be informed in advance.

Definition A.2.1 *For $\operatorname{Re} z > 0$, we define the **gamma function** by setting*

$$\Gamma(z) = \int_0^\infty t^{z-1} e^{-t}\, dt. \tag{A.5}$$

It is easy to see that the integral in (A.5), called an **Eulerian integral of the second kind**, converges absolutely for the range of z specified in the definition. By Morera's theorem, Γ is a holomorphic function on the right half plane. One checks that

$$\Gamma(1) = 1, \tag{A.6}$$

and, by integration by parts, that the functional equation

$$x\Gamma(x) = \Gamma(x+1) \tag{A.7}$$

holds. As a result of (A.6) and (A.7),

$$\Gamma(k+1) = k! \tag{A.8}$$

holds for positive integers k.

The following technical fact concerning the so-called **Gaussian integral** will occur repeatedly:

Lemma A.2.2 *It holds that*

$$\int_{\mathbb{R}^N} e^{-|x|^2}\, dx = \pi^{N/2}.$$

Proof: By Fubini's theorem, it suffices to treat the case $N = 1$. Let

$$S = \int_{-\infty}^{\infty} e^{-x^2}\, dx.$$

Then

$$\begin{aligned} S^2 &= \int_{-\infty}^{\infty} e^{-x^2}\, dx \int_{-\infty}^{\infty} e^{-y^2}\, dy \\ &= \int_{\mathbb{R}^2} e^{-x^2-y^2}\, dxdy & (A.9) \\ &= \int_0^{2\pi} \int_0^{\infty} e^{-r^2} r\, drd\theta & (A.10) \\ &= \pi, \end{aligned}$$

where we have changed to polar coordinates to go from (A.9) to (A.10). Therefore $S = \sqrt{\pi}$, as desired. ∎

Now we may evaluate $\Gamma(1/2)$. Using the change of variable $s^2 = t$, we see that

$$\Gamma(1/2) = \int_0^{\infty} t^{-1/2} e^{-t}\, dt = 2\int_0^{\infty} e^{-s^2}\, ds = \sqrt{\pi}. \qquad (A.11)$$

Using integration by parts, we may analytically continue the gamma function to $\mathbb{C} \setminus \{0, -1, -2, \ldots\}$. The function is meromorphic, and has a simple pole at each of $0, -1, -2$, etc. Good sources of information on the gamma function are Carrier, Crook, and Pearson [1] and Greene and Krantz [1].

Next we have

Definition A.2.3 *For $\operatorname{Re} z > 0$ and $\operatorname{Re} w > 0$, we define the* **beta function** *by setting*

$$\beta(z, w) = \int_0^1 s^{z-1}(1-s)^{w-1}\, ds. \qquad (A.12)$$

The integral in (A.12) is called an **Eulerian integral of the first kind**. The beta function is holomorphic in each variable separately. We have the following connection between the gamma and beta functions:

Proposition A.2.4 *If $\operatorname{Re} z > 0$ and $\operatorname{Re} w > 0$, then it holds that*

$$\beta(z, w) = \frac{\Gamma(z)\Gamma(w)}{\Gamma(z+w)}. \qquad (A.13)$$

A.2. THE CONSTANTS OF EUCLIDEAN SPACE 283

Proof: By analytic continuation, it suffices to verify the identity for z, w real and positive. Now we have

$$\begin{aligned}
\Gamma(z)\Gamma(w) &= \int_0^\infty s^{z-1}e^{-s}\,ds \int_0^\infty t^{w-1}e^{-t}\,dt \\
&= \int_0^\infty e^{-s}s^{z-1} \int_s^\infty e^{-t+s}(t-s)^{w-1}\,dt\,ds \\
&= \int_0^\infty s^{z+w-2} \int_s^\infty e^{-t}\left(\frac{t}{s}-1\right)^{w-1}\,dt\,ds \\
&= \int_0^\infty s^{z+w-1} \int_1^\infty e^{-ts}(t-1)^{w-1}\,dt\,ds \\
&= \int_1^\infty \int_0^\infty s^{z+w-1}e^{-s}t^{-z-w}(t-1)^{w-1}\,ds\,dt \\
&= \int_1^\infty (t-1)^{w-1}t^{-z-w}\,dt \cdot \Gamma(z+w) \\
&= \int_0^1 t^{z-1}(1-t)^{w-1}\,dt \cdot \Gamma(z+w) \\
&= \beta(z,w) \cdot \Gamma(z+w). \qquad\blacksquare
\end{aligned}$$

Next we will calculate the surface area of the unit ball in $\mathbb{R}^N$:

Proposition A.2.5 *Let $B = \{x \in \mathbb{R}^N : |x| < 1\}$ and $\Sigma = \partial B$ its boundary. Then the $N-1$ dimensional Hausdorff measure of Σ (i.e. the rotationally invariant area) is*

$$\omega_{N-1} \equiv \sigma(\Sigma) = \frac{2\pi^{N/2}}{\Gamma(N/2)}. \tag{A.14}$$

Proof: By Lemma A.2.2 and using spherical coordinates, we have

$$1 = \int_{\mathbb{R}^N} e^{-\pi|x|^2}\,dx = \int_\Sigma d\sigma \int_0^\infty e^{-\pi r^2} r^{N-1}\,dr$$

and therefore

$$\frac{1}{\omega_{N-1}} = \int_0^\infty e^{-\pi r^2} r^{N-1}\,dr.$$

Using the change of variable $s = r^2$, we find that

$$\begin{aligned}
\frac{1}{\omega_{N-1}} &= \int_0^\infty e^{-\pi s} s^{N/2}\frac{ds}{2s} \\
&= \frac{1}{2}\pi^{-N/2} \int_0^\infty e^{-s} s^{N/2} \frac{ds}{s} \\
&= \frac{1}{2}\pi^{-N/2}\Gamma(N/2).
\end{aligned}$$

This is the desired formula. ■

Using our elementary facts about the gamma function (A.6), (A.7), and (A.11), the reader may check that the formula in Proposition A.2.5 gives the expected value for the area of the sphere in dimensions 1, 2, 3.

Now we will calculate the volume of the unit ball in $\mathbb{R}^N$.

Proposition A.2.6 *Let $B = \{x \in \mathbb{R}^N : |x| < 1\}$. Then the Euclidean volume of B is*

$$\Upsilon_N \equiv \text{vol}(B) = \frac{\pi^{N/2}}{\Gamma((N+2)/2)}.$$

Proof: Using spherical coordinates, we calculate that

$$\Upsilon_N = \omega_{N-1} \int_0^1 r^{N-1}\, dr = \frac{\omega_{N-1}}{N}.$$

But then Proposition A.2.5 tells us that

$$\Upsilon_N = \frac{2\pi^{N/2}}{N \cdot \Gamma(N/2)} = \frac{\pi^{N/2}}{\Gamma((N+2)/2)}. \qquad \blacksquare$$

Using our elementary facts about the gamma function (A.6), (A.7), and (A.11), the reader may check that the formula in Proposition A.2.6 gives the expected value for the volume of the ball in dimensions 1, 2, 3.

An immediate but interesting corollary of our calculations is the following:

Proposition A.2.7 *Let B_N be the unit ball in $\mathbb{R}^N$ and Σ_{N-1} its boundary. Let Υ_N denote the volume of B_N and ω_{N-1} the surface area of Σ_{N-1}. Then*

$$\lim_{N \to \infty} \Upsilon_N = \lim_{N \to \infty} \omega_{N-1} = 0.$$

We next note an alternative recursion formula for the volume of the ball. It gives rise to the beta function in a natural way. In $\mathbb{R}^N$, let us write $x' = (x_1, \ldots, x_{N-1})$ and $x = (x', x_N)$. Then

$$\begin{aligned}
\Upsilon_N &= \int_{B_N} 1\, dV_N(x) \\
&= \int_{x' \in B_{N-1}} \int_{|x_N| < \sqrt{1-|x'|^2}} 1\, dV_1(x_N)\, dV_{N-1}(x') \\
&= \int_{x' \in B_{N-1}} (1 - |x'|^2)\, dV_{N-1}(x') \\
&= \omega_{N-2} \int_0^1 (1 - r^2) r^{N-2}\, dr. \qquad \text{(A.15)}
\end{aligned}$$

A.2. THE CONSTANTS OF EUCLIDEAN SPACE

The change of variable $s = r^2$ converts (A.15) to a beta integral, and Proposition A.2.4 turns that into an expression in terms of gamma functions. We leave the details for the interested reader.

A similar technique may be used to calculate the value of

$$\int_{B_N} |x^\alpha|^2 \, dV_N(x)$$

for any multi-index α. These numbers are useful in various spherical harmonic expansions (see, for instance, Krantz [4]).

Guide to Notation

DESCRIPTION	NOTATION
Area of the $(N-1)$-dimensional unit sphere in $\mathbb{R}^N$	ω_{N-1}
Boundary of the set A	∂A
Cardinality of the set S	$\mathrm{card}(S)$
Characteristic function of the set A	χ_A
Closed ball of radius r centered at x	$\overline{\mathbb{B}}(x,r)$
Closure of A is a compact subset of the interior of B	$A \subset\subset B$
Closure of the set A	$\overline{A}$
Compactly supported functions of class C^α	C_c^α
Complement of the set A	${}^c A$
Complex numbers	$\mathbb{C}$
Determinant of the matrix M	$\det[M]$
Diameter of the set S	$\mathrm{diam}(S)$
Differentiability classes	
$\quad k$ times continuously differentiable	C^k
$\quad$ infinitely differentiable	C^∞
$\quad$ analytic	C^ω
$\quad k$ times continuously differentiable with k^{th} derivative Hölder continuous of order α	$C^{k,\alpha}$
Distance between sets A and B	$\mathrm{dist}(A,B)$
Divergence of the vectorfield v	$\mathbf{div}(v)$
Euclidean space	$\mathbb{R}^m$
Extended reals	$\overline{\mathbb{R}}$
Fourier transform	$\hat{f}(\xi) = \int f(x) e^{-2\pi i x \cdot \xi}\, dx$

Gaussian curvature	K
Gradient of the function f	$\mathbf{grad}\, f = \nabla f$
Hausdorff distance between sets A and B	$\mathrm{HD}(A, B)$
Hessian of the function f	$\mathbf{Hess}(f)$
Interior of the set A	$\mathring{A}$
Laplacian of the function u	$\Delta(u)$
Length of the arc Γ	$\mathrm{len}(\Gamma)$
Lipschitz constant of the map f	$\mathrm{Lip}(f)$
Mean curvature	$\mathbf{H}$
Multi-indices	$\Lambda(m) = (\mathbb{Z}^+)^m$
functions of	$\mu! = \mu_1! \cdot \ldots \cdot \mu_m!$
	$|\mu| = \mu_1 + \ldots + \mu_m$
	$x^\mu = x_1^{\mu_1} \cdot \ldots \cdot x_m^{\mu_m}$
	$|x| = |x_1|^{\mu_1} \cdot \ldots \cdot |x_m|^{\mu_m}$
	$(x)_\mu = (x_1)_{\mu_1} \cdot \ldots \cdot (x_m)_{\mu_m}$
	$\binom{x}{\mu} = \frac{(x)_\mu}{\mu!}$
operator based on	$\frac{\partial^\mu}{\partial x^\mu} = \frac{\partial^{\mu_1}}{\partial x^{\mu_1}} \cdots \frac{\partial^{\mu_m}}{\partial x^{\mu_m}}$
relations on	$\mu < \nu$ iff $\mu_1 < \nu_1, \ldots, \mu_m$
	$\mu \leq \nu$ iff $\mu_1 \leq \nu_1, \ldots, \mu_m$
Non-negative integers	$\mathbb{Z}^+$
Normal space to S at p	NS_p
Normal bundle over S	NS
Open ball of radius r centered at x	$\mathbb{B}(x, r)$
Positive integers (same as the natural numbers)	$\mathbb{N}$
Real numbers	$\mathbb{R}$
Scalar curvature	$\mathbf{S}$
Second fundamental form	$\mathrm{II}(v, w)$
Set difference	$A \setminus B$

GUIDE TO NOTATION

Sphere of radius r centered at x	$\mathbb{S}(x,r)$
Support of the function f	$\operatorname{supp} f$
Tangent space to S at p	TS_p
Tangent bundle over S	TS
Trace of the matrix M	$\operatorname{tr}[M]$
Transpose of the matrix M	M^{t}
Volume of the N-dimensional unit ball in $\mathbb{R}^N$	Υ_N

Bibliography

R. A. Adams

1. *Sobolev Spaces*, Academic Press, New York, 1975.

L. V. Ahlfors

1. *Lectures on Quasiconformal Mappings*, Wadsworth/Brooks-Cole, Monterey, 1987.

A. D. Aleksandrov

1. Uniqueness theorems for surfaces in the large. V, *American Mathematical Society Translations* (2) **21** (1962), 412–416.

2. A characteristic property of spheres, *Annali di Matematica Pura ed Applicata* **58** (1962), 303–315.

A. S. Besicovitch

1. On the fundamental geometrical properties of linearly measurable plane sets of points, I, II, III *Mathematische Annalen* **98** (1928), 422–464, **115** (1938), 296–329, **116** (1939), 349–357.

2. On the definition and value of the area of a surface, *Quarterly Journal of Mathematics* **16** (1945), 86–102.

G. Bianchi, A. Colesanti, and C. Pucci

1. On the second differentiability of convex surfaces, *Geometriae Dedicata* **60** (1996), 39–48.

E. D. Bloch

1. *A First Course in Geometric Topology and Differential Geometry*, Birkhäuser, Boston, 1997.

R. Boas

1. When is a C^∞ function real analytic?, *The Mathematical Intelligencer* **11** (1989), 34–37.

T. Bonnesen and W. Fenchel

1. *Theorie der konvexen Korper*, J. Springer, Berlin, 1934.

A. B. Brown

1. Functional dependence, *Transactions of the American Mathematical Society* **38** (1935), 379–394.

H. Brunn

1. *Über Ovale und Eiflächen*, Inaugural-Dissertation, München, 1887.

2. *Über Kurven ohne Wendepunkte*, Habilitationsschrift, München, 1889.

A. P. Calderón

1. Cauchy integrals on Lipschitz curves and related operators, *Proceedings of the National Academy of Sciences of the United States of America* **74** (1977), 1324–1327.

P. Caraman

1. *n-Dimensional Quasiconformal Mappings*, Abacus Press, Tunbridge Wells, Kent, England, 1974.

C. Carathéodory

1. Über das lineare Maß von Punktmengen, eine Verallgemeinerung des Längenbegriffs, *Nachrichten von der Gesellschaft der Wissenschaften zu Göttingen* 1914, 404–426.

F. Carrier, M. Krook, and C. Pearson

1. *Functions of a Complex Variable*, McGraw-Hill, New York, 1966.

I. Chavel

1. *Eigenvalues in Riemannian Geometry*, Academic Press, Orlando, 1984.

M. Christ

1. *Lectures on Singular Integral Operators*, CBMS Conference Series in Mathematics **77**, American Mathematical Society, Providence, 1990.

R. R. Coifman and G. Weiss

1. *Analyse Harmonique Non-commutative sur Certains Espaces Homogénes*, Lecture Notes in Mathematics, Volume 242, Springer-Verlag, Berlin, 1971.

2. Extensions of Hardy spaces and their use in analysis, *Bulletin of the American Mathematical Society* **83** (1977), 569–645.

R. Courant

1. *Dirichlet's Principle, Conformal Mapping, and Minimal Surfaces*, Interscience Publishers, New York, 1950. With an appendix by M. Schiffer.

R. Courant and D. Hilbert

1. *Methods of Mathematical Physics*, Volume 1, John Wiley & Sons (Interscience), New York, 1953.

M. de Guzman

1. *Differentiation of Integrals in $\mathbb{R}^N$*, Springer Verlag, Berlin, 1975.

G. de Rham

1. *Differentiable Manifolds*, Springer-Verlag, New York, 1984.

C. Delaunay

1. Sur la surface de révolution dont la coubure moyenne est constante, *Journal de Mathèmatiques Pures et Applicquées* **6** (1841), 309–320. With a note appended by M. Sturm.

B. A. Dubrovin, A. T. Fomenko, and S. P. Novikov

1. *Modern Geometry—Methods and Applications*, Part 1, Springer Verlag, Berlin, 1992.

J. Dugundji

1. *Topology*, Allyn and Bacon, Boston, 1966.

J. Eells

1. The surfaces of Delaunay, *The Mathematical Intelligencer* **9** (1987), 53–57.

A. Erdelyi, director

1. Higher transcendental functions and Tables of integral transforms, *The Bateman Manuscript Project*, McGraw-Hill, New York, 1953–55.

L. C. Evans and R. F. Gariepy

1. *Measure Theory and Fine Properties of Functions*, CRC Press, Boca Raton, 1992.

K. J. Falconer

1. *The Geometry of Fractal Sets*, Cambridge University Press, New York, 1985.

H. Federer

1. The (ϕ, k) rectifiable subsets of n space, *Transactions of the American Mathematical Society* **62** (1947), 114–192.

2. On Lebesgue area, *Annals of Mathematics* (2) **61** (1955), 289–353.

3. Curvature Measures, *Transactions of the American Mathematical Society* **93** (1959), 418–491.

4. *Geometric Measure Theory*, Springer-Verlag, New York, 1969.

W. Fenchel

1. Convexity through the ages, pp. 120–130 in *Convexity and its Applications*, edited by P. M. Gruber and J. M. Wills, Birkhḧauser, Basel, 1983.

G. B. Folland

1. *Real Analysis*, John Wiley & Sons, New York, 1984.

G. B. Folland and E. M. Stein

1. Estimates for the $\bar{\partial}_b$ complex and analysis on the Heisenberg group, *Communications on Pure and Applied Mathematics* **27** (1974), 429–522.

R. L. Foote

1. Regularity of the distance function, *Proceedings of the American Mathematical Society* **92** (1984), 153–155.

T. W. Gamelin and D. Khavinson

1. The isoperimetric inequality and rational approximation, *American Mathematical Monthly* **96** (1989), 18–30.

F. W. Gehring

1. Rings and quasiconformal mappings in space, *Transactions of the American Mathematical Society* **103** (1962), 353–393.
2. Topics in quasiconformal mappings, pp. 20–38 in *Quasiconformal Space Mappings*, edited by M. Vuorinen, Springer-Verlag, Berlin, 1992.

B. R. Gelbaum and J. M. H. Olmstead

1. *Counterexamples in Analysis*, Holden-Day, San Francisco, 1964.

D. Gilbarg and N. S. Trudinger

1. *Elliptic Partial Differential Equations of Second Order*, Second Edition, Springer-Verlag, New York, 1983.

P. Gilkey

1. *The Heat Equation, Invariant Theory, and the Index Theorem*, CRC Press, Boca Raton, 1995.

E. Giusti

1. *Minimal Surfaces and Functions of Bounded Variation*, Birkhäuser, Basel, 1984,

C. Gordon, D. Webb, S. Wolpert

1. One cannot hear the shape of a drum, *Bulletin of the American Mathematical Society* **27** (1992), 134–138.

M. L. Green

1. The moving frame, differential invariants and rigidity theorems for curves in homogeneous spaces, *Duke Mathematical Journal* **45** (1978), 735–779.

M. J. Greenberg and J. R. Harper

1. *Algebraic Topology: A First Course*, Addison-Wesley, Redwood City, 1981.

Greene and Krantz

1. *Function Theory of One Complex Variable*, John Wiley and Sons, New York, 1997.

H. Grötzsch

1. Über die Verzerrung bei schlichten nichtkonformen Abbildungen und eine damit zusammenhangende Erweiterung des Picardschen Satzes, *Berichte über die Verhandlungen der Sächsischen Akademie der Wissenschaften zu Leipzig, Mathematisch-Physische Klasse* **80** (1928), 503–507.

P. M. Gruber and J. M. Wills, editors

1. *Handbook of Convex Geometry*, North-Holland, Amsterdam, 1993.

J. Harrison and A. Norton

1. The Gauss-Green theorem for fractal boundaries *Duke Mathematical Journal* **67** (1992), 575–588.

M. Hirsch

1. *Differential Topology*, Springer-Verlag, New York, 1976.

H. Hopf

1. Über Flächen mit einer Relation zwischen den Hauptkrümmungen, *Mathematische Nachrichten* **4** (1951), 232–249.

L. Hörmander

1. *Linear Partial Differential Operators*, Springer Verlag, Berlin, 1969.

2. *Notions of Convexity*, Birkhäuser, Boston, 1994.

W. Hurewicz and H. Wallman

1. *Dimension Theory*, Princeton University Press, Princeton, 1948.

T. Jech

1. *Set Theory*, Academic Press, New York, 1978.

P. W. Jones

1. Quasiconformal mappings and extendability in Sobolev spaces, *Acta Mathematica* **147** (1981), 71–88.

M. Kac

1. Can one hear the shape of a drum?, *American Mathematical Monthly* **73** (1966), 1–23.

N. Kapouleas

1. Complete constant mean curvature surfaces in Eudlidean three-space, *Annals of Mathematics* (2) **131** (1990), 239–330.

2. Compact constant mean curvature surfaces in Euclidean three-space, *Journal of Differential Geometry* **33** (1991), 683–715.

3. Constant mean curvature surfaces constructed by fusing Wente tori, *Proceedings of the National Academy of Sciences of the United States of America* **89** (1992), 5695–5698.

M. D. Kirzbraun

1. Über die zusammenziehenden un Lipschitzschen Transformationen, *Fundamenta Mathematicae* **22** (1934), 77–108.

W. R. Knorr

1. *The Ancient Tradition of Geometric Problems*, Birkhäuser, Boston, 1986.

2. *Textual Studies in Ancient and Medieval Geometry*, Birkhäuser, Boston, 1989.

N. J. Korevaar, R. Kusner, B. Solomon

1. The structure of complete embedded surfaces with constant mean curvature, *Journal of Differential Geometry* **30** (1989), 465–503.

S. G. Krantz

1. Lipschitz spaces, smoothness of functions, and approximation theory, *Expositiones Mathematicae* **3** (1983), 193–260.

2. *Real Analysis and Foundations*, CRC Press, Boca Raton, 1991.

3. *Function Theory of Several Complex Variables*, Second Edition, Wadsworth, 1992.

4. *Partial Differential Equations and Complex Analysis*, CRC Press, Boca Raton, 1992.

5. *Geometric Analysis and Function Spaces*, CBMS Regional Conference Series in Mathematics **81**, American Mathematical Society, Providence, 1993.

6. *A Panorama of Harmonic Analysis*, manuscript.

S. G. Krantz and H. R. Parks

1. Distance to C^k hypersurfaces, *Journal of Differential Equations* **40** (1981), 116–120.

2. *A Primer of Real Analytic Functions*, Birkhäuser, Basel, 1992.

S. Krantz and T. D. Parsons

1. Antisocial subcovers of self-centered coverings, *American Mathematical Monthly* **93** (1996), 45–48.

M. Krein and D. Milman

1. On the extreme points of regularly convex sets, *Studia Mathematica* **9** (1940), 133–138.

R. E. Langer

1. Fourier series: The genesis and evolution of a theory, Herbert Ellsworth Slaught Memorial Paper 1, published as a supplement to the *American Mathematical Monthly* **54** (1947).

S. R. Lay

1. *Convex Sets and their applications*, Wiley, New York, 1982.

O. Lehto and K. I. Virtanen

1. *Quasiconformal Mappings in the Plane*, Springer-Verlag, Berlin, 1973.

J. Liouville

1. Théorème sur l'équation $dx^2 + dy^2 + dz^2 = \lambda \, (d\alpha^2 + d\beta^2 + d\gamma^2)$, *Journal de Mathématiques Pures et Appliquées* **15** (1850) 103.

L. H. Loomis and S. Sternberg

1. *Advanced Calculus*, Addison-Wesley, Reading, 1968.

K. R. Lucas

1. *Submanifolds of dimension $n-1$ in $\mathcal{E}^n$ with normals satisfying a lipschitz condition*, Studies in Eigenvalue Problems, Technical Report 18, University of Kansas, Department of Mathematics, 1957.

N. Lusin

1. *Leçons sur les ensembles analytiques et leurs applications.* Gauthier-Villars, Paris, 1930.

P. Mattila

1. *Geometry of Sets and Measures in Euclidean Spaces*, Cambridge University Press, Cambridge, 1995.

W. H. Meeks, III

1. The topology and geometry of embedded surfaces of constant mean curvature, *Journal of Differential Geometry* **27** (1988), 539–552.

N. Meyers and J. Serrin

1. $H = W$, *Proceedings of the National Academy of Sciences of the United States of America* **51** (1964), 1055–1056.

J. Milnor

1. *Topology from the Differentiable Viewpoint*, University Press of Virginia, Charlottesville, 1965.

H. Minkowski

1. Allgemeine Lehrsätze über die konvexen Polyeder, pp. 103–121 in *Gesammelte Abhandlungen von Hermann Minkowski*, Volume 2, B. G. Teubner, Leipzig and Berlin, 1911.

A. P. Morse

1. The behavior of a function on its critical set, *Annals of Mathematics* (2) **40** (1939), 62–70.

J. C. C. Nitsche

1. *Lectures on Minimal Surfaces*, Volume 1, Cambridge University Press, New York, 1989.

W. F. Osgood

1. A Jordan curve of positive area, *Transactions of the American Mathematical Society* **4** (1903), 107–112.

R. Osserman

1. The isoperimetric inequality, *Bulletin of the American Mathematical Society* **84** (1978), 1182–1238.

2. *A Survey of Minimal Surfaces*, Dover, Mineola, New York, 1986.

H. R. Parks

1. Some new constructions and estimates in the problem of least area, *Transactions of the American Mathematical Society* **248** (1979), 311–346.

H. R. Parks and R. M. Schori

1. A Jordan Arc in $\mathbb{R}^m$ with Positive m-Dimensional Lebesgue Measure, pp. 229–236 in *The Problem of Plateau: A Tribute to Jesse Douglas and Tibor Rado*, edited by T. M. Rassias, World Scientific Publishing, 1992.

G. Peano

1. Sur une courbe qui remplit toute une aire plane, *Mathematische Annalen* **36** (1890), 157–160.

G. J. Porter

1. k-volume in $\mathbb{R}^n$ and the generalized Pythagorean Theorem, *American Mathematical Monthly* **103** (1996), 252–256.

Yu. G. Reshetnyak

1. Liouville's theorem on conformal mappings under minimal regularity assumptions, *Akademiya Nauk SSSR* (transliterated *Siberian Mathematical Journal*) **8** (1967), 835–840.

W. Rudin

1. *Principles of Mathematical Analysis*, Third Edition, McGraw-Hill, New York, 1976.

2. *Real and Complex Analysis*, Second Edition, McGraw-Hill, New York, 1974.

3. *Functional Analysis*, McGraw-Hill, New York, 1973.

A. Sard

1. The measure of the critical values of differentiable maps, *Bulletin of the American Mathematical Society* **48** (1942), 883–890.

G. E. Shilov

1. *Elementary Functional Analysis*, MIT Press, Cambridge, Massachusetts, 1974.

K. T. Smith

1. *Primer of Modern Analysis*, Springer-Verlag, New York, 1983.

R. M. Solovay

1. A model of set-theory in which every set of reals is Lebesgue measurable, *Annals of Mathematics* (2) **92** (1970), 1–56.

E. M. Stein

1. *Singular Integrals and Differentiability Properties of Functions*, Princeton University Press, Princeton, 1970.

E. M. Stein and Guido Weiss

1. *Introduction to Fourier Analysis on Euclidean Spaces*, Princeton University Press, Princeton, 1971.

J. Steiner

1. Einfache Beweise der isoperimetrischen Hauptsätze, pp. 75–91 in *Jacob Steiner's Gesammelte Werke*, Volume 2, G. Reimer, Berlin, 1882.

S. Sternberg

1. *Lectures on Differential Geometry*, Prentice-Hall, New York, 1964.

D. J. Struik

1. *Lectures on Classical Differential Geometry*, Dover Publications, New York, 1988.

J. M. Sullivan

1. Sphere packings give an explicit bound for the Besicovitch covering theorem, *Journal of Geometric Analysis* **4** (1994), 219–231.

M. E. Taylor

1. *Pseudodifferential Operators*, Princeton University Press, Princeton, 1981.

O. Teichmüller

1. Extremale quasikonforme Abbildungen und quadratische Differentiale, *Abhandlungen der Preussischen Akademie der Wissenschaften, Mathematisch-Naturwissenschaftliche Klasse* **22** (1939), 1–197.

F. A. Valentine

1. *Convex Sets*, McGraw-Hill, New York, 1964.

N. Vilenkin

1. *Special Functions and the Theory of Group Representations*, American Mathematical Society, Providence, 1968.

R. Walter

1. Explicit examples to the H-problem of Heinz Hopf, *Geometriae Dedicata* **23** (1987), 187–213.

H. Wente

1. Counterexample to a conjecture of H. Hopf, *Pacific Journal of Mathematics* **121** (1986), 193–243.

H. Whitney

1. Analytic extensions of differentiable functions defined in closed sets, *Transactions of the American Mathematical Society* **36** (1934), 63–89.

2. A function not constant on a connected set of critical points, *Duke Mathematical Journal* **1** (1935), 514–517.

3. *Geometric Integration Theory*, Princeton University Press, Princeton, 1957.

H. Widom

1. *Integral Equations*, Van Nostrand, New York, 1969.

K. Yoshida

1. *Functional Analysis*, Fifth Edition, Springer-Verlag, Berlin, 1978.

W. P. Ziemer

1. *Weakly Differentiable Functions*, Springer-Verlag, New York, 1989.

A. Zygmund

1. *Trigonometric Series*, Cambridge University Press, Cambridge, England, 1968.

Index

α-dimensional
    density 72
    lower density 71
    upper density 72
adjoint 265
admissible cover 57
analytic sets 18
analytically convex 193
approximate continuity 89
approximately continuous 89
approximating measure 57
approximation of BV functions 104
arc 66
area
    minimizers 111
    of the unit ball 281
barycentric coordinates 68
Besicovitch covering Theorem 91
Bessel's inequality 263
beta function 281
blowing-up 118
Borel
    regular 17
    set 17
bounded
    exhaustion function 211
    turning 257
    variation 101
    variation on Ω 101
Brunn-Minkowski Inequality 241
C^k
    boundary 8
    topology on domains 279
C^∞
    varieties theorem 4, 5
Caccioppoli sets 108

Carathéodory
    construction 58
    measure 64
closed operator 264
co-area formula 136
compactness of $BV(\Omega)$ 107
complementary in the sense of Young 198
completeness
    of $BV(\Omega)$ 102
    of $L^p(m)$ 23
cone 31
conformal mapping 247
conjugate exponents 23
continuously differentiable 1
convex
    analytically 193
    combination 209
    function 196, 197
    hull 209
    integral geometric measure 64
    of order k 219
    strongly analytically 193
convexity
    of order k 217
    geometric 191
    with respect to a family of functions 204
convolution 24
countably
    m-rectifiable set 70
    (ϕ, m)-rectifiable set 70
critical
    point 157
    value 157
curvature 44

cycloid 50
De Rham
    cohomology 189
    theorem 189
defining function 8
Delaunay surface 46
differentiable 125
dilation factor 250
direction of curvature 39
directionally limited 93
Dirichlet
    eigenvalue problem 260
    integral 265
distance function 12
divergence 186
    theorem 109, 186
ϵ-thick set 238
eigenfrequencies 260
eigenfunction for the Laplacian 260
eigenvalue of the Laplacian 260
ends 46
energy integral 265
epicycloid 50
Eulerian integral
    of the first kind 281
    of the second kind 280
exhaustion function 211
extension theorem for Sobolev spaces 150
exterior derivative 188
extreme point 208
face of a simplex 68
Fatou's Lemma 21
finite perimeter 108
Fourier
    coefficient 262
    inversion iheorem 143
    transform 143
Frenet formulas 45
Friedrichs mollifier 25
functions whose partial derivatives are measures 103
gamma function 279
gauge 57
Gauss map 27

Gauss-Green formula 109
    generalized 122
Gaussian
    curvature 39
    integral 280
general position 67
generalized Gauss-Green Theorem 122
genus 46
geometric convexity 191
Gillespie measure 64
Gross measure 63
hard Sard 161
Hardy-Littlewood maximal function 86
Hausdorff
    dimension 62
    distance 274
    measure 61, 63
Helly's Theorem 191
Hilbert space 262
hull 204
hypocycloid 50
Hölder's Inequality 23
infinitely differentiable 1
inner regularity 18
integrable function 20
integral geometric measure 63
Isodiametric inequality 235
Isoperimetric Inequality 113, 242
    local 114
Jacobian, K-dimensional 125
k-jet 166
Kirchberger's Theorem 191
Kirzbraun's Extension Theorem 162
L^p
    norm 22
    space 22
Laplacian 40
Lebesgue
    area 68
    covering theorem 95
    differentiation theorem 87
    measure 17, 22
    space 22

INDEX

Lebesgue's
   covering Theorem 6
   dominated convergence theorem 21
   monotone convergence theorem 21
Leibniz's rule 2
length 66
Lipschitz
   constant 65
   function 65
local graph 14
locally
   finite perimeter 108
   integrable function 20
lower integral 20
lsc domination principle 5
m-rectifiable set 70
Marstrand's Theorem 73
maximal function 86
   with respect to 94
mean curvature 39
measurable
   function 19
   set 16
measure 15
minimal surface 50
Minkowski
   content, lower 74
   content, upper 74
   dimension, lower 78
   dimension, upper 78
   functional 214
   inequality 23, 25
   measure, lower 78
   measure, upper 78
mixed eigenvalue problem 260
multi-index 1
nearest point property 205
Neumann eigenvalue problem 260
normal
   bundle 27
   space 28
normality 96
order of contact 217

Orlicz space 200
outer measure 16
(ϕ, m)-rectifiable set 70
packing dimension
   lower 81
   upper 81
parametric hypersurface 15
Parseval's indentity 263
partition of unity 7
perimeter 107
Poincaré's Inequality 111, 113
positive reach 207
principal
   curvature 39
   direction 39
product measure 21
proper map 161
purely (ϕ, m)-unrectifiable 70
quasi-linear function 68
quasiconformal
   homeomorphism 258
   Jordan curve 257
   mapping 251
Rademacher's Theorem 130
Radon measure 17
rank of a critical value 161
reach 207
real analytic 2
rectifiable 66
reduced boundary 115
refinement 95
regular measure 17
Rellich's Theorem 107
roulette 50
σ-algebra 16
Sard's Theorem 158, 161
scalar curvature 39
second fundamental form 38
semicontinuity in $BV(\Omega)$ 102
separation of sets 95
signed distance function 12
simple m-extension operator 153
simplex 68
simplicial complex 68

Sobolev
- imbedding theorem 144
- space on a domain 147
- spaces 144

spherical measure 58, 63
Steiner symmetrization 224
step function 20
Stokes's Theorem 188
strictly convex 208
strongly
- analytically convex 193
- convex function 198

structure theorem 71 subitem for sets of finite perimeter 122
summatory functions 261
support 1
supporting hyperplane 201
surface 46
Suslin set 18, 227
symmetrized Hausdorff measure 64
tangent
- bundle 28
- cone 31
- vector 8

Taylor
- formula 165
- polynomial 166

Taylor's Theorem 166
Tietze's Extension Theorem 162
topology
- compact-open C^κ 2
- fine 2
- strong 2
- weak 2
- Whitney 2

total
- curvature 39
- extension operator 154

trace
- of order K 39
- theorem for Sobolev spaces 148

triangulable 68
triangulation 68
upper integral 20
Urysohn Lemma 4, 6

valence 6
vector sum of sets 235
vertices 68
Vitali Covering Theorem 99
volume of the unit ball 282
weak
- analytic convexity 193
- derivative 264
- gradient 264
- type $(1,1)$ 86

Weingarten map 35
Whitney
- decomposition 168, 170
- Extension Theorem 167
- family 170
- formula 163

Wiener Covering Lemma 84

Birkhäuser Advanced Texts
Basler Lehrbücher

Managing Editors:
H. Amann, Department of Mathematics, University of Zürich, Switzerland
R. Brylinski, Penn State University, University Park, PA, USA

This series presents, at an advanced level, introductions to some of the fields of current interest in mathematics. Starting with basic concepts, fundamental results and techniques are covered, and important applications and new developments discussed. The textbooks are suitable as an introduction for students and non-specialists, and they can also be used as background material for advanced courses and seminars.

We encourage preparation of manuscripts in TeX for delivery as camera-ready copy which leads to rapid publication, or in electronic form for interfacing with laser printers or typesetters. Proposals should be send directly to the editors or to: Birkhäuser Boston, 675 Massachusetts Avenue, Cambridge, MA 02139 or to Birkhäuser Verlag, P.O. Box 133, CH-4010 Basel, Switzerland.

M. Brodmann, **Algebraische Geometrie**
1989, 296 Seiten. Gebunden. ISBN 3-7643-1779-5

E.B. Vinberg, **Linear Representations of Groups**
1989. 152 pages. Hardcover. ISBN 3-7643-2288-8

K. Jacobs, **Discrete Stochastics**
1991. 296 pages. Hardcover. ISBN 3-7643-2591-7

S.G. Krantz / H.R. Parks, **A Primer of Real Analytic Functions**
1992. 194 pages. Hardcover. ISBN 3-7643-2768-5

L. Conlon, **Differentiable Manifolds: A First Course**
1992. 369 pages. Hardcover. ISBN 3-7643-3626-9

H. Hofer / E. Zehnder, **Symplectic Invariants and Hamiltonian Dynamics**
1994. 342 pages. Hardcover. ISBN 3-7643-5066-0

M. Rosenblum / J. Rovnyak, **Topics in Hardy Classes and Univalent Functions**
1994. 264 pages. Hardcover. ISBN 3-7643-5111-X

P. Gabriel, **Matrizen, Geometrie, Linear Algebra**
1996. 648 Seiten. Gebunden. ISBN 3-7643-5376-7

M. Artin, **Algebra**
1998. ca. 723 Seiten. Broschur. ISBN 3-7643-5938-2

V.S. Sunder, **Functional Analysis Spectral Theory**
1998. 256 pages. Hardcover. ISBN 3-7643-5892-0

QA 300 .K644 1999
Krantz, Steven G. 1951-
The geometry of domains in space